Perspectives in Neural Computing

Springer

London
Berlin
Heidelberg
New York
Barcelona
Hong Kong
Milan
Paris
Singapore
Tokyo

H. Malmgren, M. Borga and L. Niklasson (Eds)

Artificial Neural Networks in Medicine and Biology

Proceedings of the ANNIMAB-1 Conference, Göteborg, Sweden, 13-16 May 2000

 Springer

Helge Malmgren, BA, PhD, MD
Department of Philosophy, Göteborg University, Box 200, SE-405 30 Göteborg, Sweden

Magnus Borga, MSc, PhD
Department of Electrical Engineering, Linköping University, SE-585 83 Linköping, Sweden

Lars Niklasson, BSc, MSc, PhD
Department of Computer Science, University of Skövde, PO Box 408, SE-541 28 Skövde, Sweden

Series Editor
J.G. Taylor, BA, BSc, MA, PhD, FInstP
Centre for Neural Networks, Department of Mathematics, King's College,
Strand, London WC2R 2LS, UK

ISSN 1431-6854

ISBN 1-85233-289-1 Springer-Verlag London Berlin Heidelberg

British Library Cataloguing in Publication Data
Artificial neural networks in medicine and biology :
   Proceedings of the ANNIMAB-1 Conference, Göteborg, Sweden,
   13-16 May 2000. – (Perspectives in neural computing)
   1. Neural networks (Computer science)   2. Artificial
   intelligence – Medical applications   3. Artificial
   intelligence – Biological applications
   I. Malmgren, H.   II. Borga, M.   III. Niklasson, Lars F.
   IV. ANNIMAB-1 (2000 : Goteborg, Sweden)
   610.2'8563
ISBN 1852332891

Library of Congress Cataloging-in-Publication Data
A catalog record for this book is available from the Library of Congress

Typesetting: Camera ready by editors
Printed and bound at the Athenæum Press Ltd., Gateshead, Tyne & Wear
34/3830-543210 Printed on acid-free paper SPIN 10761446

Preface

This book contains the proceedings of the conference ANNIMAB-1, held 13-16 May 2000 in Göteborg, Sweden. The conference was organized by the Society for Artificial Neural Networks in Medicine and Biology (ANNIMAB-S), which was established to promote research within a new and genuinely cross-disciplinary field. Forty-two contributions were accepted for presentation; in addition to these, 8 invited papers are also included.

Research within medicine and biology has often been characterised by application of statistical methods for evaluating domain specific data. The growing interest in Artificial Neural Networks has not only introduced new methods for data analysis, but also opened up for development of new models of biological and ecological systems. The ANNIMAB-1 conference is focusing on some of the many uses of artificial neural networks with relevance for medicine and biology, specifically:

- Medical applications of artificial neural networks: for better diagnoses and outcome predictions from clinical and laboratory data, in the processing of ECG and EEG signals, in medical image analysis, etc. More than half of the contributions address such clinically oriented issues.

- Uses of ANNs in biology outside clinical medicine: for example, in models of ecology and evolution, for data analysis in molecular biology, and (of course) in models of animal and human nervous systems and their capabilities.

- Theoretical aspects: recent developments in learning algorithms, ANNs in relation to expert systems and to traditional statistical procedures, hybrid systems and integrative approaches.

We would like to acknowledge the cooperation of a number of individuals and organisations, who have contributed to making the ANNIMAB conference and this publication possible.

- The Department of Philosophy, Göteborg University, has graciously accepted to take the financial risk for the entire project. The Foundation for Knowledge and Competence Development (KK-stiftelsen) is financially supporting both the ANNIMAB-S and the conference. Economic support for the conference has also been obtained from the Faculty of Arts, Göteborg University, and from the Swedish Council for Research in the Humanities and Social Sciences (HSFR).

- The organising committee has through numerous meetings made an invaluable effort to create an interesting and well-balanced program with several related topics, well covered by invited and submitted papers.

- We are of course indebted to the program committee and all the individual reviewers for guaranteeing the scientific quality of the conference.

- A special acknowledgement is in order to Daniel Ruhe for all his technical support, formatting checks and (maybe most noticeable) design of web pages and the cover of the proceedings. We also want to thank ARC Science Simulations for allowing us to use the copyrighted image of the earth on this cover.

We trust that readers interested in artificial neural networks, medicine or biology will find this book to be very exciting and of high quality.

Göteborg, February 4, 2000

Helge Malmgren
Magnus Borga
Lars Niklasson

ANNIMAB-1 Organising Committee

Magnus Borga
Björn Haglund
Noél Holmgren
Lars Lindström
Helge Malmgren
Filip Radovic
Martin Rydmark

Torben Kling-Petersen
Lars Niklasson
Erik Olsson
Daniel Ruhe
Christer Svennerlind
Holger Wigström

Program Committee

Rod Adams
Charles W. Anderson
Soren Brunak
Bo Cartling
Rich Caruana
Richard Dybowski
Peter Erdi
Laurene V. Fausett
Bernard Fertil
Wulfram Gerstner
Roy Goodacre
Stephen Grossberg
Torgny Groth

Juha Karhunen
Samuel Kaski
Hans Knutsson
Paulo Lisboa
Giovanni Magenes
Rainer Malaka
Hanspeter A. Mallot
Pietro G. Morasso
Ian Nabney
Mattias Ohlsson
Steve Phelps
Tomaso Poggio
John G Taylor

Contents

Invited Presentations

Medical Image Analysis

Signal Processing in Medicine

Clinical Diagnosis and Medical Decision Support

Biomolecular Applications and Biological Modelling

Learning Methods and Hybrid Algorithms

Invited Presentations

Protein β-Sheet Partner Prediction by Neural Networks

Pierre Baldi* and Gianluca Pollastri

Department of Information and Computer Science, University of California Irvine
Irvine, CA 92697-3425

Claus A. F. Andersen and Søren Brunak

Center for Biological Sequence Analysis, The Technical University of Denmark
DK-2800 Lyngby, Denmark

Abstract

Predicting the secondary structure (α-helices, β-sheets, coils) of proteins is
an important step towards understanding their three dimensional conformations.
Unlike α-helices that are built up from one contiguous region of the polypeptide
chain, β-sheets are more complex resulting from a combination several disjoint regions. The exact nature of these long distance interactions remains unclear. Here
we introduce a neural-network based method for the prediction of amino acid partners in parallel as well as anti-parallel β-sheets. The neural architecture predicts
whether two residues located at the center of two distant windows are paired or
not in a β-sheet structure. The distance between the windows is a third essential
input into the architecture. Variations on this architecture are trained using a large
corpus of curated data. Prediction on both coupled and non-coupled residues currently exceeds 83% accuracy, well above any previously reported method. Unlike
standard secondary structure prediction methods, the use of multiple alignment
(profiles) in our case seems to degrade the performance, probably as a result of
intra-chain correlation effects.

1 Background

Predicting the secondary structure (α-helices, β-sheets, coils) of proteins is an important step towards understanding their three dimensional conformations. Unlike
α-helices that are built up from one contiguous region of the polypeptide chain, β-
sheets are built up from a combination of several disjoint regions. These regions, or
β strands are typically 5-10 residues long. In the folded protein, these strands are
aligned adjacent to each other in parallel or anti-parallel fashion. Hydrogen bonds can
form between C'O groups of one strand and NH groups on the adjacent strand and
vice versa with C_α atoms successively a little above or below the plane of the sheet.
Hydrogen bonds between parallel and anti-parallel strands have distinctive patterns,
but the exact nature and behavior of β-sheet long-ranged interactions is not clear.

While the majority of sheets seems to consist of either parallel or antiparallel
strands, mixed sheets are not uncommon. A β-strand can have 1 or 2 partner strands,

---

*and Department of Biological Chemistry, College of Medicine, University of California, Irvine. To
whom all correspondence should be addressed (*pfbaldi@ics.uci.edu*).

and an individual amino acid can have 0,1 or 2 hydrogen bonds with one or two residues in a partner strand. Sometimes one or several partner-less residues are found in a strand, giving rise to the so-called β-bulges. Finally, β-strand partners are often located on a different protein chain. How amino acids located far apart in the sequence find one another to form β-sheets is still poorly understood, as is the degree of specificity between side-chain/side-chain interactions between residues on neighboring strands, which seems to be very weak [13]. The presence of a turn between strands is also an essential ingredient.

Partly as a result of the exponentially growing amount of available 3D data, machine learning methods have in general been among the most successful in secondary structure prediction [2]. The best existing methods for predicting protein secondary structure, i.e. for classifying amino acids in a chain in one of the three classes, achieve prediction accuracy in the 75-77% range [3, 4, 8]. Therefore any improvement in β-sheet prediction is significant as a stand-alone result, but also in relation to secondary and tertiary structure prediction methods in general. Here we design and train a neural network architecture for the prediction of amino acid partners in β-sheets (see also [7, 15]).

2 Data Preparation

2.1 Selecting the Data

As always the case in machine learning approaches, the starting point is the construction of a well-curated data set. The data set used here consists of 826 protein chains from the PDB select list of June 1998 [5] (several chains were removed since DSSP could not run on them). All the selected chains have less than 25% sequence identity using the Abagyan-function [1]. The selection has been performed by applying the all against all Huang-Miller sequence alignment using the "sim" algorithm [6], where the chains had been sorted according to their quality (i.e. resolution plus R-factor/20 for X-ray and 99 for NMR).

2.2 Assigning β-sheet Partners

The β-sheets are assigned using Kabsch and Sander's DSSP program [9], which specifies where the extended β-sheets are situated and how they are connected. This is based on the intra-backbone H-bonds forming the sheet according to the Pauling pairing rules [11]. An H-bond is assigned if the Coulomb binding energy is below -0.5 kcal/mol. In wild-type proteins there are many deviations from Paulings ideal binding pattern, so Kabsch and Sander have implemented the following rules: a β-sheet ('E') amino acid is defined when it forms two H-bonds in the sheet or is surrounded by two H-bonds in the sheet. The minimal sheet is two amino acids long; if only one amino acid fulfills the criteria, then it is called β-bridge ('B'). Bulges in sheets are also assigned 'E' if they are surrounded by normal sheet residues of the same type (parallel or anti-parallel) and comprise at most 4 and 1 residue(s) in the two backbone partner segments, respectively.

A standard example of how the partner assignments are made is shown in figure 1. In the case of β-bridges the same rules are followed, while in the special case of β-bulge residues then no partner is assigned.

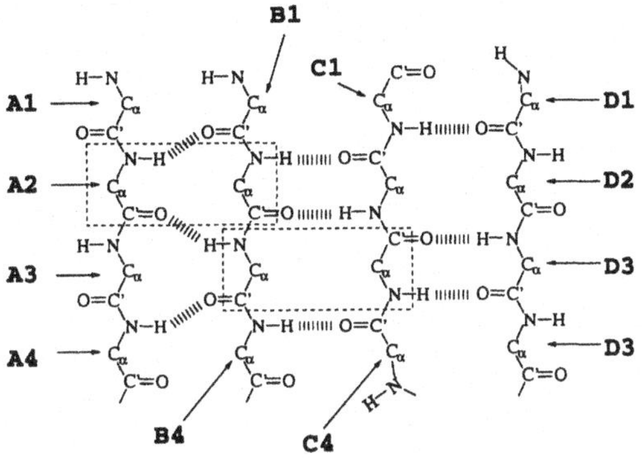

Figure 1: The assignment criteria for sheet partners are shown for two examples by the dashed boxes. That is the A sheet segment binds to the B sheet segment with a parallel sheet and residue A2 is the partner of B2. The other dashed box shows that B3 is the partner of C3, even though none of them has H-bonds in the anti-parallel B-C sheet. The other sheet partners in the example shown are: A3-B3, B2-C2, C2-D2 and C3-D3. Note that the residues A1,A4,B1,B4,C1,C4,D1,C4 are not sheet residues.

3 Neural Network Architecture

A number of different artificial neural network approaches can be considered. Because of the long-ranged interactions involved in beta-sheets, neural architectures must have either very large input windows or distant shorter windows. Very large input windows lead to architectures with many parameters which are potentially prone to overfitting, especially with sparse amino acid input encoding. Overfitting, however, is not necessarily the main obstacle because data is becoming abundant and techniques, such as weight sharing, can be used to mitigate the risk. Perhaps the main obstacle associated with large input windows is that they tend to dilute sparse information present in the input that is really relevant for the prediction [10].

Here we have used a basic two-windows approach. Since the distance between the windows plays a key role in the prediction, one can either provide the distance information as a third input to the system or one can train a different architecture for each distance type. Here, we use the first strategy with the neural network architecture depicted in Figure 2 (see also [12]). The architecture has two input windows of length W corresponding to two amino acid substrings in a given chain. The goal of the architecture is to output a probability reflecting whether the two amino acids located at the center of each window are partners or not. The sequence separation between the windows, measured by the number D of amino acids, is essential for the prediction and is also given as an input unit to the architecture with scaled activity $D/100$. As in other

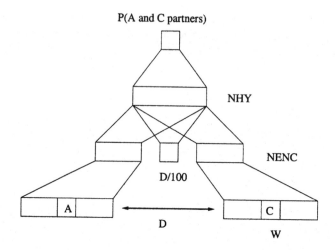

Figure 2: Neural network architecture for amino acid β-partner prediction.

standard secondary structure prediction architectures, we use sparse encoding for the 20 amino acids. Each input window is post-processed by a hidden layer comprising a number NENC of hidden units. Information coming from the input windows and the distance between the windows are combined in a fully interconnected hidden layer of size NHY. This layer is finally connected to a single logistic output unit that estimates the partnership probability. The architecture is trained by back-propagation on the relative entropy between the output and target probability distributions.

4 Experiments and Results

For training, we randomly split the data 2/3 for training and 1/3 for testing purposes. A typical split gives:

Table 1: Training set statistics, with number of sequences, amino acids, and positive and negative examples.

	Training set	Test set
Sequences	551	275
Amino acids	129119	64017
Positive ex.	37008	18198
Negative ex.	44,032,700	22,920,100

The number of negative examples (pairs of amino acids that are not partners) is of course much higher. In order to have balanced training, at each epoch we present all the 37008 positive examples, together with 37008 randomly selected negative examples at each epoch. We use a hybrid between on-line and batch training, with 50 batch blocks, i.e. weights are updated 50 times per epoch. The training set is also shuffled at each epoch, so the error is not decreasing monotonically. The learning rate per block is set at 3.8×10^{-5} at the beginning and is progressively reduced. There is no momentum term or weight decay. When there is no error decrease for more than 100 epochs, the learning rate is divided by 2. Training stops after 8 or more reductions, corresponding to a learning rate that is 256 times smaller than the initial one. Typical performances are given below for different architectural variations. The percentages are computed on the entire test set, including all the negative examples it contains.

The results of several training experiments using different variants of the same architecture are summarized in Table 2

Table 2: Performance results expressed in percentages of correct prediction. W=input window length, NENC=number of hidden units in the post-processing layers of each window, NHY=number of hidden units in the output hidden layer. The second experiment with the 10/11/7 architecture involves multiple alignments (see text). Overall percentage is the simple average of the percentage on each class.

NHY	NENC	W	beta	non-beta	total
8	7	3	83.00	79.29	81.15
8	7	4	83.00	79.80	81.40
8	7	5	82.92	80.05	81.43
8	7	6	83.27	80.37	81.87
8	7	7	83.55	80.28	81.91
10	9	6	83.25	80.60	81.93
10	9	7	83.38	83.84	83.61
10	9	8	83.49	80.84	82.16
10	11	7	83.93	83.34	83.64
10	11	7	76.32	87.77	82.04
10	12	7	82.31	84.36	83.33
12	11	7	83.41	82.30	82.86

The best overall results (83.64%) are obtained with an architecture with a window length of $W = 7$ and hidden unit layers with $NENC = 11$ and $NHY = 10$. This architecture achieves similar accuracy on both partner and non-partner classes (83.93% and 83.34% respectively). It is worthwhile to notice that a small network with three hidden units trained using the distance between the amino acids alone as input achieves an average performance of 75.39% (80.35% on beta-sheet partners and 70.43% on non-partners).

It is well known that evolutionary information in the form of multiple alignments and profiles significantly improves the accuracy of secondary structure prediction methods. This is because the secondary structure of a family is more conserved than the primary amino acid structure. Notice, however, that in the case of beta-sheet partners, intra-sequence correlations may be essential and these are lost in a profile approach where the distributions associated with each column of a multiple alignment are considered independent. To test these effects, we used the BLAST program with standard default parameters (such as BLOSUM matrix 62) to create multiple alignments of our sequences and retrained the optimal architecture found with the corresponding profiles. As can be seen in the table, the overall performance appears to slightly degrade to 82.04%. More interestingly, however, the performance on the non-partner class is improved (87.77 %), whereas the performance on the partner class is degraded (76.32%). This is consistent with a selective improvement of secondary structure prediction resulting from multiple alignments, which does not extend to beta sheet partners as a result of important intra-sequence correlations that are lost in multiple alignments. This may imply that the actual correlation between sheet sequences is higher than previously thought.

5 Discussion

Perfect prediction of protein secondary structures is probably impossible for a variety of reasons including the fact that a significant fraction of proteins may not fold spontaneously [14], that beta-strand partners may be located on a different chain, and that conformation may also depend on other environmental variables, related to solvent, acidity, and so forth. It is however comforting to observe that steady progress is being made in this area, with an increasing number of folds being solved in the structural data bases, and steady improvement of classification and machine learning methods. Here we have developed a neural network architecture that predicts beta-sheet amino acid partners with a performance of almost 84% correct prediction.

There are several directions in which this work can be extended which are currently in progress. These include:

- The development of secondary structure prediction methods for beta sheets based on sequences rather than profiles, to be combined with the profile-based methods which work better with α-helices and coils.

- The use of the present architecture as a beta-sheet predictor rather than a partner predictor, possibly in combination with another neural network.

- Various combinations of the present architectures with existing secondary structure predictor to improve beta-sheet prediction performance.

- The prediction of beta-strand partners rather than amino-acid partners.

- The combination of alignments with partner prediction in order to better predict beta-strands. In particular, a neural network could be trained to predict for each β-strand an ideal partner strand based on amino acid pairing statistics. True partner strands could then be searched by looking for regions in the sequence that have high parallel or antiparallel alignment scores with the putative ideal partner sequence.

- The use of additional information, such as amino acid properties (hydrophobicity, etc.) to improve prediction accuracy.

Acknowledgements

The work of PB is supported by a Laurel Wilkening Faculty Innovation award at UCI. The work of SB and CA is supported by a grant from the Danish National Research Foundation.

References

[1] R.A. Abagyan and S. Batalov. Do aligned sequences share the same fold? *J. Mol. Biol.*, 273:355–368, 1997.

[2] P. Baldi and S. Brunak. *Bioinformatics: The Machine Learning Approach*. MIT Press, Cambridge, MA, 1998.

[3] P. Baldi, S. Brunak, P. Frasconi, G. Soda, and G. Pollastri. Exploiting the past and the future in protein secondary structure prediction. *Bioinformatics*, 15:937–946, 1999.

[4] J. A. Cuff and G. J. Barton. Evaluation and improvement of multiple sequence methods for protein secondary structure prediction. *Proteins*, 34:508–519, 1999.

[5] U. Hobohm and C. Sander. Enlarged representative set of protein structures. *Protein Sci.*, 3:522, 1994.

[6] X. Huang and W. Miller. *Adv. Appl. Math.*, 12:337–357, 1991.

[7] T. J. Hubbard. Use of b-strand interaction pseudo-potentials in protein structure prediction and modelling. In R. H. Lathrop, editor, *In: Proceedings of the Biotechnology Computing Track, Protein Structure Prediction Minitrack of 27th HICSS*, pages 336–354. IEEE Computer Society Press, 1994.

[8] D. T. Jones. Protein secondary structure prediction based on position-specific scoring matrices. *J. Mol. Biol.*, 292:195–202, 1999.

[9] W. Kabsch and C. Sander. Dictionary of protein secondary structure: Pattern recognition of hydrogen–bonded and geometrical features. *Biopolymers*, 22:2577–2637, 1983.

[10] O. Lund, K. Frimand, J. Gorodkin, H. Bohr, J. Bohr, J. Hansen, and S. Brunak. Protein distance constraints predicted by neural networks and probability density functions. *Prot. Eng.*, 10:1241–1248, 1997.

[11] L. Pauling and R.B. Corey. Configurations of polypeptide chains with favored orientations around single bonds: Two new pleated sheets. *Proc. Natl. Acad. Sci. USA*, 37:729–740, 1951.

[12] S. K. Riis and A. Krogh. Improving prediction of protein secondary structure using structured neural networks and multiple sequence alignments. *J. Comput. Biol.*, 3:163–183, 1996.

[13] M.A. Wouters and P.M.G. Curmi. An analysis of side chain interaction and pair correlation within anti-parallel beta-sheets: The difference between backbone hydrogen-bonded and non-hydrogen-bonded residue pairs. *Proteins*, 22:119–131, 1995.

[14] P. E. Wright and H. J. Dyson. Intrinsically unstructured proteins: re-assessing the protein structure-function paradigm. *Journal of Molecular Biology*, 293:321–331, 1999.

[15] H. Zhu and W. Braun. Sequence specificity, statistical potentials, and three-dimensional structure prediction with self-correcting distance geometry calculations of beta-sheet formation in proteins. *Protein Science*, 8:326–342, 1999.

ART Neural Networks for Medical Data Analysis and Fast Distributed Learning

Gail A. Carpenter

Department of Cognitive and Neural Systems, Boston University
Boston, Massachusetts, USA
gail@cns.bu.edu

Boriana L. Milenova

Department of Cognitive and Neural Systems, Boston University
Boston, Massachusetts, USA
boriana@cns.bu.edu

Abstract

ART (Adaptive Resonance Theory) neural networks for fast, stable learning and prediction have been applied in a variety of areas. Applications include airplane design and manufacturing, automatic target recognition, financial forecasting, machine tool monitoring, digital circuit design, chemical analysis, and robot vision. Supervised ART architectures, called ARTMAP systems, feature internal control mechanisms that create stable recognition categories of optimal size by maximizing code compression while minimizing predictive error in an on-line setting. Special-purpose requirements of various application domains have led to a number of ARTMAP variants, including fuzzy ARTMAP, ART-EMAP, Gaussian ARTMAP, and distributed ARTMAP. The talk at the ANNIMAB-1 conference (Gothenburg, Sweden, May, 2000) will outline some ARTMAP applications, including computer-assisted medical diagnosis. Medical databases present many of the challenges found in general information management settings where speed, efficiency, ease of use, and accuracy are at a premium. A direct goal of improved computer-assisted medicine is to help deliver quality emergency care in situations that may be less than ideal. Working with these problems has stimulated a number of ART architecture developments, including ARTMAP-IC [1]. A recent collaborative effort, using a new cardiac care database for system development, has brought together medical statisticians and clinicians at the New England Medical Center with researchers developing expert systems and neural networks, in order to create a hybrid method for medical diagnosis. The talk will also consider new neural network architectures, including distributed ART (dART), a real-time model of parallel distributed pattern learning that permits fast as well as slow adaptation, without catastrophic forgetting. Local synaptic computations in the dART model quantitatively match the paradoxical phenomenon of Markram-Tsodyks [2] redistribution of synaptic efficacy, as a consequence of global system hypotheses.

1 ART and ARTMAP neural networks

Adaptive resonance theory originated from an analysis of human cognitive information processing and stable coding in a complex input environment [3,4]. An evolving series of ART neural network models have added new principles to the early theory and have realized these principles as quantitative systems that can be applied to problems of category learning, recognition, and prediction. Each ART network forms stable recognition categories in response to arbitrary input sequences with either fast or slow learning regimes. The first ART model, ART 1 [5], was an unsupervised learning system to categorize binary input patterns. ART 2 [6] and fuzzy ART [7] extend the ART 1 domain to categorize analog as well as binary input patterns.

Supervised ART architectures, called ARTMAP systems, self-organize arbitrary mappings from input vectors, representing features such as spectral values and terrain variables, to output vectors, representing predictions such as vegetation classes in a remote sensing application. Internal ARTMAP control mechanisms create stable recognition categories of optimal size by maximizing code compression while minimizing predictive error in an on-line setting. Binary ART 1 computations are the foundation of the first ARTMAP network [8], which therefore learns binary maps. When fuzzy ART replaces ART 1 in an ARTMAP system, the resulting fuzzy ARTMAP architecture [9] rapidly learns stable mappings between analog or binary input and output vectors.

2 Match-based learning, error-based learning, and fast learning

The central feature of all ART systems is a pattern matching process that compares the current input with a learned expectation produced by an active code, or hypothesis. ART matching leads either to a resonant state, which focuses attention and triggers learning, or to a self–regulating parallel memory search, which eventually leads to a resonant state, unless the network's memory capacity is exceeded. If the search ends at an established code, the memory representation may stay the same or may be refined to incorporate information from attended portions of the current input. If the search ends at a new code, the code's memory representation begins by learning the current input itself. This *match–based learning* process is the foundation of ART code stability. Match–based learning allows memories to change only when input from the external world is close enough to internal expectations, or when something completely new occurs. This feature makes ART and ARTMAP well suited to problems that require on–line learning of large and evolving databases.

Match–based learning is contrasted with *error–based learning*, which responds to a mismatch by sending the difference between a target output and an actual output toward zero, rather than by initiating a search for a better match. Error–based learning is naturally suited to problems such as adaptive control and the learning of sensory–motor maps, which require ongoing adaptation to present statistics. Neural networks that employ error–based learning include back propagation [10] and other multilayer perceptrons (MLPs).

Many ART applications use fast learning, whereby adaptive weights fully converge to equilibrium values in response to each input pattern. Fast learning enables a system to adapt quickly to inputs that occur only rarely but that may require immediate accurate recall. Remembering many details of an exciting movie is a typical example of fast learning. When the difference between actual output and target output defines "error," present inputs would drive out past learning, since fast learning zeroes the error on each input trial. Therefore fast learning destabilizes the memories of error–based models like back propagation. This feature restricts the domain of most MLPs to off–line applications with a slow learning rate.

3 Distributed coding

In ART and ARTMAP networks, winner–take–all competitive activation supports stable coding, but this limiting case of competition may cause category proliferation when noisy inputs are trained with fast learning. In contrast, MLPs feature distributed McCulloch–Pitts activation, which promotes noise tolerance and code compression, but which causes catastrophic forgetting with fast learning. A recently introduced family of networks called *distributed ART* models combine the best of these two worlds: distributed activation enhances noise tolerance and code compression while new system dynamics retain the stable fast learning capabilities of winner–take–all (WTA) ART systems [11]. With WTA coding, the unsupervised distributed ART model (dART) reduces to fuzzy ART and the supervised distributed ARTMAP model (dARTMAP) reduces to fuzzy ARTMAP. With distributed coding, these networks automatically apportion learned changes according to the degree of activation of each node, which permits fast as well as slow learning without catastrophic forgetting. A parallel distributed match–reset–search process also helps stabilize memory.

Distributed ART models replace the traditional neural network path weight with a *dynamic weight* equal to the rectified difference between coding node activation and an adaptive threshold. Thresholds increase monotonically during learning according to a principle of atrophy due to disuse. However, monotonic change at the synaptic level manifests itself as bidirectional change at the dynamic level, where the result of adaptation resembles long–term potentiation (LTP) for single pulse or low frequency test inputs but can resemble long–term depression (LTD) for higher frequencies. This dynamic is traced to dual computational properties of phasic and tonic components of the coding signal. During learning, the tonic component increases nonspecifically, for all inputs, while the phasic component becomes more selective, maximally favoring the current input. Seemingly paradoxical, the disappearance of LTP enhancement for high–frequency test inputs has been observed by Markram and Tsodyks [2] in the neocortex. Analysis of the dART learning system indicated how these dynamics are precisely the computational components needed to support stable coding in a real–time neural network.

4 Rules, applications, and biological substrates

ART principles have also been used to explain challenging behavioral and brain data in the areas of visual perception, visual object recognition, auditory source identification, variable–rate speech and word recognition, and adaptive sensory–motor control (e.g., [12,13]). One area of recent progress concerns how the neocortex is organized into layers. This new work suggests how "laminar computing" leads to intelligent behavior by modeling how bottom–up, top–down, and horizontal interactions are organized within the cortical layers. These interactions have thus far been studied within the visual cortex. Here, a model has been developed to show how visual cortex (1) stably develops circuits that match environmental constraints, and continues to refine this structure through adult learning; (2) binds or groups distributed information into coherent object representations; and (3) pays attention to important events (e.g., [14]). The mechanisms that govern (1) in the infant are proposed to lead to properties (2) and (3) in the adult. These results are clarifying how ART design principles are embedded within the neocortical circuits that subserve other types of intelligent behaviors, and open the way towards designing general–purpose vision systems that can autonomously learn optimal operating parameters in response to specialized image domains.

ART and dART systems are part of a rapidly growing family of attentive self–organizing systems that have evolved from the biological theory of cognitive information processing. ART modules have found their way into such diverse applications as industrial design and manufacturing, the control of mobile robots, face recognition, remote sensing land cover classification, target recognition, medical diagnosis, electrocardiogram analysis, signature verification, tool failure monitoring, chemical analysis, circuit design, protein/DNA analysis, 3–D visual object recognition, musical analysis, and seismic, sonar, and radar recognition (e.g., [15-17]). A recent book focuses on the implementation of ART systems as VLSI microchips [18].

Applications exploit the ability of an ART system to rapidly learn to classify large databases in a stable fashion, to calibrate confidence in a classification, and to focus attention upon those featural groupings that the system deems to be important based upon experience. The learned expertise of an ARTMAP system also translates to IF–THEN "rules." Within each recognition code, the expectation, or prototype represents a rule that predicts a given outcome. With WTA coding, these prototype vectors provide a transparent set of rules that characterize the decision–making process. ARTMAP neural networks have now provided new methodologies for medical database analysis. A case study of this method, applied to a cardiac database developed at the New England Medical Center (NEMC) [19] is introduced in the following section.

5 The New England Medical Center (NEMC) modeling project

A group of physicians and statisticians from the Division of Clinical Care Research at the New England Medical Center (NEMC) have created a database of Emergency Department patients who were considered for admission to the coronary care unit. A primary goal of the NEMC project is to develop methods to support a physician's decision making process. The project specifically aims to understand the utility and limitations of established and new modeling procedures and to promote their appropriate use in medical research, health care policy, and care assessment. These goals are accomplished through systematic investigation and rigorous evaluation of the relative predictive performance of the analyzed modeling methods. The project is being carried out as a collaboration between the physicians and statisticians who designed the NEMC cardiac database and researchers from Boston University and MIT.

5.1 The NEMC database

The NEMC cardiac database consists of the records of 3,068 study subjects examined at the Emergency Departments (ED) of six participating New England hospitals. The database includes clinical features available to ED physicians, such as clinical presentation, history, physical findings, electrocardiogram, socio-demographic characteristics, and coronary-disease risk factors. Of the 3,068 subjects in the database, 15.7% were diagnosed with cardiac problems requiring hospitalization. These positive outcome cases fall into three categories: arrhythmia, hemodynamic condition, and ischemia. These positive categories are not mutually exclusive; for example, arrhythmia and ischemia are usually accompanied by a hemodynamic condition. The identity of subcategories among the positive outcome variable was not used during model development. That is, the dichotomous output in the NEMC database codes only whether a patient required hospitalization, without specifying particular medical conditions.

The NEMC database includes records for each patient that represent 32 clinical variables, 199 raw ECG variables, and 78 derived ECG variables. Clinical variables quantify features such as medical complaints at arrival to the hospital, age, gender, body-mass index, history of past disease, and medication. The 199 raw ECG variables for the NEMC database were handpicked from a large pool of describing an ECG cycle. Chosen variables describe the amplitude, duration, slope, and area for segments of interest from each of the 12 leads (e.g., Q wave amplitude and duration, QRS area and duration, ST slope). Several variables describing general aspects of the ECG cycle were also included (e.g., mean ventricular rate, mean QRS duration, mean QT interval).

The 78 derived ECG variables in the NEMC database were created to separate clinically important aspects from irrelevant features of the raw signal. These variables are thought to be less sensitive to random fluctuations than the raw signal. Derived ECG variables consist of four groups:

- Five (5) derived ECG variables qualitatively describe the presence or absence of the following abnormalities in the cardiac cycle: Q waves, ST elevation, ST depression, T wave elevation, and T wave inversion. The derivation was based on recordings from all 12 leads.

- Fifty five (55) derived lead-by-lead ECG variables describe whether the five abnormalities enumerated above appear in individual leads. If an abnormality is detected in a raw lead, the same type of abnormality should be registered in one of its contiguous leads. Otherwise, the abnormality is attributed to noise and the variable is set to zero.

- Fifteen (15) summary regional ECG variables describe whether the locations of each of the five abnormalities, in the anterior, inferior, or lateral regions. For each region, at least two raw leads should have registered the abnormalities. Otherwise, the abnormality was ignored and the variable was set to zero.

- Three (3) summary dichotomous ECG variables describe the presence or absence of right bundle branch block, left bundle branch block, and left ventricular hypertrophy. According to the NEMC researchers, these variables have been used in previous regression models to override the effect of ST elevation, although it is not clear what their direct predictive value might be.

5.2 ARTMAP in the NEMC project

ARTMAP-IC [1], an extension of fuzzy ARTMAP [9] and ART-EMAP [20], was initially designed to solve a computational problems commonly encountered in medical modeling, including how to encode inconsistent cases, where identical patient records are associated with different outcomes. When the ARTMAP-IC system was introduced, its performance was evaluated on four benchmark medical databases. One of these databases was the Cleveland heart disease database from the UCI repository [21]. This database contained the records of 303 cardiology patients, 45.9% of which were diagnosed with heart disease. Each record had 13 attributes, including age, gender, heart rate, angina, ST depression, and ST slope. These initial simulation results demonstrated ARTMAP-IC's potential value for cardiac diagnosis.

Exploratory studies of the NEMC database indicated that ARTMAP-IC was not well suited for this problem. In particular, the low prevalence of positive outcomes (15.7%) rendered the system's instance counting feature counterproductive. The final successful ARTMAP algorithm for the NEMC project did, on the other hand, incorporate variations of the basic network that had been developed for other applications. This experience is typical: a given application usually benefits from certain model features but not others. A new model extension was also designed to improve the probability estimation capabilities of the ARTMAP system. Finally, although ARTMAP can handle an unlimited number of inputs, the significance of individual variables is clinically important. With 309 input variables in the NEMC database, variable selection and dimensionality reduction were important for purposes of interpretation. The project therefore included a novel method for estimation of the impact of individual input variables within the framework of ARTMAP networks. With this system, ARTMAP's generalization capabilities were seen to compare favorably to those of logistic regression and decision trees.

The logistic regression approach, on the other hand, appears to offer a simple model with reasonable discrimination and calibration capabilities. This claim of model simplicity is somewhat misleading because some of the derived variables, used as inputs to the regression, were laboriously handcrafted and have a great deal of complexity embedded in them. Still, one may argue that explicit variable derivation rules have certain advantages over complex self-organizing systems, such as ARTMAP or decision trees, because one can create rules encoding physicians' knowledge and diagnostic techniques. A counterargument is that self-organizing systems can discover unnoted patterns in the data and thus offer new diagnostic insights. A crucial test for broad acceptance of data-driven modeling approaches such as ARTMAP in the medical community would be the possibility of unraveling the information encoded in their complex structure. While an ARTMAP network is much easier to analyze and interpret than a standard backpropagation network, the high input dimensionality and the large number of category nodes pose ongoing challenges to structure visualization and rule extraction.

Acknowledgements

This research was supported in part by the Office of Naval Research (ONR N00014-95-1-0409 and ONR N00014-95-1-0657).

Technical Report CAS/CNS TR-2000-002, Boston, MA: Boston University.

References

[1] Carpenter GA, Markuzon N. ARTMAP-IC and medical diagnosis: Instance counting and inconsistent cases. Neural Networks 1998; 11:323-336

[2] Markram H, Tsodyks M. Redistribution of synaptic efficacy between neocortical pyramidal neurons. Nature 1996; 382:807-810

[3] Grossberg S. Adaptive pattern classification and universal recoding, I: Parallel development and coding of neural feature detectors & II: Feedback, expectation, olfaction, and illusions. Biological Cybernetics 1976; 23:121-134 & 187-202

[4] Grossberg S. How does a brain build a cognitive code? Psychological Review 1980; 87:1-51

[5] Carpenter GA, Grossberg S. A massively parallel architecture for a self-organizing neural pattern recognition machine. Computer Vision, Graphics, and Image Processing 1987a; 37:54-115

[6] Carpenter GA, Grossberg S. ART 2: Self-organization of stable category recognition codes for analog input patterns. Applied Optics 1987b; 26:4919-4930

[7] Carpenter GA, Grossberg S, Rosen DB. Fuzzy ART: Fast stable learning and categorization of analog patterns by an adaptive resonance system. Neural Networks 1991; 4:759-771

[8] Carpenter GA, Grossberg S, Reynolds JH. ARTMAP: Supervised real-time learning and classification of nonstationary data by a self-organizing neural network. Neural Networks 1991; 4:565-588

[9] Carpenter GA, Grossberg S, Markuzon N, et al. Fuzzy ARTMAP: A neural network architecture for incremental supervised learning of analog multidimensional maps. IEEE Transactions on Neural Networks 1992; 3:698-713

[10] Werbos P. Beyond regression: New tools for prediction and analysis in the behavioral sciences. PhD thesis, Harvard University, Cambridge, Massachusetts, 1974

[11] Carpenter GA. Distributed learning, recognition, and prediction by ART and ARTMAP neural networks. Neural Networks 1997; 10:1473-1494

[12] Desimone R. Neural circuits for visual attention in the primate brain. In: Carpenter GA, Grossberg S (eds) Neural networks for vision and image processing. MIT Press, Cambridge, Massachusetts, 1992, pp 343-364

[13] Grossberg S. How does the cerebral cortex work? Learning, attention, and grouping by the laminar circuits of visual cortex. Spatial Vision 1999a; 12:163-186

[14] Grossberg S. The link between brain learning, attention, and consciousness. Consciousness and Cognition 1999b; 8:1-44

[15] Caudell TP, Smith SDG, Escobedo R, et al. NIRS: Large scale ART–1 neural architectures for engineering design retrieval. Neural Networks 1994; 7:1339-1350

[16] Christodoulou CG, Huang J, Georgiopoulos M, et al. Design of gratings and frequency selective surfaces using fuzzy ARTMAP neural networks. Journal of Electromagnetic Waves and Applications 1995; 9:17-36

[17] Soliz P, Donohoe, GW. Adaptive resonance theory neural network for fundus image segmentation. In Proceedings of the World Congress on Neural Networks (WCNN'96), Erlbaum, Hillsdale, New Jersey, 1996, pp 1180-1183

[18] Serrano–Gotarredona T, Linares–Barranco B, Andreou, AG. Adaptive resonance theory microchips: Circuit design techniques. Kluwer Academic Publishers, Boston, 1998

[19] Carpenter GA, Kopco N, Milenova BL, et al. A neural network method for supervised learning and medical database analysis. Statistics in Medicine, to appear

[20] Carpenter GA, Ross WD. ART-EMAP: A neural network architecture for object recognition by evidence accumulation. IEEE Transactions on Neural Networks 1995; 6:805-818

[21] Murphy PM, Aha DW. UCI repository of machine learning databases. University of California, Department of Information and Computer Science, Irvine, California. http://www.ics.uci.edu/~mlearn/MLRepository.html, 1992

Modelling Uncertainty in Biomedical Applications of Neural Networks

Georg Dorffner[1], Peter Sykacek, Christian Schittenkopf

Austrian Research Institute for Artificial Intelligence, Vienna, Austria

Abstract

In this paper we argue that the explicit account of uncertainty in data modeling is particularly important for biomedical applications of neural networks and related techniques. There are several sources of uncertainty of a model, including noise, bias and variance. Unless one attempts to identify or minimize the sources that contribute to errors of a particular application, one only has a sub-optimal solution. If, on the other hand, one does attempt to model uncertainty, one gets several major advantages. We discuss several methods for modeling uncertainty, including density estimation, Bayesian inference and complex noise models, in the context of several sample applications – most notably in the domain of biosignal processing.

1 Neural Computation for Biomedical Applications

Neural computation is a widespread and interdisciplinary field, ranging from advanced data analysis via computational neuroscience to cognitive science. In biomedical applications mainly the use of methods from neural computation for flexible nonlinear data analysis is of great interest. In this realm, neural computation has long gone beyond the application of traditional "neural network" models (like multilayer perceptrons and selforganizing maps) deeply penetrating other disciplines like statistics, signal or time series processing. Methods include recent advances such as independent component analysis [1][2], hidden Markov models with extended observation densities [2], or support vector machines [4]. What unites these seemingly diverging methodologies is their strive for flexible, nonlinear and adaptive, models of stochastic data.

2 A Model's Uncertainty

Viewed in this sense, neural computation provides methods and methodologies for the estimation and application of stochastic data models. As such, all methods have one thing in common: they commit an *error* when providing output estimates or predictions. In neural network literature this aspect has long been reduced to the observation that the goal of training is to minimize this error while avoiding

---

[1] Also with Dept. of Medical Cybernetics and Artificial Intelligence, University of Vienna, Austria

overfitting to training data. With recent authors (e.g. [5], [6] and others) we argue that much more can be said about the possible error a model commits. As we will make clear below, a model's error can be viewed in the more global context of its "uncertainty" in providing an estimate at the output, such as the probability for a class, or the approximation of a continuous value. When succeeding in providing an estimate of this uncertainty and its possible sources – in addition to the estimate of the target – the overall application becomes more valuable in that

- it is more robust in that it can tell the user when to trust it and when not

- it is therefore much more likely to be accepted by the user as an automated tool, e.g. for decision support

- it can generally perform better if one excludes those cases where uncertainty exceeds a certain threshold

- it simply extracts more information about the underlying data than a model which does not account for its uncertainty

- it becomes more amenable to be integrated in larger systems together with other models expressing their own estimates and uncertainty

All of these reasons for uncertainty estimation become even stronger when considered in a biomedical context (replace 'user' with 'clinician' or 'physician'). In particular, many biomedical applications are "safety-critical" in that an inappropriate use of automated systems can endanger a person's health or life. Examples are monitoring systems in intensive care, diagnostic devices used as an aid in deciding upon the proper treatment of a disease, or automated tools to extract information from medical data such as X-ray or ultrasound images. But even if an application is not safety-critical in the above sense, knowing about its uncertainty can greatly increase its usefulness and acceptance as a valuable tool. An example is sleep medicine, where the community has been looking for an automated tool for all-night sleep analysis which is robust enough to deliver trustworthy results [7].

3   The Sources of Uncertainty

[8] distinguishes between three major sources for errors: *noise* in the data, a model's *bias* and a model's *variance*. Following [6], an even more detailed distinction can be made:

- **Target noise[2]:** Stochasticity of data brings with it that, even if the optimal data model can be found, the target variable fluctuates around the estimate

---

[2] Another source of noise is input noise due to insufficient or erroneous measurement. Usually this type of noise is subsumed into target noise, although a separate estimation would be possible.

given by the model. This reflects the fact that the dependency to be estimated is not completely visible in the data.

- **A model's bias due to model complexity:** It is well known that a model with too simple a complexity (e.g. a linear one) naturally has a systematic deviation from the optimal output when the theoretically optimal model is more complex (e.g. nonlinear). A complex model like a neural network has a low bias (which is defined as the expectation over the limit of infinitely many model estimations).

- **A model's bias due to its training algorithm:** Much less attention is usually paid to the fact that a model estimation ("training") procedure can create an inherent bias by not being able to reach the theoretical optimum (e.g. a gradient-based algorithm getting stuck in local optima).

- **A model's variance due to insufficient data**: Models are much more prone to overfit wherever there are not sufficient amounts of data to support a particular model over others. A single neural network usually has a large error due to variance.

- **A model's variance due to training:** Again, another – often unnoticed – source of variance comes from the fact that estimation algorithms usually do not find the optimal solution. Variance is created, for instance, if different training runs end up in different local optima.

In short, the main ways of either modeling or minimizing the resulting uncertainty are the following:

- Using a complex enough model can minimize a model's bias. Usually it is impossible to estimate the remaining bias. The bias due to training can be minimized by either using a global optimization procedure or by performing multiple runs.

- Using a *committee* of models estimated on different training sets can minimize a model's variance. Through a resampling technique like n-fold cross-validation it can also be estimated. Another, more principled way of accounting for variance in the data is *density estimation* of the input data. Variance due to the training can be minimized by again performing multiple training runs.

- Noise can be modeled using more sophisticated noise models. Classical neural network training – as was pointed out by [5] – implicitly assumes constant Gaussian noise, thereby collapsing both input and target noise into one noise model. Based on so-called mixture-density networks [9], this assumption can be relaxed resulting in a more appropriate noise model for many applications.

4 Bayesian Inference

A framework that nicely combines many of the above-mentioned aspects is that of Bayesian inference in model estimation [10][11]. The Bayesian approach describes all parameters of a model (e.g. the weights in a neural network) by means of a probability distribution. Instead of finding *the* optimal model, the aim is to estimate the distribution of probable models based on the observed data. By doing this, one captures the following sources of uncertainty:

- Averaging over the posterior density of model outputs – which is done to obtain the output of the system – is akin to applying a committee of models, therefore minimizing a model's variance.
- The estimation of the model's variance is done in a very principled and therefore powerful way by viewing the model in a probabilistic sense from the beginning.
- The Bayesian model evidence (in short, the factor by how much the probability distribution can be narrowed down based on the data) can be used as another measure of uncertainty, by expressing how well a model describes the data. Model selection based on evidence ensures finding the model with the smallest bias (given the prior assumptions).

In regression problems, model uncertainty thus can be measured by the width of the posterior distribution of models ("error bar"). In classification problems, the uncertainty is implicitly contained in the derived probability estimates (which are called to be "moderated" through the estimation process. Only noise modeling is outside the scope of Bayesian inference and must still be accounted for explicitly.

5 Examples

5.1 Cross-Validation and Committees

The first is a typical biomedical application in the area of diagnostic support. In [12] we report about the application of multilayer perceptrons (MLP) to the task of recognizing the presence of coronary artery disease based on features extracted from stress-ECG examinations. The results of a ten-fold cross-validation indicate the mean (around 80% sensitivity for 75% specificity) and variance (8% for sensitivity vs. 12% for specificity) of the performance. While this is straight-forward and should be standard in any application (although it often is not – see [13]), it demonstrates that by virtue of the variance obtained from the cross-validation one can get an overall estimate of the uncertainty due to variance for a *single* network classifier. To minimize this source of error, the final application should be a *committee* built from all ten networks from cross-validation. A subsequent cross-validation on a larger data set (which was not available in this particular application) should then yield an estimate of the committee's variance. This example demonstrates that even a single cross-validation might not be enough to optimize a model's performance.

5.2 Density Estimation

While these applications of cross-validation and committees are good at estimating the variance of the *entire* model, the technique can be taken even further to estimate the variance depending on the input. Real-world data usually is not evenly distributed across the space of all possible input vectors. As mentioned above, uncertainty due to variance depends on the density of data points in each area of the input space. This is illustrated by a simple toy example – taken from the area of sleep EEG analysis – using just one input (for visualization purposes). In Fig. 2, two features of an all-night sleep-EEG – namely stochastic complexity [14] and the power in the delta band – are plotted against each other.[3] It becomes clear that in this regression problem the dependency is highly nonlinear, demanding a nonlinear model for optimal performance. The results of five runs using a multilayer perceptron are shown vis-à-vis the results of 5 runs of a linear regression. The bias of the linear method is clearly visible. The results show that in the area of input values around -1.7 and above 1.5 – where relatively few data points are available – the variance between the runs of the MLP is much larger than elsewhere.

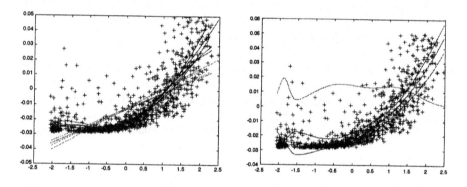

Figure 1: Left: 5 runs each of linear and non-linear regression for estimating depth-of-sleep. Right: Confidence values and error bars obtained by density estimation.

A powerful alternative to this empirical derivation of the variance is to estimate the density of the input data directly (e.g. using a Gaussian mixture model – compare [5]). The results of such density estimation are also plotted in Fig. 1. The inverse of data density can be used directly as an error bar to indicate that the model is much more uncertain for inputs around -1.7 and above 1.5.

[15] gives a more realistic example of this use of density estimation. There we have shown that density estimation can be used to increase performance, if the resulting measure of uncertainty is used to reject inputs that are uncertain, and that Bayesian

---

[3] This could be taken as a simple example of estimating depth-of-sleep – expressed by delta power – from complexity. Of course, in this simple instantiation this is not a realistic application, but it can be taken as a prototype for continuous sleep analysis.

model estimation can further improve the results. The goal was to predict manually scored sleep stages based on a number of features extracted from EEG and EOG signals.

5.3 Noise Modelling

Fig. 2 demonstrates the capabilities of mixture density networks to model noise processes that are not Gaussian. On the same regression problem as in Fig. 1, it becomes clear that the fluctuations – especially for negative input values – do not appear to follow a normal distribution. An MDN with two Gaussian kernels can estimate the skewness of the distribution (the centers of each kernel are depicted by solid lines, while the widths of the kernels are indicated by the dotted lines). This provides useful additional information about the possible deviations of observations from the estimated output.

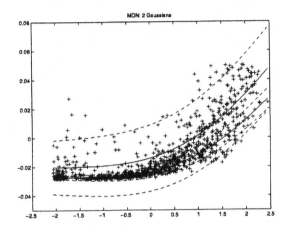

Figure 2: Estimation of the non-Gaussian noise with 2 kernels.

5.4 The SIESTA Project

The SIESTA project[4] is another example of how to incorporate uncertainty estimation into a biomedical application, in particular, in sleep medicine. The major difference to [15] is that SIESTA aims at producing continuous profiles of sleep based on only partially labeled data. The final system is designed as a "semi-Bayesian" model accounting for uncertainty both on a feature level and in the resulting output (see [16] and [17] for first results).

An example of how Bayesian inference can be used for estimating uncertainty is shown in Fig 3. An EEG signal is plotted, which is contaminated by muscle

---

[4] SIESTA ("A New Standard for Integrating Polygraphic Sleep Recordings into a Comprehensive Model of Human Sleep and Its Validation in Sleep Disorders") is Biomed-2 project no. BMH4-CT97-2040, sponsored by the European Commission, DG XII.

artifacts, as they frequently occur during sleep when the person is moving. A common method for extracting features from the signal is the use of autoregressive models, estimated over a window of quasi-stationary signal. When the model estimation is done in the Bayesian framework, the so-called Bayesian evidence (as described above) can be directly used as a measure of uncertainty, as depicted in Fig. 3 (solid line below the signal). Since the artifact resembles noise much more than the regular signal, and therefore the autoregressive model is – in a sense – much more uncertain in this area, the evidence decreases, while staying high elsewhere.

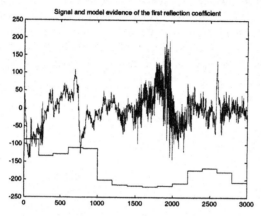

Figure 3: The Bayesian evidence of an autoregressive model of artifact-contaminated EEG

6 Summary

In summary, we would like to argue that uncertainty estimation is paramount in many biomedical applications. Many different techniques exist – from simple density estimation to a full-blown Bayesian approach – to provide a model with an indicator of its own uncertainty, thus improving both the model's performance and robustness.

Acknowledgments

This work is supported by the Austrian Federal Ministry of Science and Transport. We are deeply indebted to Stephen Roberts, Oxford University, for much valuable input in numerous discussions.

References

[1] Bell AJ, Sejnowski TJ. An information maximisation approach to blind separation and blind deconvolution. Neural Computation 1995; 7:1129-1159

[2] Oja E. From neural learning to independent components. Neurocomputing 1998; 22:187-200

[3]   Penny WD, Roberts SJ. Hidden Markov Models with Extended Observation Densities, Imperial College, Neural Systems Research Group, Technical Report TR-98-15, 1998

[4]   Vapnik V. The Nature of Statistical Learning Theory. Springer-Verlag, New York, 1995

[5]   Bishop CM. Neural Networks for Pattern Recognition. Clarendon Press, Oxford, 1995

[6]   Penny WD, Roberts SJ. Neural network predictions with error bars. Imperial College, Neural Systems Research Group, Research Report TR-97-1, 1997

[7]   Hasan J. Automatic analysis of sleep recordings: a critical review. Annals-Clin-Res. 1985; 17: 280-287

[8]   Breiman L. Bias, variance, and arcing classifiers., Statistics Department, University of California, Berkeley, CA, USA, Tech. Rep. 460, 1996

[9]   Bishop CM. Mixture Density Networks, Department of Computer Science and Applied Mathematics, Aston University, Birmingham, UK, NCRG/94/004, 1994

[10] Neal RM. Bayesian Learning for Neural Networks. Springer, Berlin/Heidelberg/New York/Tokyo, 1996

[11] MacKay DJC. Bayesian Methods for Adaptive Models. California Institute of Technology, Pasadena, CA, 1992

[12] Dorffner G, Leitgeb E, Koller H. A comparison of linear and non-linear classifiers for the detection of coronary artery disease in stress-ECG, in Horn W. et al.(eds.), Artificial Intelligence in Medicine, Springer, Berlin, LNAI 1620, 1999, pp.227-231

[13] Flexer A. Statistical Evaluation of Neural Network Experiments: Minimum Requirements and Current Practice. in Trappl R.(ed.): Cybernetics and Systems '96, Oesterreichische Studiengesellschaft fuer Kybernetik, Wien, 2 vols, 1996., pp.1005-1008

[14] Rezek IA, Roberts SJ. Stochastic Complexity Measures for Physiological Signal Analysis. IEEE Transactions on Biomedical Engineering 1998; 44:1186-1191

[15] Sykacek P, Dorffner G, Rappelsberger P, Zeitlhofer J. Experiences with bayesian learning in a real world application. In Jordan M.I., et al.: Advances in Neural Information Processing Systems 10, MIT Press/Bradford Books, Cambridge/London, 1998, pp.964-970

[16] Flexer A, Sykacek P, Rezek IA, Dorffner G. Using Hidden Markov Models to build an automatic, continuous and probabilistic sleep stager for the SIESTA project (extended abstract), Medical and Biological Engineering and Computing 1999; Supplement 2, Proceedings of EMBEC'99, p.1658-165

[17] Sykacek P, Roberts SJ, Rezek IA, Flexer A, Dorffner G. Reliability in preprocessing – Bayes rules SIESTA (extended abstract), Medical & Biological Engineering & Computing 1999; suppl. EMBEC'99

Neural Computation in Medicine: Perspectives and Prospects

Richard Dybowski

Medical Informatics Laboratory, Division of Medicine,
King's College London (St Thomas' Campus), London, UK

Abstract

In 1998, over 400 papers on artificial neural networks (ANNs) were published in the context of medicine, but why is there this interest in ANNs? And how do ANNs compare with traditional statistical methods? We propose some answers to these questions, and go on to consider the 'black box' issue. Finally, we briefly look at two directions in which ANNs are likely to develop, namely the use of Bayesian statistics and knowledge-data fusion.

1 Cognitive and data modelling

In the wake of publication of the back-propagation algorithm in 1986, the number of ANN-oriented articles featured in the Medline database has grown from 2 in 1990 to 473 in 1998. But why is there this interest in ANNs within medicine?

The design of ANNs was originally motivated by the phenomena of learning and recognition, and the desire to model these cognitive processes. The *cognitive-modelling* branch of ANN research is still active, and it is relevant to medicine in providing models of psychological and cerebral dysfunction [1, 2, 3]. However, in the late 1980s, a more pragmatic stance emerged, and ANNs became to be seen also as tools for *data modelling*, primarily for classification.

Medicine involves decision making, and classification is an integral part of that process, but medical classification tasks, such as diagnosis, can be far from straightforward. At least two sources of difficulty can be identified.

- Clinical laboratories are being subjected to an ever-increasing *workload*. Much of the data received by these laboratories consists of complex figures, such as cytological specimens – objects traditionally interpreted by experts – but experts are a limited resource.

- The *complexity* of patient-related data can be such that even an expert can overlook important details. With the production of high volumes of possibly high-dimensional data from instruments such as flow-cytographs, and the integration of disparate data sources through *data fusion*, this is becoming an increasing problem.

Presented with such problems, it is natural that the medical fraternity should turn to ANNs in the hope that these adaptable models could alleviate at least some of the problems; however, the use of ANNs in medicine has raised a number of issues. One of these is the relationship between ANNs and statistics; another is the use of 'black box' systems in medicine. We look at both these issues and then consider two directions in which ANNs are likely to develop.

2   A statistical perspective

2.1   Multilayer perceptrons

Multilayer perceptrons (MLPs) with sigmoidal hidden-node functions are the ANNs most commonly used in medicine. Medical data are diverse and, at times, highly complex [4], but MLPs are universal approximators. The flexibility of MLPs has enabled them to be applied to a wide variety of medical fields [5, 6, 7, 8, 9, 10, 11], including oncology [12], dentistry [13], bacteriology [14], and sleep research [15]. For example, Tarrasenko [16] describes how an MLP was used to associate an electroencaphalograph (EEG) with deep sleep, REM sleep or wakefulness. EEGs from nine patients were digitized and, in order to reduce the dimensionality of feature space, the EEGs were characterized by the ten coefficients of an autoregressive time-series model. These coefficients were used as the input vectors for an MLP with three outputs. The number of hidden units was varied from 4 to 20, and five initially random weight vectors were used. The best-performing MLP had a misclassification rate of 10.6%. In comparison, nearest-neighbour classification gave a misclassification rate of 18.3%.

The genesis and renaissance of ANNs took place within various communities, and papers published during these periods reflected the disciplines involved: biology and cognition; statistical physics; and computer science. But it was not until the early 1990s that a probability-theoretic perspective emerged, which regarded ANNs as being within the framework of statistics [17, 18, 19, 20].

A recurring theme of this literature is that many ANNs are analogous, or identical to, existing statistical techniques. For example, a popular statistical method for modelling the relationship between a binary response variable and a vector of covariates is *logistic regression*, but a single-layer perceptron with a logistic output function and trained by a cross-entropy error function will be functionally identical to a main-effects logistic regression model. Furthermore, an MLP can be regarded as a both a non-linear extension of logistic regression and as a particular type of projection pursuit regression model. However, Ripley & Ripley [21] point out that the statistical algorithms for fitting projection pursuit regression are not as effective as those for fitting MLPs.

One may ask whether the apparent similarity between ANNs and existing statistical methods means that ANNs are redundant. One answer to this is given by Ripley [22]:

> The traditional methods of statistics and pattern recognition are either *parametric* based on a family of models with a small number of parameters, or *non-parametric* in which the models used are totally flexible. One of the impacts of neural network methods on pattern

recognition has been to emphasize the need in large-scale practical problems for something in between, families of models with large but not unlimited flexibility given by a large number of parameters.

Another response is to point out that the widespread fascination for ANNs has attracted many talented researchers and potential users into the realm of data modelling. It is true that the neural-computing community re-discovered some statistical concepts already in existence, but this influx of participants has created new ideas and refined existing ones. These benefits include the *learning of sequences* by time delay and partial recurrence [23], and the creation of powerful visualization techniques, such as *generative topographic mapping* [24]. Thus the ANN movement has resulted in statisticians having available to them a collection of techniques to add to their repertoire. Furthermore, the placement of ANNs within a statistical framework has provided a firmer theoretical foundation for neural computation. This has led to new developments such as the Bayesian approach to ANNs [25], and to improvements to existing neural methods [26].

ANNs can be used jointly with conventional statistical methods. In the scheme suggested by Goodman & Harrell [27], the performance metrics of a generalized linear model are compared with those from an MLP. If the two approaches are not found to differ with statistical significance, the linear model is chosen since it simpler with respect to computation and interpretation. The authors demonstrate this approach in the context of a coronary artery bypass dataset.

MLPs are trained using exemplars from each class of interest, but many pathologies are uncommon. In such situations a balanced training set should be used for the under-represented classes [28]. However, this may not be feasible when abnormalities are very rare, in which case *novelty detection* should be considered [29].

It should be emphasized that, even with correct training, an ANN will not necessarily be the best choice for a classification task in terms of accuracy. This has been highlighted by Wyatt [30], who wrote:

> Neural net advocates claim accuracy as the major advantage. However, when a large European research project, StatLog, examined the accuracy of five ANN and 19 traditional statistical or decision-tree methods for classifying 22 sets of data, including three medical datasets [31], a neural technique was the most accurate in only one dataset, on DNA sequences. For 15 (68%) of the 22 sets, traditional statistical methods were the most accurate, and those 15 included all three medical datasets.

But one should add the comment made by Michie et al [31] on the results of the StatLog project:

> With care, neural networks perform very well as measured by error rate. They seem to provide either the best or near best predictive performance in nearly all cases ...

Nevertheless, in order to justify the implementation of an ANN, its ease of implementation and performance should be compared with that obtained from one or more appropriate standard statistical techniques. Furthermore, one should also be aware of emerging non-neural alternatives, such as *support vector machines* for classification and *independent component analysis* for visualization, and also make comparisons with these, if appropriate.

2.2 Self-organizing feature maps

The most common neural system for unsupervised training is Kohonen's *self-organizing feature maps* (SOFMs) [32]. The aim of SOFMs is to map an input vector to one of a set of neurons arranged in a lattice, and to do so in such a way that positions in input space are topologically ordered with locations on the lattice. Hertz et al [33] liken this process to an elastic net, existing in input space, which wants to come as close as possible to the input vectors of a training set.

Amongst the interesting medical applications of SOFMs are their use for classification of craniofacial growth patterns [34], the extraction of information from electromyographic signals with regard to motor unit action potentials [35], and magnetic-resonance image segmentation [36]. With regard to the first example, an SOFM was used to extract the most relevant information from mandibular growth data. The position of a patient on the resulting map could be used to aid orthodontic diagnosis and treatment.

The example by Glass & Reddick [36] demonstrates how an SOFM can be used for image segmentation. An SOFM was trained with pixels from viable tumors and necrotic tissue, as visualized by magnetic resonance images. An MLP was then trained to distinguish between these two types of tissue on the basis of the SOFMs final input-node weights. Consequently, the SOFM-MLP combination characterized the type of tissue represented by a new pixel. This provided a non-invasive technique to assess an osteosarcoma patient's response to chemotherapy.

The so-called Growing Cell Structure [37] is related to SOFMs, and Walker et al [38] used this technique for the cytodiagnosis of breast carcinoma. In addition to performing classification, the method also enabled them to visualize how the input variables were involved in the classification.

Although the SOFM algorithm provides a means of visualizing the distribution of data points in input space, Bishop [39] points out that this can be weak if the data do not lie within a two-dimensional subspace of the higher-dimensional space containing the data. Another observation, made by Balakrishnan et al [40] is that Kohonen feature maps are similar to the statistical technique of k-means clustering, yet it has been our observation that many papers describing an SOFM do not compare the efficacy of it with another visualization technique applied to the same dataset.

2.3   Radial-basis function networks

The second most commonly used ANNs in medicine are the *radial-basis function networks* (RBFNs). An RBFN can be regarded as a type of generalized linear discriminant function, a linear function of functions that permits the construction of non-linear decision surfaces. The basis functions of an RBFN define *local responses* (*receptive fields*). Typically, only some of the hidden units (basis functions) produce significant values for the final layers. This is why RBNFs are sometimes referred to as *localized receptive field networks*. In contrast, all the hidden units of an MLP are involved in determining the output from the network (they are said to form a *distributed representation*). The receptive-field approach can be advantageous when the distribution of the data in the space of input values is multimodal. Furthermore, RBFNs can be trained more quickly than MLPs [41], but the number of basis functions required grows exponentially with the number of input nodes, and an increase in the number of basis functions increases the time taken, and amount of data required, to train an RBFN adequately.

Construction of an RBFN is a two-step process. The first step is typically performed using an unsupervised method, such as k-means clustering or an SOFM, to define the basis functions. This exploits the distribution of the classes in feature space. The second step uses supervised learning for the output-layer weights via linear optimization, which is faster than the typical training algorithms used for MLPs. The potential advantages of these two steps have been investigated for a number of medical applications. For example, an RBFN approximated well the nonlinearity of heart dynamics [42]. Each basis function provided a local reconstruction of the dynamics in the space spanned by the basis functions. Other examples include the classification of cervical-tissue fluorescent spectra [43], the identification of bacteria (clustering with respect to mass spectra) [44] and spinal disorders (clustering with respect to motion data) [45]. However, even when an SOFM suggests good discrimination between the classes of interest, an RBFN may not perform as well as an MLP. For example, when the EEG classification mentioned in Section 2.1 was repeated using an RBFN in place of the MLP, the misclassification rate increased slightly to 11.6% [16].

When conditions are such that an RBFN can act as a classifier [46, 47], an advantage of the local nature of RBFNs compared with MLP classifiers is that a new set of input values that falls outside all the localized receptor fields could be flagged as not belonging to any of the classes represented. In other words, the set of input values is novel. This is a more cautious approach than the resolute classification that can occur with MLPs, in which a set of input values is always assigned to a class, irrespective of the values.

2.4   Adaptive Resonance Theory Networks

Although not a common ANN, an *adaptive resonance theory* (ART) *network* is noteworthy. The ART process [48] can be regarded as a type of hypothesis test [49]. A pattern presented at an input layer is passed to a second layer, which is interconnected to the first. The second layer makes a guess about the category to which the original pattern belongs, and this hypothetical identity is passed back to the first layer. The hypothesis is compared with the original pattern and, if found to be a close match, the hypothesis and original pattern reinforce each other (*resonance* is said to take place). But if the hypothesis is incorrect, the second layer produces another guess. If the second layer cannot eventually provide a good match with the pattern, the original pattern is learned as the first example of a new category. Spencer et al [50] incorporated ART into a system capable of discovering temporal patterns in ICU data and predicting the onset of haemodynamic disorders.

Although ART provides unsupervised learning, an extension called ARTMAP [51] combines two ART modules to enable supervised learning to take place. Harrison et al [52] describe how ARTMAP can be used to update a knowledge base. They do so in the context of ECG diagnosis of myocardial infarction and the cytopathological diagnosis of breast lesions.

In spite of resolving the stability/plasticity dilemma [53], the ART algorithms can be sensitive to noise [54]. Furthermore, Ripley [22] questions the virtue of the ART algorithms over adaptive k-means clustering [55].

3 The "black-box" issue

A criticism leveled against neural networks is that they are 'black-box' systems [30, 56] (although this can be less of a problem for the mathematically minded [57]). By this it is meant that the manner in which a neural network derives an output value from a given feature vector is not comprehensible to the non-specialist, and that this lack of comprehension makes the output from neural networks unacceptable.

There are a number of properties that we desire in a model, two of which are accuracy (the 'closeness' of a model's estimated value to the true value) and interpretability. By *interpretability*, we mean the type of input-output relationships that can be extracted from a model and which are comprehensible to the intended users of the model.

An interpretable model is advantageous for several reasons. Firstly, It could be educational by supplying a previously unknown but useful input-output summary. This, in turn, can lead to new areas of research. Secondly, it could disclose an error in the model when an input-output summary or explanation contradicts known facts.

Does the lack of interpretability, as defined above, make a model unacceptable? That depends on the purpose of the model. Suppose that the choice of a statistical model for a given problem is reasonable (on theoretical or heuristical grounds), and an extensive empirical assessment of the model (for example, by cross-validation and prospective evaluation) shows that its parameters provide an acceptable degree of accuracy over a wide range of input vectors. The use of such a model for prediction would generally be approved, subject to a performance-monitoring policy. Why not apply the same reasoning to neural networks, which are, after all, non-standard statistical models?

But suppose that we are interested in *knowledge discovery*; by this we mean the extraction of previously unknown but useful information from data. With a trained MLP, it is difficult to interpret the mass of weights and connections within the network, and the interactions implied by these. However, the goal of *rule extraction* [58] is to map the (possibly complex) associations encoded by the functions and parameters of a trained ANN to a set of comprehensible if-then rules. If successful, such a mapping would lead to an interpretable collection of statements describing the associations discovered by the ANN, which, in turn, may lead to clinical insight.

Another approach to providing interpretability is to use a *hybrid neuro-fuzzy system* [59]. These are feed-forward networks built from if-then rules containing linguistic terms based on domain knowledge, and the membership functions associated with the fuzzy rules can be tuned to data by means of a training algorithm. Both rule extraction and hybrid neuro-fuzzy systems are responses to those clinicians unwilling to use a predictive model lacking interpretability, even when the model is highly accurate.

4   Future directions

4.1   Bayesian neural computation

Whereas classical statistics attempts to draw inferences from data alone, *Bayesian statistics* goes further by allowing data to update prior beliefs via Bayes' theorem. Advantages to neural computation provided by the Bayesian framework include

- a principled approach to fitting an ANN to data via regularization,

- allowance for more than one possible MLP architecture by *model averaging*, and for multiple solutions to the training of a single MLP architecture by using a *committee of networks*,

- automatic selection of features to be used as input to an MLP from a set of candidate features (*automatic relevance determination*).

However, there are disadvantages to the Bayesian approach, namely that it tends to require more computation than the maximum likelihood approach, and it requires explicit descriptions of the prior probability distributions, which are often subjective.

Bayesian ANNs have not yet found their way into general use, but, given their capabilities, we expect them to take a prominent role in medical neural computation.

4.2 Knowledge-data fusion

If both data and domain knowledge are available, how can these two resources be used together? This is the focus of *knowledge-data fusion*. One approach is use networks in which nodes and their connections have an interpretation in the domain of interest. An example of this is seen with hybrid fuzzy-neuro systems. Another example is the concept of *probabilistic networks*; however, unlike an MLP, the nodes of a probabilistic network are related stochastically, and inference can take place in different directions along the network.

The stochastic relationships of a probabilistic network are defined by conditional probabilities at the non-terminal nodes, and for a conditional probability with a complex likelihood, an MLP may be an effective model. This idea of using MLPs at the nodes of a probabilistic networks is one approach to implementing *data fusion*, such as the integration of electrophysiology signals with both radiological images and patient records for decision support. By using knowledge-data fusion to integrate data from disparate sources, with ANNs acting as modules for the framework, virtual patient records within a hospital-wide intranet could be updated in a principled manner.

References

[1] L. Aakerlund and R. Hemmingsen. Neural networks as models of psychopathology. *Biological Psychiatry*, 43(7):471–82, 1998.

[2] W.W. Tryon. A neural network explanation of posttraumatic stress. *Journal of Anxiety Disorders*, 12(4):373–85, 1998.

[3] C.A. Galletly, C.R. Clark, and A.C. McFarlane. Artificial neural networks: a prospective tool for the analysis of psychiatric disorders. *Journal of Psychiatry & Neuroscience*, 21(4):239–47, 1996.

[4] R. Dybowski and V. Gant. Artificial neural networks in pathology and medical laboratories. *Lancet*, 346:1203–1207, 1995.

[5] J.T. Wei, Z. Zhang, S.D. Barnhill, K.R. Madyastha, H. Zhang, and J.E. Oesterling. Understanding artificial neural networks and exploring their potential applications for the practicing urologist. *Urology*, 52(2):161–172, 1998.

[6] W. el Deredy. Pattern recognition approaches in biomedical and clinical magnetic resonance spectroscopy: A review. *NMR in Biomedicine*, 10(3):99–124, 1997.

[7] A. Tewari. Artificial intelligence and neural networks: Concept, applications and future in urology. *British Journal of Urology*, 80 Supplement 3:53–8, 1997.

[8] J.P. Hogge, D.S. Artz, and M.T. Freedman. Update in digital mammography. *Critical Reviews in Diagnostic Imaging*, 38(1):89–113, 1997.

[9] P.Y. Ktonas. Computer-based recognition of EEG patterns. *Electroencephalography & Clinical Neurophysiology – Supplement*, 45:23–35, 1996.

[10] W. Penny and D. Frost. Neural networks in clinical medicine. *Medical Decision Making*, 16(4):386–389, 1996.

[11] D. Itchhaporia, P.B. Snow, R.J. Almassy, and W.J. Oetgen. Artificial neural networks: Current status in cardiovascular medicine. *Journal of the American College of Cardiology*, 28(2):515–521, 1996.

[12] R.N. Naguib and G.V. Sherbet. Artificial neural networks in cancer research. *Pathobiology*, 65(3):129–139, 1997.

[13] M.R. Brickley, J.P. Shepherd, and R.A. Armstrong. Neural networks: A new technique for development of decision support systems in dentistry. *Journal of Dentistry*, 26(4):305–309, 1998.

[14] R. Goodacre, M.J. Neal, and D.B. Kell. Quantitative analysis of multivariate data using artificial neural networks: A tutorial review and applications to the deconvolution of pyrolysis mass spectra. *Zentralblatt fur Bakteriologie*, 284(4):516–539, 1996.

[15] C. Robert, C. Guilpin, and A. Limoge. Review of neural network applications in sleep research. *Journal of Neuroscience Methods*, 79(2):187–193, 1998.

[16] L. Tarassenko. *A Guide to Neural Computing Applications*. Arnold, London, 1998.

[17] J.S. Bridle. Probabilistic interpretation of feedforward classification network outputs, with relationships to statistical pattern recognition. In F. Fogleman-Soulie and J. Herault, editors, *Neurocomputing: Algorithms, Architectures, and Applications*, pages 227–236. Springer-Verlag, Berlin, 1991.

[18] B.D. Ripley. Statistical aspects of neural networks. In O.E. Barndorff-Nielsen, J.L.Jensen, and W.S. Kendell, editors, *Networks and Chaos — Statistical and Probabilistic Aspects*, pages 40–123. Chapman & Hall, London, 1993.

[19] S.-I. Amari. Mathematical methods of neurocomputing. In O.E. Barndorff-Nielsen, J.L.Jensen, and W.S. Kendell, editors, *Networks and Chaos — Statistical and Probabilistic Aspects*, pages 1–39. Chapman & Hall, London, 1993.

[20] B. Cheng and D.M. Titterington. Neural networks: A review from a statistical perspective. *Statistical Science*, 9:2–54, 1994.

[21] B.D. Ripley and R.M. Ripley. Neural networks as statistical methods in survival analysis. In R. Dybowski and V. Gant, editors, *Clinical Applications of Artificial Neural Networks*, chapter 9. Cambridge University Press, Cambridge. In press.

[22] B.D. Ripley. *Pattern Recognition and Neural Networks*. Cambridge University Press, Cambridge, 1996.

[23] S. Haykin. *Neural Networks: A Comprehensive Foundation*. Macmillan, New York, 1994.

[24] C.M. Bishop, M. Svensen, and C.K.I. Williams. GTM: the generative topographic mapping. *Neural Computation*, 10(1):215–234, 1997.

[25] D.J.C. MacKay. A practical Bayesian framework for back-propagation networks. *Neural Computation*, 4(3):448–472, 1992.

[26] A. Flexer. Connectionists and statisticians, friends or foes? In J. Mira and F. Sandoval, editors, *Proceedings of the International Workshop on Artificial Neural Networks*, LNCS 930, pages 454–461. Springer, 1995.

[27] P.H. Goodman and F.E. Harrell. Neural networks: Advantages and limitations for biostatistical modelling. Technical report, Washoe Medical Center, University of Nevada, October 1998.

[28] L. Tarassenko. Neural networks. *Lancet*, 346:1712, 1995.

[29] L. Tarassenko, P. Hayton, N. Cerneaz, and M. Brady. Novelty detection for the identification of masses in mammograms. In *Proceedings of the 4th IEE International Conference on Artificial Neural Networks*, pages 442–447, Cambridge, 1995. Cambrdige Univeristy Press.

[30] J. Wyatt. Nervous about artificial neural networks? *The Lancet*, 346:1175–1177, 1995.

[31] D. Michie, D.J. Spiegelhalter, and C.C. Taylor, editors. *Machine Learning, Neural and Statistical Classification*. Ellis Horwood, New York, 1994.

[32] T. Kohonen. Self-organizing formation of topologically correct feature maps. *Biological Cybernetics*, 43:59–69, 1982.

[33] J. Hertz, A. Krogh, and R.G. Palmer. *Introduction to the Theory of Neural Computation*. Addison-Wesley, Reading, MA, 1991.

[34] C.J. Lux, A. Stellzig, D. Volz, W. Jager, A. Richardson, and G. Komposch. A neural network approach to the analysis and classification of human craniofacial growth. *Growth, Development, & Aging*, 62(3):95–106, 1998.

[35] C.I. Christodoulou and C.S. Pattichis. Unsupervised pattern recognition for the classification of EMG signals. *IEEE Transactions on Biomedical Engineering*, 46(2):169–178, 1999.

[36] J.O. Glass and W.E. Reddick. Hybrid artificial neural network segmentation and classification of dynamic contrast-enhanced MR imaging (DEMRI) of osteosarcoma. *Magnetic Resonance Imaging*, 16(9):1075–1083, 1998.

[37] B. Fritzke. Growing cell structures – a self-organising network for unsupervised and supervised learning. *Neural Networks*, 7:1441–1460, 1994.

[38] A.J. Walker, S.S. Cross, and R.F. Harrison. Visualisation of biomedical datasets by use of growing cell structure networks: A novel diagnostic classification technique. *Lancet*, 354:1518–1521, 1999.

[39] C.M. Bishop. *Neural Networks for Pattern Recognition*. Clarendon Press, Oxford, 1995.

[40] P.V. Balakrishnan, M.C. Cooper, V.S. Jacob, and P.A. Lewis. A study of the classification capabilites of neural networks using unsupervised learning: A comparison with k-means clustering. *Psychometrika*, 59:509–525, 1994.

[41] J. Moody and C. Darken. Fast learning in networks of locally-tuned processing units. *Neural Computation*, 1:281–294, 1989.

[42] A. Bezerianos, S. Papadimitriou, and D. Alexopoulos. Radial basis function neural networks for the characterization of heart rate variability dynamics. *Artificial Intelligence in Medicine*, 15(3):215–234, 1999.

[43] K. Tumer, N. Ramanujam, J. Ghosh, and R. Richards-Kortum. Ensembles of radial basis function networks for spectroscopic detection of cervical precancer. *IEEE Transactions on Biomedical Engineering*, 45(8):953–961, 1998.

[44] R. Goodacre, E.M. Timmins, R. Burton, N. Kaderbhai, A.M. Woodward D.B. Kell, and P.J. Rooney. Rapid identification of urinary tract infection bacteria using hyperspectral whole–organism fingerprinting and artificial neural networks. *Microbiology*, 144(5):1157–1170, 1998.

[45] J.B. Bishop, M. Szpalski, and S.K. Ananthraman D.R. McIntyre M.H. Pope. Classification of low back pain from dynamic motion characteristics using an artificial neural network. *Spine*, 22(24):2991–2998, 1997.

[46] R. Dybowski. Classification of incomplete feature vectors by radial basis function networks. *Pattern Recognition Letters*, 19:1257–1264, 1998.

[47] I.T. Nabney. Efficient training of RBF networks for classification. Technical report NCRG/99/002, Neural Computing Research Group, Aston University, 1999.

[48] G.A. Carpenter and S. Grossberg. A massively parallel architecture for a self-organizing neural pattern recognition machine. *Computer Vision, Graphics, and Image Processing*, 37:54–115, 1987.

[49] M. Caudill and C. Butler. *Naturally Intelligent Systems*. MIT Press, Cambridge, MA, 1990.

[50] R.G. Spencer, C.S. Lessard, F. Davila, and B. Etter. Self-organising discovery, recognition and prediction of haemodynamic patterns in the intensive care unit. *Medical & Biological Engineering & Computing*, 35(2):117–123, 1997.

[51] G. Carpenter, S. Grossberg, and J. Reynolds. ARTMAP: Supervised real-time learning and classification of nonstationary data by a self-organizing neural network. *Neural Networks*, 4:565–588, 1991.

[52] R.F. Harrison, S.S. Cross, R.L. Kennedy, C.P. Lim, and J. Downs. Adaptive resonance theory: A foundation for 'apprentice' systems in clinical decision support? In R. Dybowski and V. Gant, editors, *Clinical Applications of Artificial Neural Networks*, chapter 13. Cambridge University Press, Cambridge. In press.

[53] S. Grossberg. Adaptive pattern classification and universal recording: I. parallel development and coding of neural feature detectors. *Cybernetics*, 23:121–134, 1976.

[54] B. Moore. ART1 and pattern clustering. In D. Touretzky, G. Hinton, and T. Sejnowski, editors, *Proceedings of the 1988 Connectionist Models Summer School*, pages 174–185, Sa Mateo, CA, 1989. Morgan Kaufmann.

[55] D.J. Hall and D. Khanna. The ISODATA method of computation for relative perception of similarities and differences in complex and real computers. In K. Enslein, A. Ralston, and H.S. Wilf, editors, *Statistical Methods for Digital Computers*, pages 340–373. Wiley, New York, 1977.

[56] D. Sharp. From "black-box" to bedside, one day? *The Lancet*, 346:1050, 1995.

[57] R.A. Kemp, C. MacAulay, and B. Palcic. Opening the black box: the relationship between neural networks and linear discriminant functions. *Analytical Cellular Pathology. 14(1):19-30, 1997*, 14(1):19–30, 1997.

[58] R. Andrews, A.B. Tickle, and J. Diederich. A review of techniques for extracting rules from trained artificial neural networks. In R. Dybowski and V. Gant, editors, *Clinical Applications of Artificial Neural Networks*, chapter 12. Cambridge University Press, Cambridge. In press.

[59] D. Nauck, F. Klawonn, and R. Kruse. *Foundations of Neuro-Fuzzy Systems*. John Wiley, Chichester, 1997.

Discriminating Gourmets, Lovers, and Enophiles? Neural Nets Tell All About Locusts, Toads, and Roaches

Wayne M. Getz

William C. Lemon

Division of Insect Biology, University of California Berkeley, California, USA

getz@nature.berkeley.edu

Abstract

Here we consider the issue of choice and how neural systems can be used to investigate the processes of discrimination, as well as the evolution of different kinds of choice-related behavior in animals. We develop these ideas in the context of three studies, among others. The first study is on the evolution of specialization in animals using locust feeding behavior as the leitmotif, where decision making in individuals is modeled by a 3-layer-perceptron. In this study the fitness of individuals depends on their response to signals from plants and the density of individuals using those plants [1]. The second is a study that investigates the evolution of species recognition in sympatric taxa using female mate choice in frogs as the leitmotif [2]. Here individuals are modeled by Elman nets (3-layered perceptrons with feedback) and their fitness is determined by their ability to discriminate conspecifics from heterospecifics. The third is a study of the response characteristics of a recurrent Hopfield-type neural network to input that represents olfactory stimuli. The connectivity of this net reflects the basic architectural features of the neuron in the insect antennal lobe, as typified by cockroaches or bees [3].

1 Introduction

Discrimination is the essence of behavior, no matter how rudimentary. For example, the sea hair (*Aplysia*, spp.), a naked mollusk, needs to discriminate between various chemical and mechanical types of stimuli to appropriately and reflexively withdraw its siphon into its mantel at times of danger. We may reasonably believe that neural network models, which can be viewed as devices for categorizing inputs into classes, provide us with insight into how *Aplysia* and other organisms use reflexive behavior in discriminatory contexts. Is it too far fetched to believe that neural nets can also provide insights into more complex discriminatory behaviors such as food selection by locusts, mate choice in frogs, or cockroaches using odor plumes to find food? In a sense, this is the same dilemma that ecologists face when asking whether aggregated species models can be used to address questions in community ecology.

In both cases, the plausibility of the models as representations of real processes is stretched to the limit, and we are reduced to convincing ourselves that playing with such models generates knowledge that otherwise would be unavailable. That models can generate hypotheses about how real processes work, is true to be sure. The success of the modeling enterprise, however, rests on whether these hypotheses ultimately lead to uncovering some truths that would, in the absence of models, remain buried forever. In this mini-expose, focusing on our own work and that of our collaborators, we review how neural nets provide some insights into three processes that are very crude abstractions of systems found in nature. We hope to demonstrate that the value of neural network models depends on how the hypotheses they generate are translated into knowledge through a dialectical interplay of theory and empiricism.

2 Locust Gourmets

The first application we review is that of neural networks used to investigate the evolution of generalist versus specialist strategies in consumers competing for resources in a limited space. The leitmotif used to carry out simulations in this study [1] was how guilds of closely related species of herbivorous insects are able to coexist while competing for several sympatric species of plants (e.g. see reviews [4,5]) In the simulations, the decision making processes of individuals were emulated by computations of 100 3-layer perceptrons (each consisting of 2 input, 3 hidden, and 1 output unit—see Fig. 1) which then produce 6 different scalar outputs in response to one of 6 pairs of input stimuli. Each pair of inputs represents one of four plant resource types (Fig. 1). The relative fitness of each of the 100 individuals (i.e., 100 perceptrons each with a different set of synaptic weights) was determined by the following three factors: 1.) the size of its response to each of the four resources, 2.) an intrinisic value for each of the four resources, 3). the decline in the value of resources with the number of individuals competing to use them.

In general, neural networks can adapt to their environments through evolution or through learning, although the two processes are only distinguishable in the way networks are updated. In this study [1], an evolutionary (e.g. genetic algorithm), rather then a learning process (e.g. back-propagation), was used to update the networks [6,7]. Specifically, this model addressed the question of why there are more host plant specialist than generalist insect consumers when laboratory studies indicate that generalists are able to use a much broader spectrum of plants than is observed in nature. It has been speculated that perceptual constraints coupled with considerations of foraging efficiency play a role in the evolution of specialization [4,5].

The results of these simulations [1] indicated that the probability of species evolving particular host plant preferences depends on the different signals produced by plants of different nutritive values. Networks could evolve to exploit a single plant type or to generalize across two or more plant types. Evolutionary equilibria typically involved guilds of complementary species that together constitute an 'ideal free distribution' in terms of productivity of the different plants.

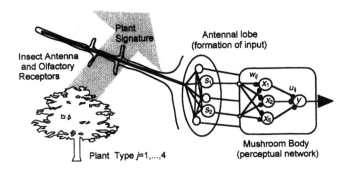

Figure 1. Each of four plant types is assumed to produce a unique response in the projection neurons of the antennal lobe of an insect brain. This response, modeled as a two unit input layer of a 3-layer preceptron, is normalized so that it effectively produces the ratio of the two inputs (S1/S2), which progressively increases from 1/4 in plant type 1, to 2/3 in plant type 2, 3/2 in plant type 3 and 4 in plant type 4. These inputs S_1 and S_2 are then processesed in a perceptual network in the mushroom bodies, modeled by three hidden units and one output unit in our three-layer perceptron. The input signals are propagated through synaptic weights w_{ij} to the three-unit hidden layer x_j and from there through synaptic weights u_{ij} to the output signal y. The output y is a measure of the insect's ability to discriminate the plant represented by the corresponding input and is used in conjunction with a fitness value, based on the number of insects using that host plant type, to generate a population index for the phenotype represented by the synaptic weights w_{ij} and u_{ij}. Modified from [1].

The mix in phenotypes within these guilds depends critically on the order of appearance of the various combinations of specialist and generalist phenotypes, and this order depends on the difficulty of the perceptual task faced by the phenotypes (Fig. 2). Any differences in the relative utilization by a generalist of different plant species will lead to the emergence of one or more specialists that exploit the plants most under-utilized by the generalist. Evolutionary changes in guild structure are less frequent than mutational rates would suggest but are saltatory when they occur. The strategy to specialize may dominate for two reasons: specialization appears to evolve more readily in complex environments and the ideal free distribution is more easily matched in a population density context by a group of specialists or by generalists in concert with specialists than by a generalist alone.

Earlier studies [8,9] used simple three-layer feed-forward perceptrons to test whether the neural networks respond to stimuli in a similar way as animals. These models have been very accurate at predicting stimulus-response characteristics, including generalization, supernormality (peak shift) and stimulus intensity effects. Such neural networks exhibit behavior predictive of the behavior of real consumers. Leow [10] describes a series of increasingly complex neural networks each of which allows a simulated creature to search for food and to evade danger by using olfactory cues. Behaviors such as obstacle avoidance and risk-taking emerge naturally from the networks' interaction with the environment. More complex systems of networks [11] have been used to model foraging behavior of egrets asked to choose between feeding in a flock or foraging individually. Local enhancement and flock foraging was preferred when resources were patchy, while individual foraging was preferred when resources were evenly distributed. Other simple neural nets have been developed that learn to avoid aversive stimuli by trial and error or by imitation [12].

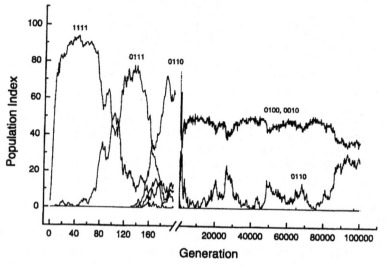

Figure 2. Population indices of six insect phenotypes (designated by ones and zeros to indicate which plant types the network perceives maximally—e.g. 1111 is a generalist not distinguishing between any of the plant types, 0100 is a specialist on plant type 2 able to discriminate it from 1 and the lumped combination of 3 and 4, and 0110 is less specialized able to discriminate the lumped combination of 2 and 3 from 1 and from 4) are plotted for one of the simulations of a population evolving in an environment in which plant types 2 and 3 each have a carrying capacity for 50 individuals while plant types 1 and 4 are toxic and have 0 carrying capacity. The values plotted are the relative population sizes of the six phenotypes every generation up to 1000 and every 100 generations thereafter. Modified from [1].

3 Toad Lovers

Neural networks have been used to model the response of animals to signals sent by conspecifics or heterospecifics. In most of these models the evolution of the phenotypes of the senders and receivers depends on the choices that the networks make. One such example is the study of the coevolution of a mimicry complex [13] consisting of a palatable prey species that evolves to avoid predation by mimicking the appearance of an unpalatable species. The effectiveness of the mimicry is determined by how difficult it becomes for the predator to discriminate between the phenotypes of the two potential prey species. Each of the individuals in the two prey populations of senders and the one predator population of receivers was represented by a 3-layer perceptron. In this simulation it benefited both types of senders, the original (i.e., the distasteful species) and the mimic (the tasty species), to elicit the same response from the receiver (predator) and it benefited the receiver to respond to only one sender (discriminating between original and mimic). The receiver was thus in conflict with one of the senders (as in Batesian mimicry). The most common response was for mimicry to develop due to the mimic evolving toward the original faster than the original moved away. As the two sender phenotypes converged there was selection on the receiver to discriminate along a stimulus dimension where the two senders were still distinguishable. Even after mimicry was established the model and the mimic were constantly changing in appearance.

Another example of where network choice was important to an evolutionary outcome used recurrent Elman nets (a three-layer perceptron with feedback from the third layer to the second) to investigate the evolution of species recognition in sympatric taxa [2]. In this study, mate selection by female frogs, based on the way they process the calls of conspecific and heterospecific male frogs, was the leitmotif for the specifics of the analysis. The simulations addressed the effectiveness of recognition mechanisms based on recognition of self only (Paterson's specific mate recognition model) versus those based on discrimination of self versus others (Dobzhansky's character displacement model). They were also designed to help determine the influence of interactions with other species (sound environment hypothesis) and the relative variation of signals within the species (feature invariance hypothesis) on the evolution of mating signals, as well as addressing the controversial hypothesis that selection for species recognition generates sexual selection.

Results presented in this study [2] indicated that call decoding strategies based only on self-recognition do not result in accurate species recognition while those based on discrimination of self versus others are more effective. The neural network weighted signal features in a manner suggesting that the total sound environment as opposed to the relative variation of signals within a species is more important in the evolution of recognition mechanisms. Finally, selection for species recognition generated substantial variation in the relative attractiveness of signals within the species and thus could result in sexual selection.

A similar study [14] showed that Elman nets selected to recognize or discriminate simple patterns may possess emergent biases towards pattern size or symmetry components, preferences often exhibited by real females, and investigated how these biases shape signal evolution. This study also induced the Elman nets to evolve toward responses to an actual mate recognition signal, the call of the tungara frog *Physalaemus pustulosus*. The Elman net was capable of recognizing the call of the tungara frog and made remarkably accurate quantitative predictions about how well females generalize to novel calls. These predictions were stable over several network architectures. The authors concluded that the degree to which female tungara frogs respond to a call may be an incidental by-product of a sensory system selected simply for species recognition.

Other recent studies have shown that neural networks predict signal phenotypes that have been observed in real animals. Examinations of the conflicts between sender and receiver confirm that a receiver bias for costly signals insures honest senders, a concept well supported by studies of behavioral ecology [15,16]. Network based models often predict the evolution of preference for symmetrical signals [17,18]. Gradient interaction models, yet another form of network based species recognition models, demonstrate clearly how generalization can generate a preference for symmetrical variants of a display [19, but see 20].

4 Roach Enophiles

One of the most common uses of neural networks is in modeling the processing of sensory information in the brain. The scope of these models ranges from investigations of neural coding of simple stimuli to experiments generating realistic behavior in artificial organisms. One recent study investigated how odors are represented in the olfactory sensory system of insects [3]. The relatively simple neural network model was inspired by our current knowledge of the first synaptic layer of the cockroach olfactory system, the antennal lobes (Fig. 3). This network was used to investigate how elements of the network, such as synaptic strengths, feedback circuits and neural activation functions, influence the formation of an olfactory code in neurons that project from the antennal lobes to the mushroom bodies, the higher association area of the insect brain.

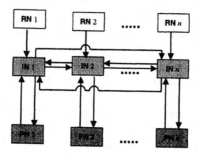

Figure 3. A schematic of the architecture underlying the neural system modeled by Getz and Lutz [3]. The arrows represent the four types of synapses: receptor neuron (RN) excitatory feed-forward onto interneurons; interneuron (IN) inhibitory feed-forward onto other interneurons; IN inhibitory feed-forward onto projection neurons (PN); and PN excitatory feedback onto IN. Projection neurons fire through disinhibition. Modified from [3].

Simulation studies of the antennal lobe network demonstrated that the network is able to produce codes independent or virtually independent of odor concentration over a given range. This concentration range is moderately dependent on the time required for voltage to decay to its resting potential in an activated neuron, strongly dependent on the strength of excitatory feedback from projection neurons onto antennal lobe intrinsic interneurons, and overwhelmingly dependent on the slope of the activation function that transforms the voltage of depolarized neurons into the rate at which spikes are produced. When excitatory feedback from the projection neurons to the intrinsic interneurons is strong, the activity in the projection neurons undergoes transitions from initial states to stimulus specific equilibrium states that are maintained once the stimulus is removed. When the projection neuron-intrinsic interneuron feedback is weak the projection neurons are more likely to relax back to a stimulus independent equilibrium state in which case the code is not maintained beyond the application of the stimulus.

Neural networks have been used to model other sensory systems such as a three-layer, two-pathway version of on-off cells of the optic chiasm of the fly visual system [21]. This network generated a realistic three component response to on and off stimulation. At the other end of the spectrum, a phototactic network and a robot used to simulate chemotaxis by the flatworm *Caenorhabditis elegans* produced reliable phototaxis regardless of the locomotory parameters of the robot [22]. An interesting model of sensory processing and behavioral control examined prey orientation behavior of waterstriders [23]. This model had six input units representing the legs of the waterstriders detecting water vibrations and two output units controlling left turns and right turns. The model accurately predicted orientation toward disturbances in the water caused by prey. Simulated lesions in the input units produced results matching those observed in real animals.

5   Conclusions

From the studies discussed above, it is clear that neural networks models provided a powerful tool for elucidating the dynamic complexities of discriminatory processing in behavioral and evolutionary contexts. In some studies, the plausibility of existing hypotheses about choice and discrimination were examined. In most studies, new hypotheses were generated and these are now open for testing. The intellectual payoff of any particular study can only be assessed in the course of time, but the neural network paradigm has certainly become pervasive in our analyses of discriminatory processes in biology. As with all modeling paradigms in population biology, the value of applying neural networks to problems in ethology and evolution depends more on the craftsman than on the tool.

References

[1] Holmgren NMA, Getz WM. Evolution of host plant selection in insects under perceptual constraints: a simulation study. Evol Ecol Res 2000; 2:1-25

[2] Ryan MJ, Getz WM. Signal decoding and receiver evolution: an analysis using a neural network. In preparation

[3] Getz WM, Lutz A. A neural network model of general olfactory coding in the insect antennal lobe. Chem Senses 1999; 24:351-372

[4] Bernays EA, Wcislo WT Sensory capabilities, information-processing, and resource specialization. Q. Rev. Biol. 1994: 69: (2) 187-204

[5] Bernays EA. The value of being a resource specialist: Behavioral support for a neural hypothesis. American Naturalist 1998: 151: (5) 451-464

44

[6] Nolfi S, Parisi D. Learning to adapt to changing environments in evolving neural networks. Adaptive Behavior 1996; 5:75-98

[7] Nolfi S, Parisi D, Elman JL. Learning and evolution in neural networks. Adaptive Behavior 1994; 3:5-28

[8] Ghirlanda S, Enquist M. Artificial neural networks as models of stimulus control. Anim Behav 1998; 56:1383-1389

[9] Bolhuis JJ. The development of animal behavior: from Lorenz to neural nets. Naturwissenschaften 1999; 86:101-111

[10] Leow WK. Computational studies of exploration by smell. Adaptive Behavior 1998; 6:411-434

[11] Toquenaga Y, Kajitani I, Hoshino T. Egrets of a feather flock together. Artificial Life 1994; 1:391-411

[12] Schmajuk NA, Zanutto BS. Escape, avoidance and imitation: a neural network approach. Adaptive Behavior 1997; 6:63-129

[13] Holmgren NA, Enquist M. Dynamics of mimicry evolution. Biol J Linn Soc 1999; 66: 145-158

[14] Phelps SM, Ryan MJ. Neural networks predict response biases of female tungara frogs. Proc Royal Soc Lond B 1998; 265:279-285

[15] Arak A, Enquist M. Conflict, receiver bias and the evolution of signal form. Phil Trans Royal Soc Lond B 1995; 349:337-344

[16] Krakauer DC, Johnstone RA. The evolution of exploitation and honesty in animal communication: a model using artificial neural networks. Phil Trans Royal Soc Lond B 1995; 348:355-361

[17] Enquist M, Arak A. Symmetry, beauty and evolution. Nature 1994; 372:169-172

[18] Johnstone RA. Female preference for symmetrical males is a by-product of selection for mate recognition. Nature 1994; 372:172-175

[19] Enquist M, Johnstone RA. Generalization and the evolution of symmetry preferences. Proc Royal Soc Lond B 1997; 264:1345-1348

[20] Dawkins MS, Guilford T. An exaggerated preference for simple neural network models of signal evolution? Proc Royal Soc Lond B 1995; 261:357-360

[21] Sarikaya M, Wang W, Ogmen H. Neural network model of on-off units in the fly visual system: simulations of dynamic behavior. Biological Cybernetics 1998; 78:399-412

[22] Morse TM, Ferree TC, Lockery SR. Robust spatial navigation in a robot inspired by chemotaxis in *Caenorhabditis elegans*. Adaptive Behavior 1998; 6:393-410

[23] Snyder MR. A functionally equivalent artificial neural network model of the prey orientation behavior of waterstriders (Gerridae). Ethology 1998; 104:285-297

An Unsupervised Learning Method that Produces Organized Representations from Real Information

Teuvo Kohonen

Neural Networks Research Centre, Helsinki University of Technology
P.O. Box 5400, FIN-02015 HUT, Finland
teuvo.kohonen@hut.fi

Abstract

The neural-network theories aim at two goals in medicine and biology: modeling of the neural structures and functions, and development of computational methods for the analysis of the experimental data. The Self-Organizing Map (SOM) was originally intended for the explanation of certain brain functions and organizations, but it has later been accepted as a new statistical analysis method to many fields of science and technology. At least 3700 scientific works on the SOM have been published. In its basic form, the SOM forms illustrative nonlinear projections of high-dimensional data manifolds, and these projections, usually produced on a two-dimensional display grid, help in the visualization and understanding of the relationships between complex data sets.

1 Introduction

One may coarsely divide the neural-network models into the following categories: 1. Feedforward networks, such as the multilayer Perceptrons. 2. Feedback networks, such as the Hopfield network. 3. Competitive-learning networks.

The *feedforward networks* are meant to transform input representations into reduced and simplified output representations. Sometimes the purpose may be to generate continuous-valued control signals in response to input events, such as in robotics, but the outputs can also be discriminated to elicit active responses at particular output ports selectively. The feedforward neural networks are closely related to the functional expansions of the classical mathematical approximation theory.

In *feedback networks*, the output state ensues from the input state after numerous recursions provided by the abundant feedbacks, whereby the saturating nonlinearities of the transfer relations play a central role. These networks can be embedded into the general theory of nonlinear feedback systems.

Competitive learning in general means that the analyzing system consists of a great number of structurally similar, although parametrically different processing units, all of which study the same input event. A *model* of some observation is associated with each unit. The processing units communicate mutually and compare their responses to the same input. By means of a specific kind of lateral interaction, or by some more straightforward maximum selector mechanism, the processing unit with the highest response can be singled out as the "winner" in the competition, and it suppress the responses of all the other processing units. This system then directly operates as a decoder of input. In this way the simple competitive-learning scheme is formally similar to the classical *vector quantization (VQ)* method [1].

The classifications carried out in any of the previous neural-networks models, however, do not yet reflect any *structures* or *organization* of the input data sets. Nonetheless, the relations between the data samples should somehow be mapped to the medium on which the information is represented, like in the biological brain that the artificial neural networks are supposed to model. Indeed, there exists one category of competitive-learning models with this ability: the *Self-Organizing Map (SOM)*. The objective of the present paper is to describe the SOM and to point out its applications in biology and medicine.

2 The Self-Organizing Map (SOM)

2.1 General

The SOM usually consists of a two-dimensional regular grid of nodes. A model of some observation is associated with each node. The SOM algorithm computes the collection of the models so that it optimally describes the domain of (discrete or continuously distributed) observations. The models are automatically modified such that similar models become closer to each other on the grid than the more dissimilar ones. In this sense the SOM is also a similarity graph, and a clustering diagram. Its computation is a nonparametric, recursive regression process. By far the majority of SOM works use real vectors, similar to the codebook vectors in the classical vector quantization, as the models associated with the nodes. Like in VQ, the model vectors then partition the signal space into Voronoi sets. A Voronoi set is the set of all input vectors that have a common model vector as the nearest neighbor in the input space.

However, it is possible to construct an indefinite number of VQ-type Self-Organizing Maps, by defining different matching criteria between the input and model vectors. The following types of models have been used in SOM architectures: 1. Vectorial models with various distance measures. 2. Linear subspaces defined by sets of basic vectors. 3. Operators and parametrized filters such as the LPC estimator. 4. Symbol strings. 5. Fuzzy set expressions. 6. Genetic-algorithm parameters.

2.2 The original SOM algorithm

Around 1981–82, the original SOM algorithm was derived heuristically, trying to implement a stepwise regression of the models into the distribution of vectorial input data [2]. Let the $\mathbf{x}(t) \in R^n$ denote real vectorial input samples, with $t = 1, 2, \ldots$ the sample index, and let the models be real vectors of the same dimensionality: $\mathbf{m}_i \in R^n$ is the ith model. In a stepwise regression, a subset of the $\mathbf{m}_i$, $i \in N_c$ is updated at each step, whereby the models must be regarded as time-dependent, $\mathbf{m}_i = \mathbf{m}_i(t)$. The subset N_c is determined in such a way that $\mathbf{m}_c(t)$ is the model closest to $\mathbf{x}(t)$ in the Euclidean metric:

$$c = \arg\min_i \{\|\mathbf{x}(t) - \mathbf{m}_i(t)\|\} . \tag{1}$$

The subset N_c consists of all the models around c. The regression step is then written as

$$
\begin{aligned}
\mathbf{m}_i(t+1) &= \mathbf{m}_i(t) + \alpha(t)[\mathbf{x}(t) - \mathbf{m}_i(t)] \text{ for } i \in N_c , \\
\mathbf{m}_i(t+1) &= \mathbf{m}_i(t) \text{ otherwise} ,
\end{aligned} \tag{2}
$$

where $\alpha(t)$ is the scalar-valued "learning gain factor." It is also possible to weight the corrections within N_c as a function of the distance of model i from model c over the grid, whereby (2) can be written

$$
\mathbf{m}_i(t+1) = \mathbf{m}_i(t) + h_{ci}(t)[\mathbf{x}(t) - \mathbf{m}_i(t)] , \tag{3}
$$

with $h_{ci}(t)$ denoting the weighting within N_c: it is customary to call $h_{ci}(t)$ the *neighborhood function*, and it can be regarded as a time-variable smoothing kernel.

The theory of the SOM is very complicated and its discussion must be abandoned here.

3 Applications of the SOM in medicine and biology

At least 3700 scientific publications on the SOM have been written [3]. Its main application areas are: 1. Statistical analysis at large, in particular data mining and knowledge discovery in databases. 2. Analysis and control of industrial processes and machines. 3. New methods in telecommunications, especially optimization of telephone traffic and demodulation of digital signals. 4. Medical and biological applications.

Very coarsely, the medical and biological applications can be divided as follows: 1. Medical imaging. 2. Medical signal analysis. 3. Diagnostic methods. 4. Biochemical analyses. 5. Monitoring of medical equipment. 6. Patient profiling. 7. Modeling of biological functions for their better understanding.

The examples below are real cases but the list may not be exhaustive.

Consider first the *medical imaging* applications. With the SOM, the following problems have been handled: medical image understanding [4], image inspection [5], segmentation of medical images [6], [7], [8], [9], [10], contour estimation in ultrasonic images of the human heart [11], detection of coronary artery disease [12], morphological analysis of muscle biopsies [13], cardio-angiographic sequence coding [14], liver tissue classification [15], foreing object detection [16], classification of brain tumors [17], chromosome location and feature extraction [18], [19], [20], [21], multiparameter image visualization [22], [23], [24], multispectral image visualization [25], and compression of x-ray images [26].

From *medical signal analysis* the following applications can be mentioned: electroencephalography [27], [28], [29], [30], [31], [32], [33], [34], [35], [36], [37], [38], [39], [40], [41], [42], [43], [44], [45], [46], [47], [48], magnetoencephalography [49], [50], [51], electrocardiography [52], [53], [54], [55], [56], [57], [58], [59], electromyography [60], [61], [62], [63], [64], [65], [66], classification of motor cortical discharge patterns [67], [68], [69], sleep cycle recognition [70], [71], analysis of blood pressure time series [72], [73], [74], classification of lung sounds [75], and gait pattern analysis [76], [77], [78].

The automated *diagnosis* could benefit from the artificial neural networks and in particular from the SOM in many ways. Unfortunately the bias in the development of the medical expert systems has been on the traditional artificial intelligence. Nonetheless one can mention the following studies, which have a clear diagnostic objective: general problems in medical diagnosis [79], [80], [81], diagnosis of acute appendicitis [82], [83], differentation of anaemiae [84], diagnosis of chronic hepatopathies [85], glaucomatosis testing [86], [87], and diagnosis of dysphonia and voice quality [88], [89], [90], [91], [92], [93], [94].

Exploration of novice-expert performance is indirectly related to medical diagnoses [95].

Medicine and biology rely on *chemical*, especially *biochemical analyses*. General problems of the use of the SOM in chemistry have been discussed in [96]. Molecular analysis has been performed in [97], [98], [99]. Protein classification by the SOM is a major application field [100], [101], [102], [103], [104], [105], [106], [107], [108], [109], [110], [111], [112]. Functional analysis of nuclei acids has been made in [113]. Other areas are analysis of microbial systems using pyrolysis mass spectrometry [114], [115], [116], [117], analysis of electrophoresis images [118], and analysis of biological crystals [119]. One also has to mention the in vitro studies of chemotherapeutic agents [120] and drug discovery for cancer [121].

There could be plenty of applications of the SOM for the *monitoring of medical equipment*. A study of the instrumentation of an anaesthesia system [122], [123] can be mentioned as an example.

Another natural application area for the SOM would be the *profiling of patients* according to laboratory tests and anamneses in order to direct the patients to proper treatment. This author, however, is only aware of two graduate studies carried out at Helsinki University of Technology.

Finally one might state that originally the "artificial neural networks" stemmed from the early *neural models* intended to explain various brain functions. However, the more realistic biological models must include plenty of details in order to describe true phenomena quantitatively, and therefore they are computationally heavy. Neither are they able to implement much more than rather trivial information-processing functions. One additional aspect is that the biological organisms carry out a substantive amount of information processing by chemical agents, such as chemical modulators and messengers, which are usually neglected in the artificial neural-networks theory.

Notwithstanding the Self-Organizing Map algorithm seems to explain the development of various brain organizations. An overview of the biological implementation of the SOM and the possible role of the chemical agents in it has been published recently [124], [125]. The role of the SOM as a model of cortical organizations and functions has also been discussed in the following contexts: maps in the visual cortex (orientation, ocular dominance, color) [126], [127], [128], [129], [130], [131], [132], [133], [134], [135], [136], [137], [138], [139], [140], facial feature location [141], tonotopic map [142], auditory cortex of the bat [143], classification of auditory brainstem responses [144], representation of external space [145], simulation of the motor cortex [146] and the sensorimotor system [147], peripheral pre-attention [148], Braitenberg vehicles [149] and cortical-map reorganization [150].

Due to the length of the list and lack of space, the following references have been presented in compressed form. The complete publishing data can be found from [3].

References

[1] Gray RM. IEEE ASSP Mag 1984; 1:4–29

[2] Kohonen T. Self-Organizing Maps. Springer-Verlag, 2nd ed 1997

[3] Kaski S, Kangas J, Kohonen T. Bibliography of Self-Organizing Map (SOM) Papers: 1981–1997. Neural Computing Surveys 1998; 1(3&4):1–176. Available in electronic form at http://www.icsi.berkeley.edu/~jagota/NCS/: vol 1, pp 102–350

[4] Tsao ECK et al. Proc 5th Florida AI Res Symp, 1992

[5] Sardy S, Ibrahim L. Opt Eng 1996; 35:2182–7

[6] Ahmed MN, Farag AA. Proc ICNN'97

[7] Bertsch H, Dengler J. DAGM-Symp Mustererkennung, 1987

[8] Dhawan AP, Arata L. Proc ICNN'93

[9] Dhawan AP, Arata L. Computer Meth and Programs in Biomed 1993; 40:203–15

[10] Karpouzas I et al. Cybernetica 1995; 38:195–9

[11] Helbing M et al. 4th Int Workshop on Systems, Signals and Image Processing, 1997

[12] Cios KJ et al. Proc Computers in Cardiology, 1990

[13] Gabriel G et al. Proc IJCNN-93

[14] Joo CH, Choi JS. Trans Korean Inst of Electr Eng 1991; 40:374–381

[15] Pan HL, Chen YC. Patt Rec Lett 1992; 13:355–368

[16] Patel D et al. Proc WCNN'94

[17] Vercauteren L et al. Proc INNC'90

[18] Turner M et al. Image and Vision Computing 1993, 11:235–239

[19] Turner M et al. Proc ICANN'94

[20] Turner M et al. Proc British Machine Vision Assoc Conf, 1992

[21] Turner M et al. In: Artificial Neural Networks 2, 1992

[22] Manduca A. Proc ICNN'94

[23] Manduca A. In: Intelligent Engineering Systems Through Artificial Neural Networks, 1994

[24] Manduca A. Proc ICIP-94

[25] Myklebust G et al. Proc Eighth IEEE Symp on Computer-Based Medical Systems, 1995

[26] Oravec M. Proc 3rd Int Conf on Digital Signal Processing, 1997

[27] Berger A et al. Proc 5th Fuzzy Days, 1997

[28] Dorffner G et al. Proc ICANN'93

[29] Elo P al. In: Artificial Neural Networks 2, 1992

[30] Elo P. Technical Report 1-92, Tampere University of Technology, 1992

[31] Flotzinger D et al. Biomed Tech (Berlin) 1992; 37:303–309

[32] Flotzinger D et al. Proc WCNN'93

[33] Flotzinger D. Proc ICANN'93

[34] Jervis BW et al. In: IEE Colloquium on 'Intelligent Decision Support Systems and Medicine' (Digest No 143), 1992

[35] Jervis BW et al. IEE Proc Science, Measurement and Technology 1994; 141:432–40

[36] Joutsiniemi SL et al. IEEE Trans Biomed Eng 1995; 42:1062–8

[37] Kaski S, Joutsiniemi SL. Proc ICANN'93

[38] Morton PE et al. Proc IEEE 17th Annual Northeast Bioengineering Conf, 1991

[39] Peltoranta M. PhD thesis, Graz University of Technology, 1992

[40] Peltoranta M, Pfurtscheller G. Med & Biol Eng & Comput 1994; 32:189–96

[41] Pfurtscheller G et al. Electroencephal and Clin Neurophys 1992; 82:313–315

[42] Pfurtscheller G et al. Electroencephal and Clin Neurophys 1996; 99:416–25

[43] Pradhan N et al. Computers and Biomed Res 1996; 29:303–13

[44] Pregenzer M et al. In: Proc ESANN'95

[45] Roberts S, Tarassenko L. Proc 2nd Int Conf Artificial Neural Networks (IEE), 1991

[46] Roberts S, Tarassenko L. IEE Colloquium on 'Neurological Signal Processing' (Digest No. 069), 1992

[47] Roberts S, Tarassenko L. IEE Proc F [Radar and Signal Processing] 1992; 139:420–425

[48] Roberts S, Tarassenko L. Med & Biol Eng & Comput 1992; 30:509–517

[49] Portin K. PhD thesis, Helsinki University of Technology, 1998

[50] Portin K et al. Electroencephal and Clin Neurophys 1996; 98:273–80

[51] Portin K et al. Proc XXVII Ann Conf Finnish Physical Society, 1993

[52] Conde T. Proc ICNN'94

[53] Dokur Z et al. Med Eng & Phys 1997; 19:738–41

[54] Hu YH et al. Proc NNSP'95

[55] Ishikawa S et al. Trans Inst Electr, Inf & Comm Eng 1996; J79D-II:1646–9

[56] Koski A. Patt Rec Lett 1996; 17:1215–22

[57] Morabito M et al. Proc Computers in Cardiology, 1991

[58] Presedo J et al. In: Computers in Cardiology, 1996

[59] Reinhardt L et al. Ann of Noninv Electrocardiol 1997; 2:331–337

[60] Abel EW et al. Med Eng & Phys 1996; 18:12–17

[61] Bodruzzaman M et al. Proc WCNN'95

[62] Christodoulou CI, Pattichis CS. Proc 1995 IEEE Int Conf Neural Networks

[63] Graupe D, Liu R. Proc 32nd Midwest Symp on Circuits and Systems, 1990

[64] Pattichis CS et al. IEEE Trans Biomed Eng 1995; 42:486–96

[65] Pattichis CS et al. 1994 IEEE Int Conf on Neural Networks, World Congr on Computational Intelligence, 1994

[66] Schizas CN et al. In: Computer-Based Medical Systems, 1991

[67] Lin S et al. Proc ICANN'95

[68] Lin S et al. Proc SPIE 1996; 2718:540–51

[69] Lin S et al. Neural Computation 1997; 9:607–21

[70] Blanchet M et al. Proc IJCNN-93

[71] Blanchet M et al. Proc 7'th Symp Biological and Physiological Engineering, 1992

[72] Rodriguez MJ et al. Proc Computers in Cardiology, 1993

[73] Rodriguez MJ et al. Proc IWANN '93

[74] Rodriquez MJ et al. Proc WCNN'93

[75] Kallio K et al. In: Artificial Neural Networks, 1991

[76] Köhle M et al. Proc CBMS'97

[77] Köhle M, Merkl D. Proc ESANN'96

[78] Köhle M, Merkl D. Proc ACNN'96

[79] Papadourakis G et al. Math & Comp in Simul 1996; 40:623–35

[80] Schizas CN et al. Proc 1992 Int Biomedical Engineering Days

[81] Taibi G et al. Proc 4th Italian Workshop on Neural Nets, 1992

[82] Pesonen E et al. Int J Bio-Med Comp 1996; 40(3):227–33

[83] Pesonen E et al. Methods Inform Med 1998; 37:59–63

[84] Evans W et al. Proc Int Conf on Neural Networks and Expert Systems in Medicine and Healthcare, 1994

[85] Zaharia CN, Barbu C. 8th European Simulation Symposium, 1996

[86] Liu X, Cheng G, Wu J. Proc ICNN'94

[87] Liu X, Cheng G, Wu JX. Artif Intell in Med 1994; 6:401–15

[88] Leinonen L et al. Scand J Log Phon 1993; 18:159–167

[89] Leinonen L et al. Folia Phoniatrica et Logopaedica 1997; 49:9–20

[90] Leinonen L et al. Suomen Logopedis-Foniatrinen Aikakauslehti 1991; 10(2):4–9

[91] Leinonen L et al. In: Artificial Neural Networks, 1991

[92] Leinonen L et al. J Speech and Hearing Res 1992; 35:287–295

[93] Leinonen L et al. Tekniikka logopediassa ja foniatriassa 1992; (26):41–45

[94] Leinonen L et al. Suomen logopedis-foniatrinen aikakauslehti 1996; 16:89–96

[95] Stevens RH et al. Proc WCNN'95

[96] Gasteiger J, Zupan J. Angewandte Chemie (int ed in English) 1993; 32:503–527

[97] Bienfait B, Gasteiger J. J Molec Graph & Modell 1997; 15:203–215, 254–258

[98] Bienfait B. J Chem Inf & Comp Sci 1994; 34:890–8

[99] Zell A et al. Proc ICNN'94

[100] Andrade MA et al. Biol Cyb 1997; 76:441–50

[101] Andrare MA et al. Protein Eng 1993; 6:383–390

[102] Ferrán EA et al. Proc First Int Conf on Intelligent Systems for Molecular Biology, 1993

[103] Ferrán EA, Ferrara P. In: Artificial Neural Networks, 1991

[104] Ferrán EA, Ferrara P. Int J Neural Networks 1992; 3:221–226

[105] Ferrán EA et al. In: Artificial Neural Networks 2, 1992

[106] Ferrán EA, Pflugfelder B. Comput Appl Biosci 1993; 9:671–680

[107] Ferrán EA, Ferrara P. Biol Cyb 1991; 65:451–458

[108] Ferrán EA, Ferrara P. Comput Appl Biosci 1992; 8:39–44

[109] Ferrán EA, Ferrara P. Physica A 1992; 185:395–401

[110] Merelo JJ et al. Proc IWANN'91

[111] Merelo JJ et al. Proc Neuro-Nimes '91

[112] Merelo JJ et al. Neurocomputing 1994; 6:443–454

[113] Giuliano F et al. Comput Applic Biosci 1993; 9:687–693

[114] Goodacre R. Microbiol Europe 1994; 2:16–22

[115] Goodacre R et al. Zentralblatt für Bakteriologie–Int J Medical Microbiology, Virology, and Parasitology and Infectious Diseases 1996; 284:501–515

[116] Goodacre R et al. J Appl Bacteriology 1994; 76:124–134

[117] Goodacre R et al. Chemom Intell Lab Syst 1996; 34:69–83

[118] Menard F, Fogelman-Soulié F. Proc INNC'90

[119] Fernandez JJ, Carazo JM. Ultramicroscopy 1996; 65:81–93

[120] van Osdol WW et al. Proc WCNN'95

[121] Weinstein JN et al. Proc WCNN'95

[122] Vapola M et al. Proc ICANN'94

[123] Vapola M et al. Proc Conf Artificial Intelligence Research in Finland, 1994

[124] Kohonen T. Proc WCNN'94

[125] Kohonen T, Hari R. Trends Neurosci 1999; 22:135–139

[126] Bauer HU et al. Network: Computation in Neural Systems 1997; 8:17–33

[127] Bauer HU et al. Proc WSOM'97

[128] Bauer HU. Proc ICANN'94

[129] Bauer HU. Neural Computation 1995; 7:36–50

[130] Obermayer K et al. In: Artificial Neural Networks, 1991

[131] Obermayer K et al. IEICE Trans Fund Electr Comm Comp Sci 1992; E75-A:537–545

[132] Obermayer K. Proc Conf Prerational Intelligence, 1993

[133] Obermayer K et al. Phys Rev A [Statist Phys, Plasmas, Fluids & Related Interdisc Topics] 1992; 45:7568–7589

[134] Obermayer K et al. Proc Natl Acad Sci USA 1990; 87:8345–8349

[135] Obermayer K et al. Proc IJCNN-90

[136] Obermayer K et al In: Advances in Neural Information Processing Systems 3, 1991

[137] Obermayer K et al. In: Advances in Neural Information Processing Systems 4, 1992

[138] Obermayer K. Adaptive neuronale Netze und ihre Anwendung als Modelle der Entwicklung kortikaler Karten. Infix Verlag, Sankt Augustin, 1993

[139] Pomierski T et al. Proc ICANN'93

[140] Saarinen J, Kohonen T. Perception 1985; 14:711–719

[141] Luckman AJ, Allinson NM. In: Visual Search, 1992

[142] Kita H, Nishikawa Y. Proc WCNN'93

[143] Martinetz T et al. In: Connectionism in Perspective, 1989

[144] Grönfors T. Proc Conf Artificial Intelligence Research in Finland, 1994

[145] Morasso P, Sanguineti V. Proc ICANN'94

[146] Glaria-Bengoechea A, Burnod Y. In: Artificial Neural Networks, 1991

[147] Saxon JB. Master's thesis, Texas A&M University, 1991

[148] Luckman AJ, Allinson M. Proc Soc Photo-opt Instr Eng 1990; 1197:98–108

[149] Werkowitz EB. Master's thesis, Air Force Inst of Tech, 1991

[150] Sutton III GG et al. Neural Computation 1994; 6:1–13

On Forgetful Attractor Network Memories

Anders Lansner and Anders Sandberg

Numerical Analysis and Computer Science, Royal Inst. of Technology

Stockholm, Sweden

ala@nada.kth.se, asa@nada.kth.se

Karl Magnus Petersson and Martin Ingvar

Clinical Neuroscience, Karolinska Hospital

Stockholm, Sweden

karlmp@neuro.ks.se, martin@ingvar.com

Abstract

A recurrently connected attractor neural network with a Hebbian learning rule is currently our best ANN analogy for a piece cortex. Functionally biological memory operates on a spectrum of time scales with regard to induction and retention, and it is modulated in complex ways by sub-cortical neuro-modulatory systems. Moreover, biological memory networks are commonly believed to be highly distributed and engage many co-operating cortical areas.

Here we focus on the temporal aspects of induction and retention of memory in a connectionist type attractor memory model of a piece of cortex. A continuous time, forgetful Bayesian-Hebbian learning rule is described and compared to the characteristics of LTP and LTD seen experimentally. More generally, an attractor network implementing this learning rule can operate as a long-term, intermediate-term, or short-term memory. Modulation of the print-now signal of the learning rule replicates some experimental memory phenomena, like e.g. the von Restorff effect.

1   Introduction

Recurrent attractor neural networks, like for example the early binary associative memories and the more recent Hopfield net, have been proposed as models for biological associative memory [1,2]. They can be regarded as possible formalizations of Donald Hebb's original ideas of synaptic plasticity and cortical cell assemblies [3]. A number of psychological memory and Gestalt perception phenomena have been qualitatively reproduced based on such network models [4,5]. Their biological plausibility is supported by results from simulations with biophysically detailed cortical pyramidal cells and inhibitory interneurons [6,7].

The recurrent connectivity can certainly be found in the cortex. The cortical horizontal fiber systems extend over large distances and are responsible for up to 80 % of the synapses on some cortical pyramidal cells [8]. These synapses are usually situated on spines and are likely to be plastic [9] and involved in learning.

Functionally cortical memory operates on a spectrum of time scales with regard to induction (episodic–semantic) and retention (long-term memory–working memory), and it is modulated in complex ways by sub-cortical neuromodulatory systems [10]. Moreover, the underlying memory networks are commonly believed to be highly distributed and engage many co-operating cortical areas [11].

Here we focus on the temporal aspects of induction and retention of memory in a connectionist type attractor model of cortex. A continuous time, forgetful Bayesian-Hebbian learning rule is described and related to biological synaptic plasticity. Further, the functional consequences of using such a learning rule in an attractor network are examined.

2 Learning rule and network model

The Bayesian-Hebbian learning rule reinforces connections between simultaneously active units and weakens or makes connections inhibitory between anti-correlated units. It is based on a probabilistic view of learning and retrieval, with input and output unit activities representing confidence of feature detection and posterior probabilities of outcomes, respectively [12-14]. The resulting weight matrix is typically symmetric, allowing for fixed point attractor dynamics in a recurrent network.

The input space of the network is a vector of random variables

$$\mathbf{X} = (X_1, X_2, \dots X_N)$$

which generates pattern vectors $\mathbf{x}$. Each variable X_i can take on a set of M_i different values. Thus x_i will be composed of M_i binary component attributes $x_{ii'}$ (the i':th possible state of the i:th attribute) with a normalized total probability of one. Such a group of component attributes covering the input space is called a hypercolumn [13] in analogy with cortical hypercolumns.

Given a certain input $\mathbf{x}$ we want to estimate the probability of a class or set of attributes y, which can be divided into individual events $y_{jj'}$ in the same way as $\mathbf{x}$. If we condition on $\mathbf{X}$ (where unknown attributes retain their prior distribution) and assume the attributes x_i to be both independent and conditionally independent we get:

$$P(y_{jj'}|\mathbf{X}) = P(y_{jj'}) \prod_i^N \frac{P(y_{jj'}|X_{ii'})}{P(y_{jj'})}$$

$$= P(y_{jj'}) \prod_i^N \frac{\sum_{i'}^{M_i} P(y_{jj'}|x_{ii'}) P(x_{ii'}|X_i)}{P(y_{jj'})}$$

$$= P(y_{jj'}) \prod_i^N \left(\sum_{i'}^{M_i} \frac{P(y_{jj'}, x_{ii'})}{P(y_{jj'}) P(x_{ii'})} o_{ii'} \right)$$

where $o_{ii'}$ represents $P(x_{ii'}|X_i)$.

If **X** represents known or estimated information, we want to create a neural network which calculates p(**y**), from the given information. If we take the logarithm of the above formula we get:

$$\log\left(P(y_{jj'}\big|\mathbf{X})\right) = \log\left(P(y_{jj'})\right) + \sum_i^N \log\left(\sum_i^{M_i} \frac{P(y_{jj'}, x_{ii'})}{P(y_{jj'})P(x_{ii'})} o_{ii'}\right)$$

which represents a kind of π-sigma neural network. Here we identify weight and bias, and the activation function.

$$\beta_{jj'} = \log\left(P(jj') + p_0\right)$$

$$w_{ii',jj'} = \frac{P(ii', jj') + p_{00}}{\left(P(ii') + p_0\right)\left(P(jj') + p_0\right)}$$

$$f(h_{ii'}) = e^{h_{ii'}}$$

The parameter p_0 represents the smallest probability measurable and prevents taking the logarithm of zero. The value of p_{00} is typically p_0^2 but may have a different value.

Since the independence assumption is often only approximately fulfilled and we deal with approximations of probability, it is necessary to normalize to one the output within each hypercolumn.

2.1 The incremental learning rule

The standard correlation based learning rule for attractor ANN suffer from catastrophic forgetting, that is, all memories are lost as the system gets overloaded. To cope with this situation, we have introduced an incremental, continuous time version of this learning rule [15].

Here, P reflects estimates of activation rates of the pre- (P_i) and postsynaptic (P_j) unit respectively and their co-activation (P_{ij}). In the previous model these were estimated using hard counters, but the incremental version instead uses leaky counters. A pre- and a post-synaptic variable Z has been introduced to allow a softer manner of estimating the temporal aspects of activation:

$$Z_i(t) = Z_i(t-1) + (S_i - Z_i(t-1))\frac{dt}{\tau_{Z_i}}$$

$$Z_j(t) = Z_j(t-1) + (S_j - Z_j(t-1))\frac{dt}{\tau_{Z_j}}$$

S_i and S_j are the activations of the pre- and postsynaptic units given by the input (in terms of mean firing frequency). To allow for a delayed print-now signal (see below) to be effective another state variable, E, has also been included. It acts similar to the eligibility trace in reinforcement learning or as a synaptic marker:

$$E_i(t) = E_i(t-1) + (Z_i(t) - E_i(t-1))\frac{dt}{\tau_E}$$

$$E_j(t) = E_j(t-1) + (Z_j(t) - E_j(t-1))\frac{dt}{\tau_E}$$

$$E_{ij}(t) = E_{ij}(t-1) + (Z_i(t)Z_j(t) - E_{ij}(t-1))\frac{dt}{\tau_E}$$

The rates of activation P are then estimated from the E variable. In this step there is a print-now signal (κ) which regulates the degree to which the event is imprinted.

$$\hat{P}_i(t) = \hat{P}_i(t-1) + (E_i(t) - \hat{P}_i(t-1))\kappa\frac{dt}{\tau_P}$$

$$\hat{P}_j(t) = \hat{P}_j(t-1) + (E_j(t) - \hat{P}_j(t-1))\kappa\frac{dt}{\tau_P}$$

$$\hat{P}_{ij}(t) = \hat{P}_{ij}(t-1) + (E_{ij}(t) - \hat{P}_{ij}(t-1))\kappa\frac{dt}{\tau_P}$$

As can be seen, all variables implement running averages with their specific time constants. Finally, the time-dependent network weights and biases are calculated as above, but using the running average estimated P-values.

3   Results

This incremental learning rule have been studied from a number of different perspectives. First its functional correspondence to experimentally observed synaptic plasticity has been evaluated. Further, the properties of an attractor network with such a forgetful learning rule has been examined. Finally, we have looked more closely into the effect of modulating the print-now signal in this network model. In the following the most important results from these studies are summarized.

58

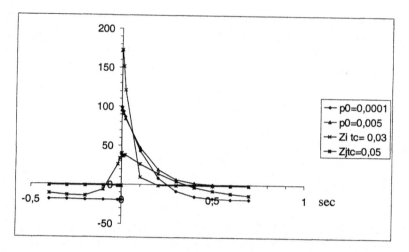

Figure 1: The dependency of sign and magnitude of the log of the synaptic weight change on the time separation between pre- and postsynaptic activation. The effect of varying the parameters p_0 and τ_z is illustrated.

3.1 The Hebbian-Bayesian learning rule and LTP

Recent studies using paired recordings from single cells, have shown quantitatively how timing between pre- and postsynaptic spike is crucial for the the type and magnitude of the synaptic plasticity [16]. We have demonstrated that our learning rule, originally derived from a computational statistical inference perspective, reproduces many aspects of synaptic plasticity like LTP and homo- ("pre not post") and hetero-synaptic ("post not pre") LTD [17].

In the simulations described here, the pre- and postsynaptic units were stimulated by trains of short single spikes (10 ms wide) with a frequency of 1 Hz and a time difference that was adjusted. Figure 1 shows one result from this study, i.e. the dependency of sign and magnitude of synaptic weight change on the time separation between pre- and postsynaptic activation. The shapes of these curves with the parameters given are in several respects semi-quantitatively similar to e.g. experimental data on LTP/LTD from Bi and Poo [16].

3.2 Attractor memory with forgetful learning

The incremental learning rule described above has been introduced in a recurrent attractor network. The learning and forgetting dynamics were studied and evaluated [15]. In these simulation experiments the network was first trained by the presentation of the training patterns, followed by a testing period where no learning took place.

A comparison between a standard network using a hard counter type learning rule and one using the incremental version demonstrates that the latter avoids catastrophic forgetting by forgetting the oldest patterns while recently learned patterns remain accessible. The forgetting does not occur immediately: for longer learning time constants the pattern is stored well until a certain number of interfering patterns have been stored, when it starts to gradually fade.

Further, as more patterns are learned, the basins of attraction of old patterns become smaller and shallower, eventually disappearing altogether. This results in a change in convergence speed such that younger patterns are completed faster than older ones, and average convergence time increases linearly as more memories are stored.

When stimulated with noise (randomly activated units with the same density as the patterns) the network typically converges to one of the learned patterns. The probability of convergence towards a certain memory decreases with its age.

3.3 Modulating the print-now signal

Endogenous processes activated by experience can modulate memory strength in terms of recall probability [18]. Hormones and neuromodulators can affect how strongly memories are retained [19]. The von Restorff effect shows up as an improved memory of a distinctly different item (the isolate), while the other items in the set are less well retained (retroactive and proactive inhibition/interference) [20-21].

Some of these factors can be interpreted in the framework of memory consolidation as a relevance modulation of the "print-now" signal by regulating memory encoding and synaptic plasticity. Such a mechanism relates closely to neuromodulation in the brain, e.g. the effect of dopamine [22] and acetylcholine in synaptic plasticity [23]. By modulating the time constant of our incremental Bayesian-Hebbian learning rule (κ) we can model the effect of such modulation of the print-now signal on encoding of information into, for example, long-term memory.

We trained a 100 unit attractor network with 10 hypercolumns of 10 units in each. Retrieval was tested by activating the units with a trained pattern in which the activity in three hypercolumns out of ten had been randomized.

Figure 2 (left) shows the mean overlap over all patterns between retrieved and memorized pattern as a function of the learning time constant τ_p. At long τ_p the network learns too slowly to learn the patterns, while at short τ_p the network learns quickly but forgets the oldest patterns. Figure 2 (right) demonstrates that the recall probability under noise input conditions can be affected by the print-now signal, such that an older pattern can take precedence over more recent ones.

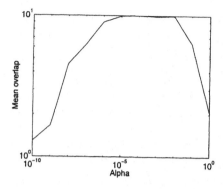 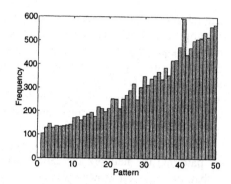

Figure 2, left: Mean overlap after convergence from noisy cue as a function of the inverse learning time constant (α). Right: Frequency of attractors with medium stimulation and plastic network. Pattern 41 is the isolate (κ double that for the other patterns).

4 Discussion and Conclusions

The incremental Bayesian-Hebbian learning rule proposed and studied here has several interesting properties. It mimics functionally important aspects of synaptic plasticity and, at the same time, when introduced into an attractor network it reproduces systems level phenomena seen in human memory research. These results add further support to the attractor network model for cortical associative memory. Yet, many important aspects of cortical memory processes are still lacking in this simple model like, for example, temporal sequence processing capability, and self-organizing feature extraction processes. Moreover, we have only considered one memory network in isolation, while biological memory systems comprise many such interacting components, each one specialized for its particular contribution to overall system performance.

Acknowledgements

Support for this work from the Swedish Research Council for Engineering Sciences (TFR, grant No. 97274) and the Swedish Medical Research Council (MFR, grant No. 12716) is gratefully acknowledged.

References

[1] Willshaw D, Buneman O, and Longuet-Higgins H. Holographic associative memory. Nature 222 (1969), 960, 1969.

[2] Hopfield J J. Neural Networks and physical systems with emergent collective computational properties, Proc. Natl. Acad. Sci. USA, 81(3088-3092), 1982.

[3] Hebb D O. The Organization of Behavior. John Wiley Inc., New York, 1949.

[4]   Freeman W. J. The physiology of perception. Scientific American 264, 78-85, 1991.

[5]   Quinlan P. Connectionism and Psychology. A Psychological Perspective on Connectionist Research. Harvester, Wheatsheaf, New York, 1991.

[6]   Haberly L B and Bower J M. Olfactory cortex: model circuit for study of associative memory? Trends Neurosci 12, 258-64, 1989.

[7]   Fransén E and Lansner A. A model of cortical associative memory based on a horizontal network of connceted columns. Network 9, 235-264, 1998.

[8]   LeVay S, and Gilbert C D. Laminar patterns of geniculocortical projections in the cat, Brain Res, 113, 1-19 , 1976.

[9]   Artola A, Bröcher S, and Singer W. Different voltage-dependent thresholds for the induction of long-term depression and long-term potentiation in slices of the rat visual cortex, Nature (London), 347, 69-72, 1990.

[10] MeGaugh J M. Memory – a century of consolidation. Science, 287, 248-251, 2000.

[11] Fletcher P C, Frith C D, and Rugg M D. The functional neuroanatomy of episodic memory. TINS 20, 213-18, 1997.

[12] Lansner A and Ekeberg Ö. A one-layer feedback artificial neural network with a Bayesian learning rule", Int. J Neural Systems, 1, 77-87, 1989.

[13] Lansner A and Holst A. A higher order Bayesian neural network with spiking units. Int J Neural Systems 7, 115-128, 1996.

[14] Holst A. The Use of a Bayesian Neural Network Model for Classification Tasks. PhD thesis, Dept of Numerical Analysis and Computing Science, Royal Institute of Technology, Stockholm, Sweden, TRITA-NA-P9708, 1997.

[15] Sandberg A, Lansner A, Petterson K-M, and Ekeberg Ö. An incremental Bayesian learning rule. Tech rep TRITA-NA-P9908, Royal Institute of Technology, NADA, 1999.

[16] Bi G-Q and Poo M-M. Synaptic Modifications in Cultured Hippocampal Neurons: Dependence on Spike Timing, Synaptic Strength, and Postsynaptic Cell Type. J Neurosci 18, 10464-72, 1998.

[17] Stanton P K. LTD, LTP, and the sliding threshold for long-term synaptic plasticity. Hippocampus 6, 35-42, 1996.

[18] Christianson S-A, editor. Handbook of Emotion and Memory: Current Research and Theory. Erlbaum, Hillsdale, NJ, 1992.

[19] Martinez J L, Schulteis G, and Weinberger S B. Learning and Memory: A Biological View, chapter 4, pages 149-198. Academic Press, second edition, 1991.

[20] von Restorff H:. Analyse von vorgängen in Spurenfeld. Psychologiche Forschung 18, 299-342, 1933.

[21] Parker A, Wilding E, and Akerman C. The von Restorff effect in visual object recognition memory in humans and monkeys: The role of frontal/perirhinal interaction. J Cognitive Neurosci, 10, 691-703, 1998.

[22] Wickens J and Kötter R. Cellular models of reinforcement. In Houk J C, Davis J L, and Beiser D Q, editors, Models of Information Processing in the Basal Ganglia, 187-214. MIT Press, 1995.

[23] Woolf N J. The critical role of cholinergic basal forebrain neurons in morphological change and memory encoding: a hypothesis. Neurobiol Learn Mem, 66, 258-66, 1996.

Outstanding Issues for Clinical Decision Support with Neural Networks

P.J.G. Lisboa, A. Vellido and H. Wong

School of Computing and Mathematical Sciences
Liverpool, UK
p.j.lisboa@livjm.ac.uk

Abstract

Neural networks are widely used in potential medical applications, going as far as their introduction into commercial products. However, these pilot studies often ignore important aspects for clinical decision support. The context for the development of new medical technology is introduced, outstanding issues of principle for clinical support are identified, and their technical implications for neural network methodology are discussed.

1 Introduction

Medical patents citing neural networks are being granted at an increasing rate, at least in the largest market for medical devices, the US. Figure 1 shows the application of Fisher and Pry's [1] model of technological innovation, which predicts an exponential expansion. It is clear that computational intelligence in patented devices has a strong component of neural networks, usually in combination with other artificial intelligence methods such as fuzzy logic and rule-based systems.

Despite the rush for patents, there are currently only a small number of commercial clinical products on the market, quite out of step with the plethora of pilot studies in almost every aspect of medical research. To what extent do these studies herald a transfer of current decision support methodologies to neural networks? In order to put this question into context, it is worth digressing briefly to the general context for innovation in medical care.

Success in medical innovation is determined by what makes financial, not just technological sense. Substantial improvements may result from improvements in clinical expertise, medical equipment or drugs. For instance, the Imperial Cancer Research Fund in the UK estimated in 1996 that an 10% increase in survival from cancer could be achieved if the consistency of care services (diagnosis, prognosis and therapy) could be uniformly raised to best practice. This amounts to 16,000 lives saved each year, in the UK alone. The reality of care in every country raises very considerable ethical questions which arise from the commodification of healthcare [2], as the economic drivers are an unavoidable reality which embraces us all [3]. If novel medical devices are to make an impact on care, they will have to directly address these pressures.

64

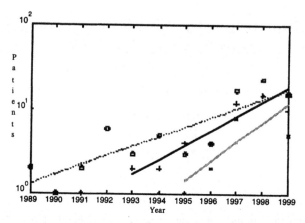

Figure 1: Medically related U.S. patents with artificial intelligence. Since 1999, 15 patents have cited neural networks as a substantial system component, and 5 of them are hybrids with expert systems.

The positive side of market forces is that there is a concerted effort to harmonise standards across the US and EU, an example of which is the Mutual Recognition Agreement signed in 1998, and expected to be fully implemented in 2001. In Europe, the Medical Devices Directives require all equipment 'placed on the market' after 30 June 2001 to carry a *CE* marking. It is likely that any form of automated decision support will qualify the device as medium or high risk, requiring assessment by a Notified Body, a design assessment and a type test. The only exceptions to this will be 'custom-made devices and devices intended for clinical investigation, which are not sold for profit'. Regulatory requirements oblige the design assessment for any novel methodology to involve benchmarking, against an appropriate 'substantially equivalent' methodology. In the case of neural networks, this could be compared with multivariate linear statistics but, among the many MEDLINE citations directly relevant to neural networks in medical care, less than one-half include a detailed performance comparison against an optimised alternative method of data analysis. New products will also have to make a fresh technical case for *best practice* in the neural network design, and carry out a detailed risk analysis. Key elements in this include the detection of *novel inputs*, which may require extrapolation from the domain of validity of the model, and a robust *validation* procedure, to determine the size of the clinical trial required to support of the product's claims – especially important as the cost of the trials can dwarf the technical costs of product development.

2 The role of decision support in medicine

Computers are everywhere present in the clinical environment, and their role is paramount. In Magnetic Resonance Imaging, for instance, the image contrast is deliberately adjusted to highlight what the radiologist considers to be particular metabolites of interest, which is just an example of how subjectivity is introduced into ostensibly objective clinical measurements. Given that the clinician already interfaces with a digitally re-constructed best estimate of the original physiological measurement, it is tempting to take the next step and use computers to extract higher-level information directly from the data. However, the transition from data collection to decision support is substantial, both in its practical implications for patient care, and in the nature of the necessary supporting technology.

There is clearly an essential role in signal processing, which is fast developing beyond Fourier methods and wavelets, with the introduction of Independent Components Analysis [4]. But what about the traditional domains of artificial intelligence, in diagnosis, prognosis and treatment advice – is there a role for neural networks in these areas?

Computer Assisted Decision Support in medicine has at least the role of enhancing the consistency of care. Secondly, it has the potential to cover rare conditions, since no clinical expert can be expected to possess encyclopaedic knowledge of all of the exceptional manifestations of diseases, even within a specialist domain. Thirdly, the expanding range of patient information that is made available in electronic form, makes it feasible to more accurately quantify important clinical indicators, such as the relative likelihood for competing diagnoses [5-7], or the clinical outcome [8-10]. Accurate quantification here may require *calibration for skewed data.*

In some cases, computer-assisted diagnoses have been claimed to be even more accurate than those by clinicians. Nevertheless, these systems have not been accepted for routine clinical use [11]. If artificial intelligence shows any signs of having 'come of age in the 90s', Papnet must figure highly, on the evidence from the large scale clinical trials using it. This system has been shown to enable the reliable detection of cancerous cells or epithelial fragments in slides which were scored negatively during routine cytological processing [12]. Given the critical importance of this test for the patient, there appears to be a solid case for introducing an element of automation into the screening process. This is arguably supported by a study that estimated a standard measure of cost effectiveness, the cost per life-year saved [13]. The addition of Papnet rescreening into a routine screening protocol on a biennial basis was costed, on 1994 prices, at $48,474 per life-year saved, against $67,918 estimated for annual mammography screening of women aged 40-49, and $113,000 for one test for Prostate Specific Antigen. Yet, even at a modest additional cost per test to the patient, cost-benefit and other considerations have caused a protracted wait to capture a sizeable proportion of the market, eventually leading the original developer Inc. into a take-over by a rival company.

However, does this really mean that neural network methods are any more promising in their practical outlook than symbolic logic? After all, high-level clinical reasoning is expected to follow predicate logic, and rely strongly on physiological models, both of which are at opposite ends of the epistemological spectrum from pure phenomenology. What are the desirable attributes that clinicians expect to see in practical decision support systems?

Learning from experience is a reasonable expectation to make of any intelligent system, quite apart from the cost and the logistics of maintaining and regularly updating automated decision support tools. Another common expectation is *model interpretation*, which can be critical to identify failure modes in phenomenological models. At a more practical level, medical data are rarely routinely acquired specially for data-based modelling, consequently they are often incomplete, although sufficient to enable the clinician to arrive at firm diagnosis on the basis of prior knowledge. An example from breast cancer survival modelling is the recording of metastic lymph nodes. If too many are observed, then parts of the remaining information may not be entered because the prognosis is then known to be poor, resulting in a record with *missing data*. Worse still, the surgeon can only excise some of the axillar nodes, resulting in substantial *input noise* in a highly predictive measurement.

3 Technical issues of neural network methodology

The previous discussion about the potential role for neural networks in medical decision support, and the regulatory framework within which they will operate, sets out a list of useful technical aspects for further consideration.

3.1 Best practice in neural network design

Neural network models must be regularised to control overfitting. This is an essential condition for reliable predictions, but the way to achieve it is open to discussion [14]. A popular regularisation scheme is to embed the neural network model, whether predictive or self-organising, within a Bayesian statistical framework [15], which can be extended to Automatic Relevance Determination (ARD). However, the variables selected are generally not stable [6], therefore this is an area in need of further research [16].

An important consideration in any decision model is the provision of a reject option, to better qualify data near the decision boundary, as well as outliers. In Bayesian models, this involves integrating over the distributions of model parameters, which is very time consuming. An expeditious approximation to this integral is the marginalisation of the network outputs, which moderates them towards the guessing line when the model parameter distributions have large variance.

3.2 Calibration for skewed data

In binary classification problems, calibration is the accuracy of the estimation of the conditional. It can be empirically measured simply by grouping the data according to the value of the network output, and measuring the proportion in each class. However, in medical applications data are frequently acquired by convenience sampling, rather than through detailed experimental design, which leads to skewed prior distributions, generally reflecting the relative prevalence of the classes.

Neural networks are expected to be well calibrated, but the Bayesian framework will always marginalised to midrange. In the case where the data are skewed, and where an estimate of the prior probability of each class is available, this can be used to marginalise the Bayesian network output to the correct guessing line, which is the value of the prior. An example of this is the diabetes data set described in [14], where the ratio of diabetics to controls is 2:1. Figure 2 illustrates the calibration curves for neural network models trained with error weighting [17] to equalise the both class priors, followed by Bayesian compensation to the assumed prior, which is taken to be the empirical prevalence of each class.

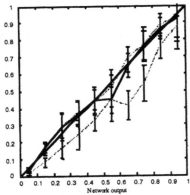

Figure 2: Calibration curve for the diabetes data, averaging over 20 cost weighted ARD networks, for balanced data (off the calibration line), with Bayesian compensation to the prior (solid line), and after marginalisation (short dash).

In survival modelling, accurate calibration is crucial since the network's estimates of the hazard in each time interval are multiplied successively to obtain the survival curve [18]. Since a monthly study with 5-year follow-up requires at least 60 time intervals, that the hazard per time interval is very low (see Figure 3), causing even small deviations from the correct calibration to build up to an unacceptable bias in the survival estimates. This is illustrated in Figure 4.

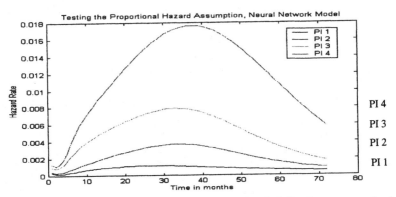

Figure 3: Neural network estimates of the hazard per month following surgery for breast cancer. The patients are grouped using a prognostic index (PI).

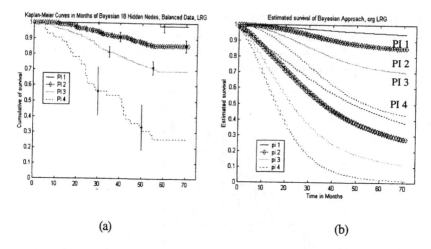

(a)   (b)

Figure 4: Survival curve for the patients in Figure 3, grouped according to a neural network derived prognostic index: a) Kaplan-Meier curves for the same patient groups and, b) network estimates with appropriate calibration (top four curves) and marginalising the hazard to 0.5 (lower four curves).

3.3 Network validation

Almost all medical applications reported in the literature either split the data into training and validation sets, or estimate performance with cross-validation. However, these methods are notoriously unreliable for estimating the misclassification error for future data. Worse still, there are no 'power calculations' for neural networks. These are central to medical statistics, but quite avoided in neural network research. Misclassification error can be parameterised using the bootstrap, and the FDA is researching this method as an estimator of the number of samples required for training, testing and validation, to demonstrate the network's performance claims, to a known a given level of confidence [19].

An additional issue here is that size of the available sample has a critical effect on the number of parameters which can be meaningfully included in the model, i.e. for which the model parameters can be reliably estimated. This, in turn, limits the maximum achievable separation. All too often, claimed performance differences lack any statistical significance, given the number of samples available. A rule of thumb to estimate the standard error of the misclassification rate is to use the binomial distribution with the expected classification accuracy. A better estimate of the significance of the difference between classifiers, and one that is very robust to distributional assumptions, can be derived from the area under the respective Receiver Operating Characteristic curves [20].

3.4 Novelty detection

Any class assignment model will gain robustness by suppressing the effect of the so-called nuisance variables, in favour of the explanatory variables with the most predictive power. This is as true of neural network models as it is of statistics, fuzzy logic, induction trees or even rule-based reasoning systems. In all cases, the model is vulnerable to making spurious predictions when presented with novel data for which the key variables have been removed from the model. It is possible to use Self-Organising Maps and ART networks for that purpose, but this is seldom done.

3.5 Incremental learning

Databases grow over time, but neural networks commonly do not. Again, ART networks form dynamic models which directly address the stability-plasticity dilemma, but their potential for medical modelling is largely untapped.

3.6 Data integrity

Clinical data commonly contain missing or mislabelled data. This may arise because the standard acquisition procedures are intended for diagnostic or prognostic purposes, and not specifically for database modelling, but also when the 'gold standard' on which the study relies is itself not fully reliable. It may even be the case that conflicts have to be resolved between expert opinion, before class labels can be established, a process which may require explicit statistical modelling [21]. Furthermore, missing data are sometimes filled-in expeditiously using feedforward models, but should more accurately be modelled to generate a self-consistent joint distribution for the present and missing data. Although this issue may be tacked within the context of Gaussian Processes, it poses severe difficulties, and is the subject of active research.

3.7 Input noise

Statistical interpretations of the neural network cost functions explicitly assume zero mean additive noise with constant variance, at the outputs [14,15], but ignore the fact that significant noise may be present in the clinical measurements themselves. This has only recently been taken-up in research [22].

3.8 Model interpretation and visualisation

Model interpretation usually requires a drastic reduction in the dimensionality of the data. Data-based methods for visualisation abound, including linear and non-linear latent variable models, unsupervised neural networks, and even the large literature on linear visualisation methods including Sammon's mapping, Fisher's Linear Discriminant, and other variants. These can be further explored in the context of neural network research [23].

It is worth noting that research trends in knowledge-based artificial intelligence, statistics and neural networks appear to be converging towards the statistical estimation of graphical models. This allows prior knowledge to be tested within a rigorous statistical framework, with heavy reliance on numerical simulation to estimate the model parameters [24], and presents a viable alternative to pure reliance on data, which is already being exploited in healthcare.

4 Conclusion

Neural networks are already replacing linear methods for advanced signal processing in image analysis, physiological measurement, and control of functional electrical stimulation. They are also being applied extensively as predictors for diagnosis, prognosis and therapy but, in most published studies, the results are not yet sufficient to confidently map-out the clinical potential of neural networks.

This paper has identified research directions in need of further research. Best practice in neural network applications includes meaningful benchmarking, accurate performance estimation, and a detailed consideration of the potential practical clinical aspects of the model. In the light of the global healthcare context, these are not optional extras, but must be included from the outset, if neural network methods are to become established as reliable and practical tools for clinical and laboratory decision support.

Acknowledgements

The authors wish to thank Ric Swindell from Christie Hospital, Manchester, for making the breast cancer data base available for this study.

References

[1] Fisher, J.C. and Pry, R.H. A simple model of technological change. Technol. Forecat. Soc. Change, 1971; 3:75-81.

[2] Kevin Wildes, S.J. More questions than answers; the commodification of health care. J. Med. and Philos. 1999, 24(3): 307-311.

[3] Khushf, G. A radical rapture in the paradigm of modern medicine: conflicts of interest, fiduciary obligations, and the scientific ideal. J. Med. and Philos. 1998, 23(1): 98-122.

[4] Cardoso, J.-F. Higher-order contrasts for Independent Components Analysis. Neur. Comp. 1999, 11: 157-192.

[5] Mango, L.J. and Valente, P.T. Neural-network-assisted nalysis and microscopi recreening in presumed negative cervical cytologic smears – a comparison. Acta Cytol. 1998, 42(1):227-232.

[6] Dreiseitl, S., Ohno-Machado, L. and Vinterbo, S. Evaluating variable selection methods for diagnosis of myocardial infarction. Proc. AMIA Symp. 1999, (1-2):246-250.

[7] Lisboa, P.J.G., Ifeachor, E.C. and Szczepaniak, P.S. (eds.) Artificial neural networks in biomedicine. Springer-Verlag, London, 2000.

[8] Wong, L.S. and Young, J.D. A comparison of ICU mortality prediction using the APACHE II scoring system and artificial neural networks. Anaesth., 45(11): 1048-54.

[9] Alonso-Betanzos A., *et al.* Applying statistical, uncertainty-based and connectionist approaches to the prediction of fetal outcome: a comparative study. Artif Intell Med 1999, 17(1):37-57.

[10] Mattfeld, T. Kestler, H.A., Hautman, R. and Gottfried, H.W. Predictio of prostatic cancer progression after radical prostectomy using artificial neural networks: a feasibility study. BJU Int. 1999, 84(3): 316-232.

[11] Shortliffe E H. The adolescence of AI in medicine: will the field come of age in the '90s? Art. Intel. Med 1993;. 5: 93-106.

[12] Mango, L.J., Tjon, R., and Herriman, J.M. An automated system for detecting missed pap smears. In: Proc. World Cong, on Neur. Netw., Wahington, D.C., II:880-883, 1996.

[13] Schechter, C.B. Cost-effectiveness of rescreening conventionally prepared cervical smears by Papnet testing. Acta Cyt. 1996, 40(6)

[14] Ripley, B.D. Pattern recognition and neural networks. Cambridge University Press, Cambridge, 1996.

[15] Bishop, C.M. Neural networks for pattern recognition. Oxford University Press, Oxford, 1995.

[16] Van de Laar, P. and Heskes, T. Pruning using parameter and neuronal metrics, Neur. Comp. 1999, 11: 977-933.

[17] Lowe, D. and Webb, A.R. Optimized feature extraction and the Bayes decision in feed-forward classifier networks. IEEE-PAMI, 13(4):355-364.

[18] Lisboa, P.J.G., Wong, H., Vellido, A., Kirby, S.P.J., Harris, P. and Swindell, R. 'Survival of Breast Cancer Patients Following Surgery: a Detailed Assessment of the Multi-Layer Perceptron and Cox's Proportional Hazard Model', IJCNN, 1998: 112-116.

[19] Chan, H.P., Sahiner, B., Wagner, R.F. and Petrick, N. Classifier design for computer-aided diagnosis: effects of finite sample size on the mean performance of classical and neural network classifiers. Med. Phys. 1999, 26(12):2654-2668.

[20] Hanley, J.A. and McNeil, B.J. The meaning and use of the area under the reciever operating characteristic (ROC) curve. Radiology 1982, 143:29-36.

[21] Hirsch, S., Shapiro, J. and Frank, P. Use of an artificial neural network in estimating prevalience and assessing underdiagnosis of asthma. Neur. Comp. App. 1997, 5(2): 124-128.

[22] Bishop, C.M. and Cazhaow, S.Q. Regression with input-dependent noise: a Bayesian treatment. *In* Adv. Neur. Inf. Proc. Syst. (NIPS) 9, 1997.

[23] Agogino, A., Ghosh, J., Perantonis, S.J., Virvilis, V. , Petridis, S. and Lisboa, P.J.G. The role of multiple, linear-projection based visualisation techniques in RBF-based classification of high-dimensional data. *To appear in* IJCNN 2000.

[24] Speigelhalter, D.J. Bayesian graphical modelling: a case-study in monitoring health outcomes. App. Statist. 1998, 47(1):115-133.

Medical Image Analysis

Cancerous Liver Tissue Differentiation Using LVQ

Kuo-Sheng Cheng

Institute of Biomedical Engineering, National Cheng Kung University
Tainan, TAIWAN, ROC.
E-mail: kscheng@mail.bme.ncku.edu.tw

Richard Sun

Institute of Biomedical Engineering, National Cheng Kung University
Tainan, TAIWAN, ROC.

Nan-Haw Chow

*Department of Pathology, National Cheng Kung University Hospital
Tainan, TAIWAN, ROC.

Abstract

Medical image processing technique provides an objective and quantitative approach for characterizing the pathological tissue images. In this paper, the fractal dimension was used to quantify the image textures for differentiating the normal and cancerous cells in the liver tissue image. Several image enhancement methods and one edge detection method were applied for accentuating the objects of interest before fractal dimension estimation. From the results, it is shown that the edge-based histogram equalization would be the best one among all these image enhancement methods. In addition, any two fractal dimensions were combined as the input features to the learning vector quantization network for tissue classification. From the results obtained from ten normal and ten cancer cases, the accuracy was demonstrated to be more than 90%. Above all, the user-friendly graphical user interface was also developed in this study.

1 Introduction

According to the survey of the public health in the Taiwan area, the liver cancer has been on the top of the ten major causes of death in the past decade [1]. This is mainly due to the high ratio of the population having the hepatitis B or C. In practice, early detection of the hepatocellular carcinoma is the only way to make treatment and cure better. Therefore, the liver cancer diagnosis is extremely important in the Taiwan area. To date, the routine examinations for the diagnosis of the liver cancer usually relies on the blood test and medical imaging, such as the ultrasonic imaging, computed tomography, or magnetic resonance imaging. However, the biopsy analysis is always used for the final confirmation. Unfortunately it heavily depends upon the pathologist's training and experience, especially in the cases with ambiguity. It is also very time consuming in mass screening. To provide the effective and quantitative parameters for biopsy analysis has been the important issue to alleviate this difficulty. Many image analysis based techniques for morphometric study have been published in the past [2-4].

With the rapid development of the computer technology, the microcomputer-based image analysis is now affordable and helpful in the microscopic image analysis. On the other hand, up to now no useful and global parameter available to effectively quantify the liver biopsy has been proposed. Besides, the parameters for nuclear morphology such as the spatial distribution, size variations, area, perimeter, etc., may be very difficult and accurate to be obtained from the microscopic image analysis. The purpose of this paper is to find the useful and global parameter for quantitatively characterizing the liver tissue image with the image processing and fractal analysis approach.

2 Materials and Methods

In this retrospective study, both the liver tissue sections surgically confirmed with and without hepatocellular carcinoma were provided by the Department of Pathology at National Cheng Kung University Hospital. The specimens were obtained from twenty subjects for 18 males and 2 females. The ages range from 37 to 77 years old with the average about 60.2. These biopsy were prepared using the routine immunoassay method [5]. The thickness for these tissue sections is about 4 μm. With the personal computer based microscopic image analysis system, the images for these sections were then observed and acquired at 400x. For the sake of simplification, the whole images containing only the normal tissues or the cancerous tissues were considered and saved for later analysis. The size for each image is 640×480. The gray level ranges from 0 to 255. In total, there were 20 cancerous and 20 normal tissue images.

2.1 Microscopic Image Analysis System

A Pentium-120 based personal computer with the image grabber was employed for image acquisition and analysis. The microscope equipped with a camera was used for visual screening the liver tissue images. All the analysis programs for the image processing and fractal dimension estimation were developed using the Microsoft Visual C++ language under the Microsoft Windows 95 environment. In addition, a menu-driven user interface was developed for easy to use. The acquired color image with RGB components is transformed into gray level (g) as $g = (R + G + B) / 3$.

2.2 Image Processing

Before estimating the fractal dimension, the liver tissue image was firstly processed using thresholding, edge detection, image enhancement methods, or their combinations. Figure 1 shows the flow chart of the image processing procedure.

In the image thresholding, the method proposed by Otsu [6] was used to determine the threshold of gray-level for separating the foreground and the background of the liver tissue image. The foreground is the interested objects such as the immunochemically stained nuclei and vessels. In the edge detection, the Sobel operator was employed in the detection of the edge points along the x and y directions.

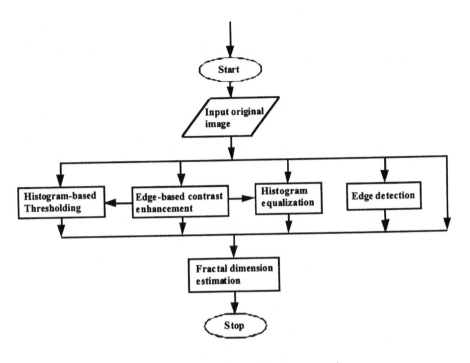

Figure 1. The procedures of image processing

Four image enhancement methods such as histogram equalization, edge-based histogram equalization [7], contrast enhancement [8], and edge-based contrast enhancement method were used for improving the contrast of the image. The third one was proposed by Negrate et al., and the fourth one was its modification. In the fourth method, the gray level of the nearest edge point replaced the mean gray level of edge points in the window mask. The range of the gray levels in our experiment would be over 1000. But, only very few pixels would be greater than 500. Therefore, the normalization procedure was needed to limit this to a practical range of 0 to 255. Four approaches were developed for the normalization. In the first approach referred as contrast enhancement or edge-based contrast enhancement in Table 1, the gray levels greater than 255 were all set to be 255. In the second and third approaches referred as modified contrast enhancement #1, #2 or modified edge-based contrast enhancement #1, #2 in Table 1, the gray levels greater than 500 and 1000 were set to be 500 and 1000, respectively. Then, the range of the gray levels were linearly transformed into 0 – 255. For the last approach, it was just to enhance the contrast for the gray levels lower than the averaged gray value, and no changes for the pixels of the higher gray levels. This is called as contrast enhancement #3 or modified edge-based contrast enhancement #3.

2.3 Fractal Dimension Estimation

To date, many methods have been proposed to estimate the fractal dimension for the gray tone images. The differential box-counting method [9] was used in this study for estimating the fractal dimension of the liver tissue image.

2.4 Learning Vector Quantization

The network as shown in Figure 2, of learning vector quantization (LVQ for short) was employed for the pattern classification in this paper. In this study, the input layer of LVQ neural network contains two nodes for the pair of fractal dimension as the input feature vector. They are the fractal dimensions for the same image processed by different image enhancement methods. The set of vectors as a codebook was initialized to be 16 vectors of weights, i.e., 8 vectors for the normal tissue and 8 vectors for the cancerous tissue.

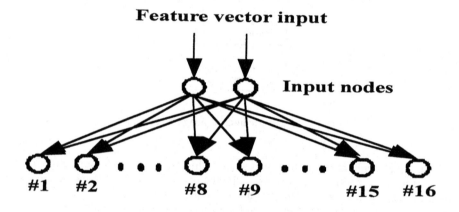

Figure 2. The structure of LVQ network

Ten images with the normal tissue and ten with the cancerous tissue were used as the training samples. For each image, ten fractal dimensions for ten randomly extracted subimages were estimated. In total, there were 200 input patterns for training. The other ten images for the normal and cancerous tissues were used as the test images. After enough iterations, the LVQ network would be typically trained with the convergence; otherwise, it would be retrained with the same input patterns but modified parameter.

3 Results

In order to investigate the effects of these image processing techniques, the combined approaches were also studied.

The fractal dimensions for ten normal and cancerous liver tissue images processed using different image enhancement methods were estimated and compared. For each image, ten subimages were randomly selected. The size of the subimage for fractal dimension estimation was 256×256. In total, there were 100 values of the fractal dimensions both for the normal tissue image and the cancerous tissue images, respectively. Using the paired t-test for investigating the significance of these fractal dimensions in cancerous tissue differentiation, these p values were all smaller than 0.001. Thus, it is shown that the tissue images without or with different image enhancement methods may be well characterized by the fractal dimension.

Table 1. The accuracy, specificity, and sensitivity of the paired fractal dimensions for liver tissue differentiation using learning vector quantization.

Paired Fractal dimensions	Carcinoma cases (True/False)	Normal cases (True/False)	Sensitivity	Specificity	Accuracy (%)
(fe^*, ia^*)	192/8	196/4	0.98	0.96	97.0
(fe, lb)	188/12	198/2	0.99	0.94	96.5
(fe, ie)	182/18	200/0	1.00	0.92	95.5
(fe, jd)	184/16	197/3	0.98	0.92	95.2
(eb, fe)	178/22	200/0	1.00	0.90	94.5
(fe, ge)	180/20	196/4	0.98	0.91	94.0
(ee, fe)	183/17	193/7	0.96	0.92	94.0
(fe, ke)	190/10	185/15	0.93	0.95	93.7
(fe, ld)	176/24	197/3	0.98	0.89	93.2
(fe, ja)	187/13	184/16	0.92	0.93	92.7

*The codes denoting the image processed by two image enhancement methods are described as follows. The first character for the first method is e: contrast enhancement; f: modified contrast enhancement #1; g: modified contrast enhancement #2; i: edge-based contrast enhancement; j: modified edge-based contrast enhancement #1; k: modified edge-based contrast enhancement #2; l: modified edge-based contrast enhancement #3. The second character for the second method is a: no processing; b: higher thresholded; d: histogram equalization; e: edge-based histogram equalization.

Four sizes of the subimage such as 32×32, 64×64, 128×128, and 256×256 used in the fractal dimension estimation for liver cancerous tissue differentiation were investigated. From the ROC curves, the subimage with the size of 256×256 would be shown to be the best choice.

In order to increase performance of differentiation, pairs of the fractal dimensions estimated from the liver tissue images with different image enhancement methods are also studied. These paired fractal dimensions were input to the LVQ network as the input feature vectors for classification. Afterwards, the sensitivity, specificity, predictive values for both the normal and cancerous tissue, and total accuracy were then evaluated. Only ten paired of fractal dimensions with the high accuracy were listed in Table 1. The total accuracy for the case of 'fe vs. ia' is highest one for about 97%. The sensitivities and specificities for these paired fractal dimensions with different image enhancement methods were all greater than 90%, except one for about 89%.

4 Discussion and Conclusion

In this paper, the liver tissue image is characterized using the fractal dimension for the cancerous liver tissue differentiation. From the results, it is shown that the fractal dimension is a very effective parameter. Different methods for image enhancement were also investigated for the feature enhancement. Among the studied methods, the edge-based histogram equalization was demonstrated to be the best one. Based upon the ROC curves for different subimages used for fractal dimension estimation, it is also found that larger image size would provide better performance. Thus, the subimage size for 256×256 was suggested in this study.

Acknowledgment

This work is supported partly by the National Science Council, ROC, under the Grant# NSC84-2213-E006-088, NSC85-2213-E006-076, NSC86-2213-E006-027.

References

[1] *Public Health in Taiwan Area*, Department of Health, The Executive Yuan, ROC, March 1995.

[2] Erler BS, Troung HM, Kim SS, Huh MH, Geller SA, Marchevsky AM. A study of hepatocellular carcinoma using morphometric and densitometric image analysis. Am J Clin Pathol 1993; 100:151-157

[3] Dougherty G. Quantitative indices for ranking the severity of hepatocellular carcinoma. Comput Med Imag & Graph 1995; 19:329-338

[4] Thiran JP, Macq B. Morphological feature extraction for the classification of digital images of cancerous tissues. IEEE Trans Biomed Eng 1996; 43:1011-1020

[5] Chow NH, Hsu PI, Lin XZ, et al. Expression of vascular endothelial growth factor (VEGF) in normal liver and hepatocellular carcinomas- An immunohistochemical study. Human Path 1997; 28:698-703

[6] Otsu N. A threshold selection method from gray-level histogram. IEEE Trans Syst Man Cybern 1979; 9:115-120

[7] Leu L. Image contrast enhancement based on the intensities of edge pixels. CVGIP: Graph Models and Image Proc 1992; 54:497-506

[8] Negrate AL, Beghdadi A, Dupoisot H. An image enhancement technique and its evaluation through bimodality analysis. CVGIP: Graph Models and Image Proc 1992; 54: 13-22

[9] Sarkar N, Chaudhuri BB. An efficient differential box-counting approach to compute fractal dimensions of image. IEEE Trans Syst Man Cybern 1994; 24:115-120

Quantification of Diabetic Retinopathy using Neural Networks and Sensitivity Analysis

Andrew Hunter, James Lowell, Jonathan Owens, Lee Kennedy

School of Computing and Engineering Technology, University of Sunderland
Sunderland, Tyne and Wear, UK.
Andrew.Hunter@sunderland.ac.uk

David Steele

Sunderland Royal Eye Infirmary
Sunderland, Tyne and Wear, UK.

Abstract

The design of neural network classifiers for the identification of diabetic retinopathy is discussed. Red-free digitised fundal images are tiled, and a neural network is trained to distinguish exudates from drusen (similar appearing lesions). By quantifying the degree of retinopathy, the approach can be used to screen diabetic patients for referral. A novel form of hierarchical feature selection using sensitivity analysis is presented. The resulting neural network is compact, and achieves 91% sensitivity and specificity on a test set.

1   Introduction

Diabetic retinopathy is the commonest cause of blindness in the working age group within developed countries [1]. Early detection and treatment allows a 50% reduction in the incidence of visual loss [2]. A comprehensive screening program needs to process 30,000 diabetes mellitus patients per million population. However, the grading of fundal images is time-consuming and requires highly skilled staff.

This paper discusses the development of an automated screening tool. The system takes a quantitative approach, identifying the proportion of the fundal image that contains exudates. This can identify new cases, monitor the progress of existing cases, and grade the stage of retinopathy. Rather than identifying individual exudates (which vary significantly) the system tiles the image, and determines whether each tile contains exudates. Hence a very high level of sensitivity and specificity per tile is not necessary, as the effects of misclassifications of individual tiles tend to cancel each other out, in the statistical sense.

A number of related systems have been reported in the literature. They each attempt to detect and quantify features in the fundal image that are characteristic of diabetic retinopathy, typically microaneurysms and exudates. Neovascularisation is also characteristic of advanced retinopathy, but we are not aware of any systems that check for this feature.

Microaneurysms are the earliest indicator, and a system that counts microaneurysms has been suggested by Phillips and Spencer [3][4]. However, this requires a fluorescein angiogram, and so is not suitable for large-scale screening.

Exudates can be identified by thresholding preceded by shade correction [5]. A drawback is that other lesions (drusen, cotton wool spots and laser scars) can be mistakenly identified. Goldbaum has suggested using the lesion colour to distinguish exudates [6]; however, this approach is not particularly robust. Gardner [7] has suggested using neural networks, with one network to distinguish exudates from each of the other lesion types. After median smoothing, the red-free values were fed directly into a large neural network (using 20x20 tiles, the network had 400 inputs).

In this paper we propose a novel approach to feature selection that radically reduces the complexity of the neural network, thus increasing performance, reliability, training and execution speed. The method is hierarchical, with individual feature selection stages feeding small feature subsets through to further stages. Each stage involves training a neural network, and analysing the contribution of features using a recently introduced form of sensitivity analysis [8]. Since each individual network is reasonably compact, and sensitivity analysis is very fast, the method can be used to select from among hundreds of features extremely quickly.

The final network designed by this method has only eleven inputs and seven hidden units. It achieved 91% correct classification on the test set, compared with 60% for a network trained directly on the red-free image.

In section 2 we describe sensitivity analysis, and its application in feature selection. In section 3 we report on our experiments, and describe our approach to hierarchical feature selection. Section 4 concludes the paper.

2　Sensitivity Analysis

In sensitivity analysis, a network is analysed and an index of importance assigned to each input. The most common approach is to explicitly perturb each input in turn [9], or to calculate the gradient of the output with respect to each input [10], (inputs that have little effect on the output when changed are unimportant). We have recently introduced an alternative method that avoids some drawbacks of the aforementioned [8]. In neural network modelling, if an input is unavailable, a missing value substitution procedure is typically used (e.g. substitution of mean). For each input in turn, we calculate the sensitivity as $s_i = e_i/e$, where e_i is the network error with mean substitution of input i, and e is the normal error.

Sensitivity analysis indicates the significance of input variables in a particular neural network. This can in turn be used to select features for use in further modelling, although caution is necessary. Variables may demonstrate both inter-dependence and mutual redundancy [11], one consequence of which is that a number of networks trained on the same data may produce different sensitivity rankings. However, experience shows that the most significant few variables are consistently identified in the majority of cases. A single training run is usually sufficient to identify key variables, as the sensitivity of variables calculated using the training and test sets can be compared for consistency. Repeated training runs can be used for increased confidence.

3   Experimental Results

We divided the red-free fundal images (simulated by using the green component) into 16x16 tiles. Fast Fourier transforms and Prewitt edge-detection filters were applied. This provided three basic feature sets (*red-free*, *fourier* and *prewitt*). The basic feature sets were each processed using four techniques: summary statistics (Mean, Standard Deviation, Skew and Kurtosis), first 16 principal components, 16-bin histogram, and principal components of the histogram.

Neural networks were built using each of the twelve methods of processing, then key features selected using sensitivity analysis. The key features were combined into three composite feature vectors, and the process repeated to determine a single feature vector. Another pass of sensitivity analysis established a feature vector for the final model.

The data were taken from sixteen images captured by a conventional fundal imaging camera, and digitised from slides. Tiles were classified by an ophthalmologist using custom software; a range of representative tiles was selected, including some boundary cases only partially occupied by the lesion. Image processing (fourier and prewitt transforms) was performed using WiT image processing software; the raw output images were converted to data files by custom software. Neural models, sensitivity analysis, ROC curve and classification results were performed using the Trajan neural network package.

The data set contained 95 tiles featuring drusen, and 116 featuring exudates. The relatively small number of cases adds to the difficulty of the feature selection process. The data was divided into a training set (44 drusen, 56 exudate) and a test set (51 drusen, 60 exudate).

Standard MLP networks were used (logistic activation function), with a small number of hidden units to reflect the low data volume available (between 5 and 8, depending on the number of inputs). Input variables were minimax normalised. Weigend weight regularisation [12] was deployed, with $\lambda = 0.005$ to help restrict model complexity. Weights were randomly initialised in the range [-1,+1].

Training was by 300 epochs of on-line back propagation (learning rate 0.1, momentum 0.3), followed by 300 epochs of conjugate gradient descent. This is an efficient combination with small data sets - if CGD alone is used, it tends to exhibit over-learning very rapidly, with training and test errors diverging rapidly. However, once BP has identified a good minimum, CGD improves terminal convergence and is much more likely to remain stable. Regularisation was extremely important - without it, over-learning invariably occurred and results were inferior.

ROC curves [13] were generated to compare results. The classification threshold was set to equalize sensitivity and specificity. Performance was assessed using the proportion of test cases correctly classified at this threshold. This statistic is highly correlated with the area under the ROC curve.

As a benchmark, we trained neural networks with 256 inputs (the red-free pixel values). Gardner et. al. [7] reported over 90% performance with a 400 input network (using 20x20 tiles) with 80 hidden units. However, they used 1,000 training cases, whereas we were limited to 100; consequently, we encountered severe over-learning when building such large networks, and performance of only 52% was achieved. With the number of hidden units reduced to five, and using weigend regularisation, 60% performance was achievable. With more training data we could probably raise this, but not to the levels reported by Gardner et. al. The major cause of the disparity is no doubt a difference in the inherent difficulty presented by the data sets.

Table 1 summarises the results of the hierarchical feature selection procedure. The number of inputs to each network, and the number of these selected by sensitivity analysis, is shown together with the network's performance. The *Red-free*, *Fourier* and *Prewitt* composite networks use the features identified by the corresponding *Summary*, *PCA*, *Histogram* and *PCA Histogram* networks. The *Overall Composite* network use the features identified by the individual composite networks; sensitivity analysis on it removes a further four variables. Thus, eleven variables are used in training the final model.

In general the test rates follow the expected pattern, with higher values appearing as more significant variables with different characteristics are united in progressively higher composite levels. A minor exception is seen in the prewitt composite nework, which has slightly lower performance than the prewitt summary statistics network. This is due to the marginal value of the non-summary features in the prewitt set, combined with the inherent variability in the training process.

Table 1: Hierarchical feature selection by sensitivity analysis

Model		Inputs	Selected	Test Rate
Red-free	Summary	4	3	0.71
	PCA	16	3	0.62
	Histogram	16	3	0.63
	PCA Histogram	16	4	0.64
	Composite	13	6	0.83
Fourier	Summary	4	3	0.73
	PCA	16	4	0.62
	Histogram	16	1	0.67
	PCA Histogram	16	2	0.59
	Composite	10	4	0.77
Prewitt	Summary	4	2	0.77
	PCA	16	1	0.73
	Histogram	16	4	0.53
	PCA Histogram	16	4	0.73
	Composite	11	5	0.76
Overall Composite		15	11	0.89
Final Model		11	-	0.90

The key stage is the selection of variables using the sensitivity analysis. This is done by eye, although it could be automated. Table 2 shows the results of one sensitivity analysis (fourier summary statistics). The sensitivity ratios are shown for both training and test sets. Kurtosis is identified as having extremely low sensitivity for both sets, and can be eliminated (a sensitivity of 1.0 indicates that a feature is entirely ignored, with progressively larger figures indicating greater significance). The sensitivities of the other three variables are low but significant, and the ranking inconsistent, so they are retained. In later processing, the composite fourier network eliminates Mean and Skew, but S.D. is retained right into the final model.

Table 2: Sensitivity analysis of fourier summary statistcs

	Mean	S.D.	Kurtosis	Skew
Training	1.067	1.034	1.010	1.092
Test	1.049	1.088	1.001	1.053

The final neural network identified has 11 input variables. This neural network has 91% performance. Sensitivity analysis confirms that all the inputs are significant. The area under the ROC curve is 96.4%.

An interesting observation is that in several of the PCA networks, the first principal component has extremely *low* sensitivity, with the second and perhaps some following components highly significant. This may indicate that the first principal component corresponds to "uninteresting" or "obvious" variation, whereas the succeeding ones capture real structure.

4 Conclusion

We have demonstrated that it is possible to build a screening system for diabetic retinopathy, using neural networks and standard image processing to distinguish a key indicator of retinopathy (exudates) from other lesions. The approach is quantitative, assessing the area of the retina containing exudates. The method achieves a high level of performance (91% correct test set classification rate).

We have introduced a hierarchical feature selection method based on sensitivity analysis, which allows key features to be selected from a very large set with minimal computational effort, and despite having a small data set. The technique is well-suited to image processing domains, where a large number of candidate features can easily be generated, but determining which to use is difficult.

4.1 Future Work

The initial study has been conducted using a relatively small number of images. Our first requirement is therefore to repeat the experiments using a larger data set.

We also need to design the other parts of the classication system: lesion detection, and discrimination of exudates from cotton wool spots and laser scars.

Alternative features will be considered, including some that are sensitive to "blobs", such as Gaussian kernel convolutions and Gabor transforms, and colour information. Our approach is of course well-suited to investigating and integrating additional features with minimal effort.

References

[1]   Williams, R. Diabetes mellitus. In: Stevens. A., Raftery, J. (eds). Health Care Needs Assessment. Oxford University Press, Oxford, 1994, pp. 31-57.

[2]   Singer, D.E., Nathan, D.M., Fogel, H.A. Schachat, A.P. Screening for diabetic retinopathy. Ann. Intern. Med. 1992; 116: 660-671.

[3]   R.P. Phillips, P.G. Ross, M. Tyska, P.F. Sharp and J.V. Forrester. Detection and quantification of hyperfluourescent leakage by computer analysis of fundus fluorescein angiograms. Graefe's Arch Clin Exp Ophthalmol 1991; 229: 329-335.

[4]   T. Spencer, J.A. Olson, K.C. McHardy, P.F. Sharp and J.V. Forrester. An Image-Processing Strategy for the Segmentation and Quantification of Microaneurysms in Fluorescein Angiograms of the Ocular Fundus. Computers and Biomedical Research 1996; 29: 284-302.

[5]   R.P. Phillips, J. Forester and P. Sharp. Automated detection and quantification of retinal exudates. Graefe's Arch Ophthalmol 1993; 231: 90-94.

[6]   M.H. Goldbaum, N.P. Katz, M.R. Nelson, L.R. Haff. The discrimination of Similarly Colored Objects in Computer Images of the Ocular Fundus. Investigative Ophthalmology and Visual Science 1990; 31 (4): 617-623.

[7]   Gardner, G.G., Keating, D. Williamson, T.H. and Elliot, A.T. Automatic detection of diabetic retinopathy using an artificial neural network: a screening tool. Br J. Ophthalmol. 1996; 80: 940-944.

[8]   Hunter, A. Application of Neural Networks and Sensitivity Analysis to improved prediction of Trauma Survival. Computer Methods and Algorithms in Biomedicine (in press).

[9]   T.D. Gedeon. Data Mining of Inputs: Analysing Magnitude and Functional Measures. Int. Journal of Neural Systems 1997; 8 (2): 209-218.

[10]   J.M. Zurada, A. Malinowski and I. Cloete, Sensitivity Analysis for Minimization of Input Data Dimension for Feedforward Neural Network, IEEE International Symposium on Circuits and Systems, London, May 30-June 3, 1994.

[11]   A. Jain and D. Zongker Feature Selection: Evaluation, Application and Small Sample Performance. IEEE Trans. Pattern Analysis and Machine Intelligence 1997; 19 (2).

[12]   Weigend, A.S., Rumelhart, D.E. and Huberman, B.A. Generalization by weight-elimination with application to forecasting. In: Lippmann, R.P., Moody, J.E. and Touretsky, D.S. (eds). Advances in Neural Information Processing Systems 1991; 3: 875-882. San Mateo, CA: Morgan Kaufmann.

[13]   M.H. Zweig and G. Campbell. Receiver-Operating Characteristic (ROC) Plots: A Fundamental Evaluation Tool in Clinical Medicine. Clin. Chem 1993; 39 (4): 561-577.

Internet Based Artificial Neural Networks for the Interpretation of Medical Images

Andreas Järund, Lars Edenbrandt

Department of Clinical Physiology, Lund University,

Lund, Sweden

andreas.järund@klinfys.lu.se

lars.edenbrandt@klinfys.lu.se

Mattias Ohlsson

Department of Clinical Physiology, Lund University,

Lund, Sweden

mattias@thep.lu.se

Erik Borälv

Department of Human-Computer Interaction, Uppsala University,

Uppsala, Sweden

erik.boralv@hci.uu.se

Abstract

This paper presents a computer-based decision support system for automated interpretation of diagnostic heart images, which is made available via the Internet. The system is based on image processing techniques, artificial neural networks, and large and well validated medical databases. The performance of the neural networks detecting infarct and ischemia in different parts of the heart, measured as areas under the receiver operating characteristic curves, was in the range 0.76-0.92. These results indicate a high potential for the tool as a clinical decision support system. The system is currently evaluated by a group of pilot users in different European countries.

1 Introduction

With the aid of telemedicine a physician can contact and transfer patient data to an experienced colleague at another hospital. Thereafter they can discuss the patient case over the phone. A problem with this technique is that experienced physicians are not always available when the advice is needed. Therefore we have developed computer-based decision support systems available via the Internet as an alternative approach (Figure 1). With this technique decision support is available 24 hours per day, 365 days per year.

Our system, which is called WeAidU (Web-based Artificial Intelligence for Diagnostic Use), makes artificial neural networks available to physicians via the Internet. The neural networks are trained to interpret diagnostic heart images. The WeAidU system works as follows: The physician who is processing the acquired images at his/her

Internet

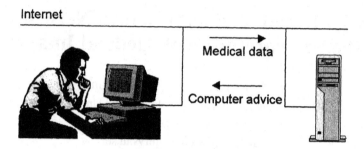

Medical data

Computer advice

Figure 1: The physician sends the medical data via the decision support system via the Internet to the intelligent server and the diagnostic advice is returnd within seconds.

workstation sends the images via the Internet to our intelligent server. Software for automated interpretation of the images is running on the server and the resulting diagnostic advice is returned within seconds via the Internet to the physician.

The system's advice is based upon advanced image processing techniques and artificial neural networks, utilizing information extracted from a large and well-validated image database. The physician-computer interface is constructed to meet the special requirements of health care professionals. Security and privacy aspects are addressed so that patient data can be transferred safely via the Internet.

2 Internet solution

The functionality of the WeAidU system is based on the client-server paradigm. The client consists of a program running on the physician's workstation, presenting the interface of the decision support. The server is a dedicated computer located at Lund University. On the server a program is running. The client and server programs are written in Java and the communication between them is handled by a concept called Remote Method Invocation (RMI). RMI is developed by SUN and is the glue in a distributed object system such as a client-server application running on two different Virtual Machines. The client and server do not use a special protocol, instead methods are called as if they were running locally. The choice of using Java as the implementation language enables multi-platform independence. Most of the potential users of the WeAidU system are connected to the Internet via a firewall. Many of these firewalls only allow for HTTP connections. The use of RMI enables the necessary information transfer between client and server via the firewall. Another advantage with RMI is the dynamic behavior of the client, which means that the clients program version always will be the latest version installed on the server. The user operating the client program can evaluate image data stored as files on the local hard disk. The WeAidU client supports Interfile (version 3.3) and DICOM files (Spec 3.10.7), two of the most commonly used image formats for nuclear medicine images. All images are stripped of pertinent information such as patient name, ID number and age. Only the data needed for the analysis is transfered between the client and the server. At present we only use the

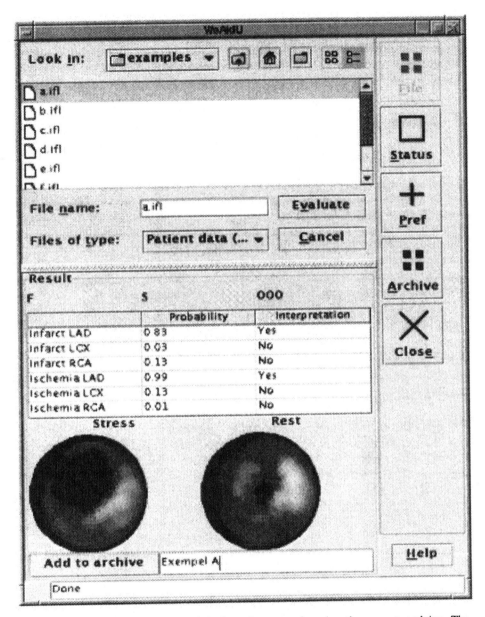

Figure 2: The interface used to send the heart images and to view the computer advice. The two circular images represents the heart in the form of so called bull's-eye images, one obtained at rest and the other after exercise.

image data. As images are sent to the server program by the client, a database stores information about the particular client request such as timestamps and raw image data.

Appropriate image processing algorithms extract the relevant neural network input parameters, which are processed by an up to date artificial neural network. After a few moments, an answer displaying the diagnostic advice will appear on the client screen. The answer is combined with an appropriate feedback form. This user feedback is of vital importance for the future enhancement of the decision support system involved.

3 Intelligent server

3.1 Image processing

Each heart image consists of 17x64 pixels and for each patient study both a rest and a stress image are analyzed. An image reduction method was used in order to extract a smaller set of features used by the neural networks. A Fourier transform technique was used as follows: Each of the two images was expanded by mirroring at row 17 and then discarding the last row, which produced 32x64 image matrices. These two matrices were input as the real and imaginary parts of a complex 32x64 matrix in a fast Fourier transform. The spatial low-frequency components, describing the rest and stress image, are found near the origin in the spatial frequency plane.

The number of components needed to reproduce the original stress and rest images where analyzed both quantitatively and qualitatively and it was found [1] that at least 15 complex low-frequency components where needed in order to recover clinically important details of the images. For more details see ref. [1].

3.2 Artificial neural networks

A 3-layer perceptron architecture with 5 hidden neurons was used. The input consisted of 30 Fourier components as described above. Six different networks were trained to diagnose infarct and ischemia in each of three different territories of the heart. In order to avoid over-training a weight elimination regularization term was used, which is controlled by a parameter α. 5-fold aaa cross validation was used in order to obtain the optimal α for each of the six networks. Optimality was measured with respect to the area under the receiver operating characteristic (ROC) curve (see section 3.4). The α that resulted in the largest ROC area was selected as the optimal α.

The six final networks classifiers were then trained on the full data set using the found α's. Each of these classifiers consisted of a committee of 20 networks, trained with different initial settings.

3.3 Medical data

All patients at the University Hospital in Lund who during the periods from November 1992 to October 1994 and from June 1995 to May 1997 had undergone both a scintigraphic examination of the heart and coronary angiography were studied retrospectively. A total of 243 patients were included. The heart images were analyzed by two experts and their interpretations were employed as gold standard. Clinical data and the results from angiography were available during the interpretation procedure.

The complete interpretation of each patient study consisted of the following six different classifications: infarct and ischemia in each of three territories corresponding to the three main coronary arteries.

3.4 Results

The performance of the artificial neural networks was evaluated using the cross validation procedure. The output values for the test cases were in the range from 0 to 1. A threshold in this interval was used above which all values were regarded as consistent with infarct or ischemia. By varying this threshold a receiver operating characteristic curve was obtained. Areas under these curves were calculated as measures of performance. Table 1 presents the areas under the receiver operating characteristic curves for the six different neural networks, one for infarct and one for ischemia for each vascular territory.

Table 1: Artificial Neural Networks results measured as ROC-areas.

Vascular Territory	Infarct	Ischemia
Left Anterior Descending Artery	0.85	0.87
Right Coronary Artery	0.88	0.76
Left Circumflex Artery	0.92	0.85

These results are in accordance with those of previous studies of automated interpretation of heart images using artificial neural networks. It has been shown that the best neural networks detected heart disease as good as or even better than experienced observers [1]. In clinical practice this type of intelligent computers will not replace, but assist the physicians by proposing an interpretation of the studies. In a recent study it was found that physicians interpreting heart images benefit from the advice of neural networks measured both as an improved performance and a decreased intra- and inter-observer variability [2]. Further it has been shown that the neural networks can maintain a high accuracy also in a hospital separate from that in which they were developed [3].

4   Future development

The system is currently evaluated by a selection of well-known European nuclear medicine departments with a widespread European representation. The evaluation process will include both a technical and a medical dimension. In the technical part the usability, acceptability, speed and safety of the system will be evaluated. In the medical part the accuracy and the clinical impact of the system will be assessed.

DICOM is not only a type of file format. It is a complete network protocol that most medical image viewing software support. Therefore, it would be possible to import image data from the physician's viewing software directly into the WeAidU client without the use of files on a hard disk. We are working on adding this feature to the system. Further, the client is written to handle different types of decision support systems and not only the interpretation of heart images. We are currently working on a similar system for the interpretation of lung images.

Acknowledgements

This study was supported by grants from the Swedish Medical Research Council (K99-14X-09893-08B), the Swedish Foundation for Strategic Research and the Swedish National Board for Industrial and Technical Development, Sweden.

References

[1] Lindahl D, Palmer J, Ohlsson M, Peterson C, Lundin A, Edenbrandt L. Automated interpretation of myocardial SPECT perfusion images using artificial neural networks. J Nucl Med 1997; 38:1870-1875

[2] Lindahl D, Lanke J, Lundin A, Palmer J, Edenbrandt L. Computer based decision support system with improved classifications of myocardial bull's-eye scintigrams. J Nucl Med 1999; 40:1-7

[3] Lindahl D, Toft J, Hesse B, Palmer J, Ali S, Lundin A, Edenbrandt L. Inter-institutional validation of an artificial neural network for classification of myocardial perfusion images. Submitted for publication

Segmentation of Magnetic Resonance Images According to Contrast Agent Uptake Kinetics Using a Competitive Neural Network

Ross J Maxwell, John Wilson, Gillian M Tozer, Paul R Barber and
Borivoj Vojnovic

Gray Laboratory Cancer Research Trust
Northwood, UK
maxwell@graylab.ac.uk

Abstract

The kinetics of the uptake of a magnetic resonance (MR) contrast agent (gadolinium–DTPA) can be monitored via the effect on signal intensity in a series of MR images. This can be used to estimate blood flow in normal tissues or tumours. We have used a competitive neural network approach to identify image pixels with similar patterns of contrast agent kinetics. This allows automatic (and un–supervised) visualisation of regions with similar blood flow. Averaging of contrast agent kinetics over self–similar pixels was useful for fitting to model functions to obtain quantitative measures of blood flow in untreated tumours and in tumours treated with the anti–vascular drug, combretastatin.

1 Introduction

Magnetic resonance imaging (MRI) is used routinely for medical diagnosis and for both pre–clinical and clinical research, e.g. in the development and testing of new drugs. The use of contrast agents for MRI (typically paramagnetic metal ion complexes) is also important in current clinical usage as it helps to highlight lesions, such as tumours, and hence aid their detection and identification. There is now considerable interest in the quantitative assessment of the rate of uptake of various MRI contrast agents as this can give a non–invasive measure of blood perfusion, blood volume and/or the vascular permeability of a diseased organ or a tumour [1].

Combretastatin A4 diphosphate (CA4–P) is a drug currently undergoing clinical trials as an anticancer agent. It has an unusual proposed mechanism of action in that it is intended to selectively destroy tumour blood vessels and hence deprive tumour cells of nutrients [2]. This approach has potential advantages over traditional anticancer drugs including the limited ability of vascular cells (unlike tumour cells) to develop drug resistance. It is important to determine the effect of CA4–P on tumour blood flow during the clinical trial phase and gadolinium–DTPA uptake MRI studies are therefore being performed. In the current study, data has been obtained from animal tumours using a similar MRI protocol to that used in patients so that the results can be related to historical blood flow measurements made in the same tumour model, using an invasive method.

2 Methods

2.1 Tumour Model

P22 carcinosarcoma tumours were grown subcutaneously in the flank of male BD9 rats. Animals were treated with saline or CA4–P (100 mg kg^{-1}) by intraperitoneal injection six hours before MRI examination. Anaesthesia was achieved with Hypnorm/Hypnovel and one tail vein was cannulated. Six animals were studied in each treatment group.

2.2 Magnetic Resonance Imaging

Animals were placed in a 4.7 Tesla horizontal bore magnet of a Varian MR spectroscopy/imaging system. A 3.0 cm surface coil was placed around the tumour as a transmitter/receiver. A sequence of 30 gradient echo images were obtained every 11.8 s; three images before administration of contrast agent followed by infusion of a dose of 0.1 mMol kg^{-1} of Gd–DTPA (Magnevist, Schering) in a volume of 1.0ml kg^{-1} given over 5 s via the cannulated tail vein attached to an infusion pump. The MRI parameters used were: TR 98ms; TE 10 ms; 2 mm slice thickness a field of view 30 x 30 mm with 256 x 148 data points (giving 0.12 x 0.20 mm in–plane resolution). The total imaging time was therefore about 6 minutes.

2.3 Data Analysis

Images were obtained by Fourier transformation of raw (time–domain) MR data to give a 256 x 256 matrix for each image. A region of interest was drawn around the tumour using 'ImageBrowser' software (Varian). Images were subsequently processed using Matlab (The Mathworks, Natick, MA, USA). A background image was calculated from the average image intensity of the second and third images. All pixels from the selected region of interest (about 10,000 to 20,000 pixels) were used as inputs to a competitive neural network with signal intensity 28 values per pixel (background image plus 27 post–gadolinium images making a 28–point input vector). Competitive learning was achieved using the Matlab Neural Network Toolbox (Version 3) functions 'newc' to create a competitive layer with eight neurons (initially with random 28–point vectors), and 'trainwb1' for training the network. At each epoch one input vector was randomly chosen for presentation to the network, the weight and bias values being updated according to a Kohonen weight learning function and a conscience bias learning function, respectively. Learning was continued for 3000 epochs. This results in clustering of time sequences in up to eight possible clusters. At the end of the learning process each pixel is assigned to the neuron with the most similar vector (time sequence). These were used to calculate segmented images. Signal intensity curves were calculated for each cluster by averaging all of the pixels in the cluster.

2.4   Estimation of Apparent Tumour Blood Fow

In order to obtain a quantitative comparison between the results for untreated and combretastatin–treated tumours, the kinetics of signal intensity change were modeled for each cluster and used to obtain average model parameters for each tumour. Under ideal conditions, such modeling should enable estimation of absolute tumour blood flow but this is not true here for the following reasons: (i) Gd–DTPA is not a freely diffusible tracer, so its uptake kinetics are influenced both by blood flow and by the permeability–surface area product (P.S) of the tumour blood vessels; (ii) signal intensity is dependent on the initial T_1 relaxation time as well as Gd–DTPA concentration and correction for the initial T_1 was not carried out in the current study; (iii) it is necessary to know the arterial concentration of Gd–DTPA during the time of tumour measurement (i.e. arterial input function, AIF). In order to estimate the AIF, we used one previously described for Gd–DTPA in rats ($C_a(t) = e^{(-0.015t)} + 0.42e^{(-0.00117t)}$, where C_a is the arterial concentration of Gd–DTPA, t is time in seconds; from Rozijn et al. [3]). This AIF was convolved with a single exponential function and used to fit the kinetics of signal intensity change averaged over each cluster. This exponential rate constant is here termed the 'apparent tumour blood flow' (in units of ml g^{-1} min^{-1}) to indicate that it depends both on blood flow and on P.S (e.g. Taylor et al. [4]).

3   Results

In general, clusters of pixels tended to be arranged in neighbouring regions of two-dimensional space (as shown in Figure 1), even though no information about pixel location was used for the clustering algorithm. Although the competitive neural network was implemented such that eight possible clusters of time sequences could have been detected, in fact between two and seven clusters were identified. In general, fewer clusters were found for the combretastatin treated tumours than for untreated tumours.

Figure 2 shows the locations and average signal intensity profiles of two main clusters for a tumour six hours after treatment with combretastatin (100 mg kg^{-1}). Cluster 1 (Fig. 2a and 2c) represents an outer region indicating uptake of Gd–DTPA. However, this is a considerably slower uptake than in most parts of untreated tumours (e.g. Fig. 1). Cluster 2 (Fig. 2b and 2d) shows that central pixels have negligible uptake of contrast agent, consistent with a dramatic shutdown of tumour blood vessels in the tumour centre.

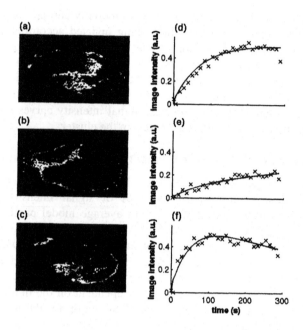

Figure 1: Clustering of contrast agent uptake kinetics for an untreated tumour. Three of the segmented regions are shown in panels (a), (b) and (c) and the corresponding image intensity changes in (d), (e) and (f), respectively. x represent mean image intensity values and solid lines show the fit to the model function.

The apparent tumour blood flow for untreated tumours was 0.35 +/− 0.06 (mean +/− SE) ml g^{-1} min^{-1}. Although this parameter is not expected to be identical to the absolute tumour blood flow determined using a freely diffusible tracer such as iodoantipyrine, the value is very similar to that of 0.34 ml g^{-1} min^{-1} obtained for untreated P22 tumours with this method [2]. A marked and significant (unpaired t–test, P < 0.01) reduction in tumour blood flow to 0.062 (+/− 0.016) ml g^{-1} min^{-1} was detected six hours after combretastatin treatment. This represents an 83% reduction in tumour blood flow, which is somewhat smaller than that previously found for this tumour model using iodoantipyrine uptake for the same dose and time after CA4–P (99% reduction, [2]). The most likely explanation for this discrepancy is that Gd–DTPA uptake is also affected by vascular permeability, which may well be modified following CA4–P induced damage to endothelial cells.

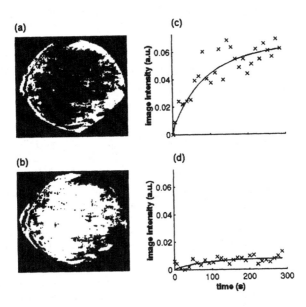

Figure 2: Clustering of contrast agent uptake kinetics for combretastatin–treated tumour. Two of the segmented regions are shown in panels (a) and (b) and the corresponding image intensity changes in (c) and (d), respectively. x represent mean image intensity values and solid lines show fit to the model function.

4   Discussion

Competitive neural networks have previously been used for unsupervised clustering of MR spectroscopy data [5]. Glass et al. [6] applied neural networks to contrast-enhance MR images of osteosarcomas and a self–organising map neural network was used for analysis of functional MRI time series data [7]. The latter approach had the advantage of utilising the two–dimensional image information. In the present   study, a similar competitive neural network to that used for MR spectroscopy data [5] was found to be useful for segmenting tumours into regions with similar gadolinium–DTPA uptake kinetics. This helps in the visualisation of spatial heterogeneity of tumour blood flow e.g. in response to anti–vascular agents. This segmentation also allows rational and unbiased selection of tumour regions for averaging so that improved fitting of pharmocokinetic models can be performed.

References

[1] Mayr NA, Hawighorst H, Yuh WT, Essig M, Magnotta VA, Knopp MV. MR microcirculation assessment in cervical cancer: correlations with histomorphological tumor markers and clinical outcome. J Magn Reson Imaging 1999; 10:267–76

[2] Tozer GM, Prise VE, Wilson J, et al. Combretastatin A–4 phosphate as a tumor vascular–targeting agent: early effects in tumors and normal tissues.. Cancer Res 1999; 59:1626–34

[3] Rozijn TH, van der Sanden BPJ, Heerschap A, Creyghton JHN, Bovee WMMJ. Influence of the pharmacokinetic model on the quantification of the Gd–DTPA uptake rate in brain tumours using direct T_1 measurements. MAGMA 1998; 6:37–43

[4] Taylor JS, Tofts PS, Port R, et al. MR imaging of tumor microcirculation: promise for the new millenium. J Magn Reson Imaging 1999; 10:903–907

[5] Maxwell RJ, Martinez–Perez I, Cerdan S, et al. Pattern recognition analysis of 1H NMR spectra from perchloric acid extracts of human brain tumor biopsies. Magn Reson Med 1998; 39:869–77

[6] Glass JO, Reddick WE. Hybrid artificial neural network segmentation and classification of dynamic contrast–enhanced MR imaging (DEMRI) of osteosarcoma. Magn Reson Imaging 1998; 16:1075–83

[7] Fischer H, Hennig J. Neural network–based analysis of MR time series. Magn Reson Med 1999; 41:124–31

Applications of Optimizing Neural Networks in Medical Image Registration

Anand Rangarajan and Haili Chui

Depts. of Diagnostic Radiology and Electrical Engineering, Yale University
New Haven, CT, USA

Abstract

We formulate feature-based medical image registration as a point matching problem. Point matching requires us to solve for the (rigid or non-rigid) spatial mapping that brings one point-set into register with the other. In order to solve for the spatial mapping, we require the point-to-point correspondences between the two point-sets. Since this information is typically unavailable, we formulate the problem w.r.t. the spatial mapping *and* the correspondences as a mixed variable objective function. Deterministic annealing and the softassign are used to obtain a robust point matching algorithm. Results are shown for 2D rigid autoradiograph alignment and 3D non-rigid mapping of cortical anatomical MR brain structures.

1 Introduction

Optimizing networks have been an important topic in neural computation for over fifteen years. Initial efforts centered around converting combinatorial optimization problems into corresponding nonlinear optimization problems by replacing binary variables by their sigmoidal counterparts [3]. The nonlinear objectives were subsequently minimized using gradient descent. While these initial approaches were beset by issues related to feasibility and constraint-satisfaction, new techniques appeared that promised to alleviate these concerns [1]. It was shown that gradually changing the gain of the sigmoidal activation function resulted in improved solutions. In addition, this approach was closely related to decreasing the temperature in mean-field annealing—a useful technique employed in statistical physics. With this connection established, some of the problems related to constraint satisfaction were also solved using winner-take-all neuronal normalization within annealing [5]. Neuronal normalization has a close parallel in Potts glass methods in statistical physics just as the sigmoidal activation function has a close parallel in naive mean-field methods.

Quadratic assignment problems (QAP) form an important subset of combinatorial optimization problems [9, 10]. This subset includes travelling salesman, graph and subgraph isomorphism, and graph partitioning (as a minor extension). QAP is characterized by a quadratic objective function with the variables obeying permutation matrix constraints. (A permutation matrix is a binary-valued matrix with all rows and columns summing to one.) When QAP is reformulated as a nonlinear optimization problem, the permutation matrix constraints become doubly stochastic matrix constraints. (A doubly stochastic matrix is a square matrix with positive numbers and all rows and columns summing to one.) As long as permutations can be obtained in the limit that the temperature goes to zero, there are no feasibility problems. In this way, minimizing the nonlinear objective function defined on doubly stochastic matrices is identical in the limit to minimizing the original QAP objective function defined on permutation matrices. However, doubly stochastic constraint satisfaction is still a difficult problem. While the neuronal normalization technique (borrowed from the Potts glass) certainly helps satisfy one of the constraints (row or column), the other constraint is still at large. Fortunately, it turns out that a new technique called Sinkhorn balancing or the *softassign* can be used to satisfy both constraints while preserving positivity [7, 8]. Sinkhorn balancing consists of alternating row and column normalizations and is guaranteed to converge to a doubly stochastic matrix. Using Sinkhorn balancing immediately improved the solutions obtained in most QAP problems. Also, it can be shown, provided certain conditions hold, that the overall dynamical system including Sinkhorn balancing (softassign) converges to a fixed point at each temperature even when Sinkhorn only approximately converges [8]. Therefore, optimizing neural networks are now capable of meeting feasibility and constraint satisfaction requirements and avoiding poor local minima when solving QAP problems.

We now move to the domain of medical image analysis. A natural place for the application of optimizing neural networks is feature-based medical image registration [4]. Here, we are given two sets of image features (in 2D or in 3D) and the task is to find an optimal mapping that brings one set of image features into register with the other. In contrast to pattern classification problems where feature vectors live in an abstract high-dimensional space, in image registration, the feature vectors live in 2D or 3D Euclidean space. Point features encoding location are the most popular kind of image feature. One of the most basic ways of formulating the image feature registration problem is: Given two sets of point features, find the best spatial mapping that brings the two point-sets into register [6].

The point matching problem requires us to solve for the spatial mapping that brings the two sets of points into register. Here, the parameterization of the spatial mapping is quite important. A simple example is when the spatial mapping is a rotation. This implies that one of the point-sets is just a rotated version of the other. A much more difficult example is when the spatial mapping is a non-rigid deformation. Here, we have to solve for the optimal non-rigid warping that can take one point-set into the other. Hidden in the formulation of point matching is the correspondence problem. If all we are given is two sets of points, finding the optimal spatial mapping requires that we find the point-to-point correspondences as well. In addition, due to the vagaries of feature extraction, it often turns out that a subset of points in either set have no corresponding homology in the other. Consequently, we have to reject these points as outliers while solving for the homologies (correspondences) between the two point-sets.

In our approach, we set up the point matching problem using an integrated correspondence and mapping objective function [2]. Our approach to solving for correspondence draws on our previous experience with optimizing neural networks. Correspondence is parameterized as a match matrix. If there are no outliers, the match matrix can be reduced to a permutation matrix. When outliers are involved, the match matrix contains a permutation matrix as a subset. However, each row and column can be all zero. In contrast to the QAP, the objective function is linear with respect to the match matrix. However, in addition to the correspondence, we also include a least-squares energy function on the spatial mapping parameters. This least-squares term is the residual error after warping one of the point-sets. The result is an integrated spatial mapping and correspondence energy function. We now describe the optimizing neural network that results from applying the earlier Sinkhorn balancing strategy to the correspondence match matrix.

2 An Integrated Correspondence and Mapping Objective Function

The point matching objective function is formulated as follows. For more details, please see [6].

$$\min_{M,W} E(M,A,w) = \sum_{i=1}^{N_1} \sum_{j=1}^{N_2} M_{ij} ||Y_j - X_i A - \phi(X_i) \cdot w||^2 + \lambda \operatorname{trace}(w^T \Phi w) \qquad (1)$$

$$\text{s. to } \sum_{i=1}^{N_1} M_{ij} \le 1, \ \sum_{j=1}^{N_2} M_{ij} \le 1, \text{ and } M_{ij} \in \{0,1\}.$$

In (1), $X_i \overset{\text{def}}{=} \{1, x_i^{(1)}, x_i^{(2)}, x_i^{(3)}\}, i \in \{1,\dots,N_1\}$ and $Y_j \overset{\text{def}}{=} \{1, y_j^{(1)}, y_j^{(2)}, y_j^{(3)}\}, j \in \{1,\dots,N_2\}$ are two 3D point-sets. The match matrix M denotes the correspondences between the two point-sets with the constraints on M signifying the binary and one-to-one nature of the correspondences. Note that the inequality constraints allow for outliers (non-homologies) in both the X and the Y point-sets. The parameter A is a 12 parameter affine spatial mapping consisting of rotation (3 parameters), translation (3 parameters), non-uniform scale and uniform shear (6 parameters). Additional non-rigid bending and stretching is accomplished via the thin-plate spline (TPS) mapping parameters. The TPS kernel is denoted by Φ. The matrix Φ (in 3D) consists of entries $\Phi_{ij} = r_{ij}$ where $r_{ij} \overset{\text{def}}{=} |X_i - X_j|$. [The ith row of Φ is denoted by $\phi(X_i)$.] The warping parameters w of size $N_1 \times 4$ taken together with the affine mapping parameters A are responsible for the non-rigid warping of point-set X onto Y. The regularization parameter λ controls the degree of the warp: larger the value of λ, smaller the warp and vice-versa.

3 The Robust Point Matching Algorithm

With the thin-plate spline mapping in place, it is reasonably straightforward to develop the deterministic annealing energy function for robust point matching. The complete form of the energy function is:

$$E(M,A,w) = \sum_{i=1}^{N_1} \sum_{j=1}^{N_2} M_{ij} ||Y_j - X_i A - \phi(X_i) \cdot w||^2 + \lambda \operatorname{trace}(w^T \Phi w)$$

$$+ \sum_{i=1}^{N_1} \mu_i (\sum_{j=1}^{N_2+1} M_{ij} - 1) + \sum_{j=1}^{N_2} v_j (\sum_{i=1}^{N_1+1} M_{ij} - 1) + T \sum_{i=1}^{N_1+1} \sum_{j=1}^{N_2+1} M_{ij} \log M_{ij}. \qquad (2)$$

Even though there are five terms, only the first two will be directly involved when we are going to solve for the transformation (A, w). The first term is the error measure. Assume for the moment that the correspondence M is known. The bending energy term depends on a parameter λ and we co-vary it with the temperature $T = \frac{1}{\beta}$. Typically, we begin with a high value of λ and quickly decrease it, with the consideration being that at first the correspondences are still far from the right answer and the transformation should not be too committed. Though this may add some complexity to the algorithm, we have found it worthwhile. The reason is that the algorithm does not seem to be very sensitive for slightly different choices of λ annealing schedules, as long as the starting value is high enough so that the transformations are not too large in the beginning and the final value small enough so that the transformations won't always be forced too close to identity. Note that in the rigid registration experiments, the spatial mapping is restricted to being a similarity transformation where there is no warping.

The Robust Point Matching (RPM) Algorithm

Initialize (A, w), β to β_0 and λ to λ_0.

Begin A: Deterministic Annealing: Repeat until $\beta > \beta_f$.

> **Begin B: Relaxation:** Repeat until convergence.
>
> Update $M_{ij} \leftarrow \exp\left(-\beta \|Y_j - X_i A - \phi(X_i) \cdot w\|^2\right)$.
>
> **Begin C: Sinkhorn Balancing:** Repeat until convergence.
>
> $$M_{ij} \leftarrow \frac{M_{ij}}{\sum_{j=1}^{N_2+1} M_{ij}}, \ \forall i \in \{1, \ldots, N_1\}.$$
>
> $$M_{ij} \leftarrow \frac{M_{ij}}{\sum_{i=1}^{N_1+1} M_{ij}}, \ \forall j \in \{1, \ldots, N_2\}.$$
>
> **End C**
>
> Least-squares solution for (A, w).
>
> **End B**
>
> $\beta \leftarrow \beta \beta_r$.
>
> $\lambda \leftarrow \lambda \lambda_r$.

End A

---

4 Results

4.1 Rigid Alignment of Primate Autoradiographs

Point-sets are obtained by Canny edge detection followed by thresholding. For more details, please see [6]. Initial conditions and the obtained registrations are shown in Figure 1. In these experiments, the following parameter values were used: initial temperature $T_0 = \frac{1}{\beta_0} = 10000$, number of iterations at each temperature $I_0 = 10$, maximum number of Sinkhorn iterations $I_1 = 30$, and annealing schedule $\beta_r = \frac{1}{0.93}$.

4.2 Non-rigid Registration of Cortical Structures

Obtaining sulcal point-set data from anatomical magnetic resonance (MR) imagery is a highly non-trivial problem. For more details please see [2]. The sulcal point-sets have about 300 points each. The RPM annealing parameters were set to be the following: $\beta_0 = \frac{1}{50}$, $\beta_{final} = 1$, $\beta_r = \frac{1}{0.95}$. The regularization parameter λ is set to force w to be small at first and then λ is decreased according to an annealing schedule. We observed that any annealing rate λ_r between 0.7 and 0.9 works quite well. We applied RPM to five sulcal point-sets and compared it with other methods which use affine transformations for brain registration. The results are shown in Fig. 2.

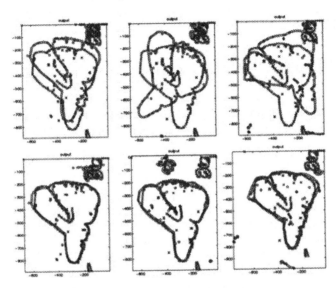

Figure 1: Top: Initial conditions. Left: Slices 360-363, Center: Slices 360-384, Right: Slices 360-415. Bottom: Final solution found by RPM. Left: Slices 360-363, Center: Slices 360-384, Right: Slices 360-415

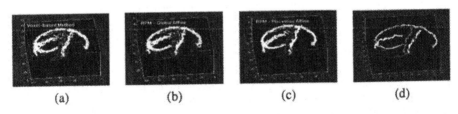

(a) (b) (c) (d)

Figure 2: Sulcal mapping: The mapping results of five brains' sulcal point-sets on the left side of the brain are shown together. (a) Voxel-based mutual information with an affine mapping, (b) RPM with an affine mapping, (c) RPM with a piecewise affine mapping, (d) RPM with a thin-plate spline mapping. Closer packing of distributions of sulcal points suggests better alignment. We clearly see the improvement of TPS-RPM over the voxel-based approach and other RPM-based approaches.

5 Conclusions

We have demonstrated the application of optimizing neural networks in two different medical image registration problems. These problems are at the opposite ends of the spatial mapping spectrum; rigid similarity transformations for autoradiograph alignment and fully non-rigid thin-plate spline transformations for brain mapping. In both applications, we used virtually the same optimizing network for establishing point-to-point correspondence (and outlier rejection). In summary, we formulated medical image registration as a mixed continuous-combinatorial optimization problem. Deterministic annealing and the softassign (using Sinkhorn balancing) were crucial in obtaining an efficient (albeit still suboptimal) robust point matching algorithm which minimized the original mixed variable objective function.

Acknowledgements

This work is partially supported by grants from the Whitaker Foundation and the National Science Foundation to A. R. We would like to thank P. Goldman-Rakic and R. Schultz for providing us with the primate autoradiograph and the MRI brain data respectively.

References

[1] G. L. Bilbro, W. E. Snyder, R. Mann, D. Van den Bout, T. Miller, and M. White. Optimization by mean field annealing. In D. S. Touretzky, editor, *Advances in Neural Information Processing Systems 1*, pages 91–98. Morgan Kaufmann, San Mateo, CA, 1988.

[2] H. Chui, J. Rambo, J. Duncan, R. Schultz, and A. Rangarajan. Registration of cortical anatomical structures via robust 3D point matching. In *Information Processing in Medical Imaging (IPMI '99)*, pages 168–181. Springer, New York, 1999.

[3] J. J. Hopfield. Neural networks and physical systems with emergent collective computational abilities. *Proceedings of the National Academy of Sciences*, 79:2554–2558, 1982.

[4] J. B. A. Maintz and M. A. Viergever. A survey of medical image registration. *Medical Image Analysis*, 2:1–36, 1998.

[5] C. Peterson and B. Soderberg. A new method for mapping optimization problems onto neural networks. *Intl. Journal of Neural Systems*, 1(1):3–22, 1989.

[6] A. Rangarajan, H. Chui, E. Mjolsness, S. Pappu, L. Davachi, P. Goldman-Rakic, and J. Duncan. A robust point matching algorithm for autoradiograph alignment. *Medical Image Analysis*, 4(1):379–398, 1997.

[7] A. Rangarajan, S. Gold, and E. Mjolsness. A novel optimizing network architecture with applications. *Neural Computation*, 8(5):1041–1060, 1996.

[8] A. Rangarajan, A. L. Yuille, and E. Mjolsness. Convergence properties of the softassign quadratic assignment algorithm. *Neural Computation*, 11:1455–1474, 1999.

[9] P. D. Simic. Constrained nets for graph matching and other quadratic assignment problems. *Neural Computation*, 3:268–281, 1991.

[10] A. L. Yuille. Generalized deformable models, statistical physics, and matching problems. *Neural Computation*, 2(1):1–24, 1990.

A Learning by Sample Approach for the Detection of Features in Medical Images.

Camille Serruys, Djamel Brahmi, Alain Giron, Nathalie Cassoux,
Raoul Triller, Phuc Le Huang and Bernard Fertil

INSERM - U 494 - CHU Pitié-Salpêtrière, 91 boulevard de l'hôpital,
75634 Paris cedex 13 - FRANCE.
Serruys@imed.jussieu.fr

Abstract

The detection of specific features in medical images is often a key support for diagnosis. Taking advantage of large bases of images where features of interest have been localized by clinicians, a modular system has been developed to spot similar features on new images. Images are scanned through a window, the size of which being previously fitted to the feature of interest. The recognition process involves a coding phase followed by a classification phase. These phases rely on unsupervised and supervised learning respectively for their implementation. Applications in Dermatology and Ophthalmology are presented.

1 Introduction

Medical images represent meaningful (although often complex) visual stimuli for experienced physicians but their characteristics are frequently hard to decipher. The development of a computational model to capture physicians' expertise is a demanding task, especially if a significant and sophisticated preprocessing of images is to be required. Learning from well-expertised images may be a more convenient approach, inasmuch a large and representative bunch of samples is available. Our approach mimics physicians' behaviour: images are parsed, looking for the presence of specific features. Modules specialised in the detection of features are built for that purpose, using a learning by sample method, which takes advantage of large data bases of typical images, carefully labelled by several clinicians. An advice is subsequently put forward on the basis of these initial exams. In this paper, the structure shared by several specialised modules is described together with some of the results obtained when dealing with the analysis of black tumors of skin and the characterisation of infected areas in retinal angiograms.

Acknowledgement: this work has been supported by a grant from the "Ligue Nationale Contre le Cancer" and by ANRS agency (Agence Nationale de Recherches sur le Sida)

2 Methodology

In order to extract medical expertise from raw images, specialised modules were designed to discriminate structures or features involved in the medical decision process (for example, a network with broad width grids is a common feature of melanoma - specificity 86% and sensitivity 35%). One module can be dedicated to the detection of that feature. In general, features can be observed in areas smaller than the whole image. By focusing on such areas -called windows thereafter- the modules classifies each of them with respect to the feature they have been taught to recognise.

2.1 The scanning of images

The recognition process has been implemented under the form of a three-stage procedure:

2.1.1 Stage 1: Image sampling

The size of regions of interest (ROIs) depends on the feature, which is search for. Image resolution and size of the observation window are therefore chosen in such a way that the feature is still detectable. In general, a window of up to 32*32 pixels can be considered as convenient. Images are subsequently sub-sampled, generating large collections of overlapping windows. These windows are given as input to the modules that, *in fine* operate convolution-like transformations of each image (next 2 stages).

2.1.2 Stage 2: Window coding

Windows to be classified are coded according to their similarities with sets of primitive windows previously obtained during specific and unsupervised learning phases (see below: the making of specialised modules). The reduction in dimensionality as well as the contextual coding which take place during that stage result in a more dedicated (and easier) handling of data.

2.1.3 Stage 3: Encoded window classification

This stage makes use of a classical feedforward network (MLP) to perform the classification of coded windows. Each window is subsequently given a probability-like coefficient, which characterises the recognition status of the feature of interest. As a consequence of this treatment, every analysed image generates several images (one per specialised module) where the features of interest are highlighted.

2.2 The making of specialised modules

Specialised modules rely on learning for their implementation. Window coding (stage 2) is achieved by means of the generalised Hebian algorithm (Principal Component-like analysis) [1] [2]. According to that procedure, eigenvectors (called here primitive windows) are obtained as a result of learning phases that involve set of windows sampled from regions of interest drawn by clinicians. Set of primitive windows fitted to various contexts, each of them being appropriate for the description of some aspects of the images (contrast, colour, texture, border...) are consequently obtained. They allow for a compact description of the windows to be analysed (Figure 1.).

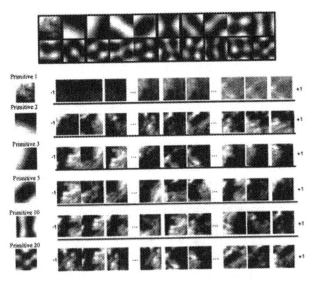

Figure 1. (upper panel) Set of 20 primitive windows generated during the learning phase. The next panel illustrates the contribution of each primitive to window analysis. Windows are ranked according to their similarity with each primitive (according to the scalar product of the window by the primitive).

Similarly the supervised learning phase, required for the realisation of the 3rd stage, also makes use of the various sets of windows. The target is set following clinician's instructions regarding the presence of the features of interest. Large sets of windows, representative of the typical situations to be characterised, are therefore required. A learning phase classically makes use of several thousand windows.

3 Applications

Some results obtained with this approach are presented below. They concern studies:
- in Dermatology, to detect some features related to the diagnosis of black tumors of skin, pigment network, pseudopods ... (a classification issue),
- in Ophthalmology, to detect virus-infected areas in retina angiograms (a classification issue), and to evaluate the accuracy of border between virus-infected areas and healthy areas (a regression issue). Accuracy of border is a valuable information that helps to detect and characterize evolution of infection.

3.1 Dermatology

Prognosis of melanoma, an invasive and malignant skin tumor, strongly relies on early detection. Unfortunately differentiating early melanomas from other less dangerous pigmented lesions is a difficult task even for trained observers since they may have near physical characteristics [3].

We are currently developing a computer-based diagnostic system designed to spot features of black tumors in digital images in order to identify melanoma. Results presented here concern detection of pseudopods, which are bulbous and often linked projections directly connected to the tumor body (Figure 2. - left panel).

Ten thousand windows of "pseudopods" extracted from 20 images of different lesions (benign and malignant) constitute the learning base. Classification is conducted with 3000 windows with pseudopods and 3000 variegated windows not containing any pseudopod-like structure, and 60 primitives are kept for the coding stage. The overall percentage of well-classified windows is about 90%.

When applied to the analysis of whole images of black tumors, a new image is obtained where areas with the relevant feature are enhanced (Figure 2. - right panel). Homogeneous areas, whether dark or light, do not generate any signal.

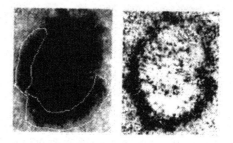

Figure 2. Image of a benign tumor with pseudopods (left panel) and its interpretation by the classifier (right panel) (pseudopod-filled areas are in dark).

3.2 Ophthalmology

Cytomegalovirus (CMV) retinitis is one the most common opportunistic infection and the first cause of blindness in patients with acquired immunodeficiency syndrome (AIDS). CMV infection follow-up is based on survey of angiograms, in order to early detect any relapse of the retinal necrosis or scar evolution. In daily clinical routine, evolution of these zones is assessed by comparison of images [4].

3.2.1 Segmentation

In a first step, our system has been used for segmenting infected area of non-infected area of retinas. The learning base is constituted by 10.000 windows taking into account representativeness of different textures extracted from 200 angiograms analyzed by ophthalmologists and 40 primitives are kept for the coding stage. The overall percentage of well-classified windows is about 85%. Promising results has been obtained on images with high variability of contrast, exposition, and texture (Figure 3.). For example, angiograms with blurred and granular (3a), irregular (3b) or overexposed (3c) infected areas and progressive (3a), fluctuant (3b) or sharp (3c) boundaries between infected and non-infected areas have been successfully analyzed.

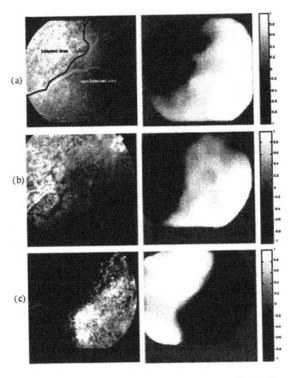

Figure 3. Segmentation of retinal lesion in angiograms. Dark pixels and light pixels label infected areas and non-infected areas respectively.

3.2.2 Boundaries accuracy

To support ophthalmologists in their daily routine and enable quantitative assessment of progression of CMV infection, a methodology taking into account the accuracy of drawing of clinicians and allowing an accurate comparison of retinal borders, has been developed [5]. This methodology, based on the analysis of multiple outlines drawn by experts (during several and independent sessions) between infected and non-infected areas, produces an envelope that characterises the accuracy of drawing (Figure 4a-b.). Clearly, the reproducibility of clinician's expertise is related to some local features of the images. Our goal is to estimate the accuracy of border from these features in order to get rid of multiple drawings.

The learning base is made of about 1500 windows underlying the boundaries and for which an accuracy value can be calculated from multiple outlines (Figure 4c.). During the learning phase, windows are given as input whereas accuracy values are given as target. Our system is able to derive useful information from the analysis of such windows and quite often provides satisfactory prediction of local accuracy of boundary (Figure 4d.).

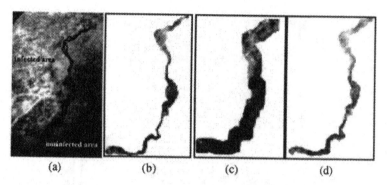

Figure 4. (a) Multiple hand-outlined segmentation, underlying boundary between infected and non-infected retina in an angiogram. (b) Corresponding accuracy envelope as calculated from multiple outlines. (c) Domain where correlation between image and accuracy is looked for (based on overlay of 64x64 pixel windows). (d) Accuracy envelope predicted by the module.

4 Conclusion

Our approach shows up effective for detection of features observable in windows and is free of translation, rotation and other transformations, which often jeopardise image analyses. It is observed that classification/regression is properly achieved as far as training sets are reasonably balanced: about 80% to 90% of correct classification of features are typically obtained (tumors or segmentation of infected area). Accuracy of border as predicted by our system tightly mimics reproducibility of hand-outlines drawn by experts (segmentation of infected area).

References

[1] Oja E. "A Simplified Neuron Model as a Principal Component Analyser.", J Math Biology 1982; 15; 267-273.

[2] Haykin S. "Neural networks: A comprehensive foundation. ", Prentice Hall, 1999.

[3] Lindelöf B., Hedblad M. A. and Sigurgeirsson B. "Melanocytic naevus or malignant melanoma? A large-scale epidemiological study of diagnostic accuracy.", Acta Dermato-Venereologica 1998; 78; 284-8.

[4] Holland G. N., Levinson R. D. and Jacobson M. A., "Dose-related difference in progression rates of cytomegalovirus retinopathy during foscarnet maintenance therapy.", Am J Ophthalmol, vol. 119, pp. 576-86, 1995.

[5] Brahmi D., Cassoux N., Giron A., Lehoang P. and Fertil B. "CAD of viral progression in CMV retinitis by the comparison of hand-outlined segmentations on series of angiograms.", International Workshop on Computer-Aided Diagnosis, 1182, Excerpta Medica International Congress, Chicago, 1998

Neural Network Based Classification of Cell Images via Estimation of Fractal Dimensions

Changjing Shang, Craig Daly, John McGrath
Division of Neuroscience and Biomedical Systems
IBLS, University of Glasgow
Glasgow G12 8QQ, UK
{c.shang, c.daly, I.McGrath}@bio.gla.ac.uk

John Barker
Department of Electronics and Electrical Engineering
University of Glasgow, Glasgow G12 8LT, UK
jbarker@elec.gla.ac.uk

Abstract

This paper[1] presents a system for classifying cells in human resistance arteries via estimating and analysing fractal dimensions of normal and abnormal cell images. The use of fractal features helps characterise and differentiate between categories of cell images. The classification task is implemented using a multi-layer feedforward neural network, which maps estimated fractal feature patterns onto their underlying cell index. This system has been applied to a large database of images (which were taken from proximal and distal areas of subcutaneous resistance arteries, of patients suffering from critical limb ischaemia, by means of laser scanning confocal microscopy), with the overall classification rate reaching 90.5%.

1 Introduction

Modelling and classification of cell and tissue images form two important tasks for the application of computer based image analysis in pathology and medicine. One of the existing advanced approaches to texture image modelling is to capture the roughness degree of the textures by estimating the fractal dimension [1, 2, 3, 4] of the image. Fractal dimension characterises the smoothness of an image surface, which is generally greater than the topological (intuitive) dimension. A number of algorithms exist for determining the fractal dimension of a surface, the approach proposed in [3, 4] is adopted in this work. It measures the fractal dimension of a given cell image via the estimation of the variogram of the image surface, thereby allowing the representation of a cell texture image with fractal features. This facilitates the classification task required.

---

[1]This work is supported by the UK Welcome Trust. The authors are grateful to Professor D. M. Titterington for his help in investigating the fractal modelling techniques.

Artificial neural networks have been widely used for pattern classification [5, 6, 7, 8]. In particular, multi-layer feedforward neural networks (MFNNs) [8] have proven very successful in classifying texture images, if representative features can be correctly identified and presented for such a network to learn. Having recognised this, this paper presents an MFNN based system to classify cell images in human resistance arteries, with the images modelled by fractal dimensions. This system has been applied to a real image database of human cells, which were taken from patients suffering critical limb ischaemia (from the areas of proximal and distal of subcutaneous resistance arteries), by the use of laser scanning confocal microscopy (LSCM).

The classical method of comparing normal and abnormal blood vessel structure relied on measurements of vessel lumen and wall thickness [9]. More recently, nuclear stains have been employed to describe the spatial arrangement of vascular cells in unfixed tissues [10, 11]. This current work provides an alternative approach, based on rigorous statistical image modelling techniques. The next section gives an outline of the method adopted for characterising cell images by computing their fractal dimensions. This is followed by a specification for the MFNN-based image classifier which maps patterns comprising fractal, and additional standard statistical, features of cell images onto their underlying cell category. Experimental results are provided in section 4, demonstrating the success of the present approach.

2 Estimation of fractal dimension

In this work, an observed image $Y = \{y(s)\}$ is assumed to be a stationary Gaussian process or a random field defined on an $M \times M$ lattice Ω, where $y(s)$ denotes the grey level of a pixel at location $s = (i, j)$, $i, j = 0, 1, \ldots, M - 1$. Given an image Y, its fractal dimension D approximately satisfies the following [1]:

$$v(d) = c\, d^{(6-2D)} = c\, d^a \qquad (1)$$

where a is termed the fractal index, c is a constant and $v(d) = E\{y(s+d) - y(s)\}^2$, denoting the variogram of the image. The relationship between fractal index a and the variogram $v(d)$ of Y can be further represented using the linear regression below, by applying log function to both sides of (1):

$$log\{v(d)\} = log(2c) + a\, log(d) \qquad (2)$$

The image variogram $v(d)$ can be estimated by

$$\hat{v}(d) = \frac{1}{N(d)} \sum_{N(d)} \{y(s+d) - y(s)\}^2 \qquad (3)$$

where $N(d)$ denotes the cardinality of the set of pairs of observations whose spatial locations are separated by a particular distance d.

Given different distances d_i, $i = 1, 2, \ldots, K$ (with K being typically a small integer), a set of $\hat{v}(d_i)$ can be obtained using (3). Applying the Least Square (LS) fitting algorithm to (2), the estimates of fractal index $\hat{a}$ and constant $\hat{c}$ can be obtained [3, 4]. Thus, the fractal dimension of Y can be estimated such that

$$\hat{D} = 3 - 0.5\hat{a} \qquad (4)$$

This reflects the fact that if the surface is very smooth, then the fractal dimension is two; if, however, the surface is extremely rough and irregular then the fractal dimension approaches three in its limit.

3 MFNN based classifier

The task of the MFNN is to accomplish classification by mapping feature patterns representing given cell images onto their underlying normal or abnormal cell category. The structure of the MFNN classifier is therefore specified as follows. The number of nodes in the input layer is set to the number of features extracted, and the number of nodes within its output layer will depend on the number of cell classes of interest in the application domain. The internal structure of the network is designed to be flexible and may contain one or two hidden layers. A suitable internal structure of the network for a particular problem can be found by experimental simulations [7].

In this work, for a given pixel location s of an image, four variograms that are respectively computed along the directions $0°, 45°, 90°$, and $135°$ are exploited. In addition, a fifth variogram is estimated, assuming the image to be isotropic. Thus, a total of five fractal features can be extracted. These, together with the mean and variance features of the image, are used as input for the MFNN to perform classification. That is, the classifier has seven input nodes.

The training of an MFNN is essential to its runtime performance. For this purpose, feature patterns that represent different cell images, coupled with their corresponding underlying class indices, are selected as the training data. The back-propagation algorithm [8] is utilised to complete such training.

4 Simulation results

4.1 Image database

The DNA stains have been used to label the various cell nuclei. Serial optical slices were taken along z axis ($1\mu m$ apart) from the first adventitial (adv.), the second smooth muscle (sm.), to the last endothelial (end.) nuclei by placing the LSCM focus on the top of the blood vessel in the x-y plan. The resultant image database consists of 286 section images, each of which is of a size 512×512 with the grey levels ranging from 0 to 255. These images were captured in different tissue samples from ischaemia patients in the regions of subcutaneous resistant arteries. Among them, 163 were taken from 4 samples in the area of proximal (normal) and 123 images from 3 samples in the area of distal (abnormal). Examples of these images are shown in Figure 1. Different images from this database are used for training and testing.

4.2 Fractal dimensions of cell images

For the given 286 cell images, 286 fractal feature patterns can be obtained using the method described in section 2. For instance, Figure 2 illustrates a scatterplot of $(log(d_i), log\hat{v}(d_i))$ and that of $(log(d_i), log(v(d_i)))$, for the image of row 2 column

1 in Figure 1(a) and that of row 1 column 2 in Figure 1(b), respectively. Where a straight line indicates the LS regression through four points at small distances d_i ($d_1 = 1; d_2 = 2^{\frac{1}{2}}; d_3 = 2; d_4 = 2^{\frac{3}{2}}$), using the model presented in (2). This result shows as an example that this modelling approach allows very good fitting of the variograms of real cell images. The estimated fractal dimensions of all the cell images shown in Figure 1 are plotted in Figure 3. Clearly, these fractal dimensions are different; the normal tissue images have a smaller fractal dimension compared to the abnormal ones. This indicates that the roughness degree of a normal tissue is smaller than that of an abnormal one. Supported by an MFNN classifier, the resultant distinctive fractal characteristics make it possible to identify automatically different cell images, normal or abnormal.

4.3 Classification performance

In the simulations carried out the structure of the MFNN classifier was further specified. It contains 6 output nodes, representing adventitial, smooth mussel and endothelial cell images of normal tissues, and the same three image categories of abnormal ones, and a single hidden layer with the number of hidden nodes set to 9, 12, 18, 24 or 30 in different simulations. 86 images that jointly cover all cell categories of normal or abnormal tissue considered were taken from the image database to train the classifier. After training, the remaining 200 different images which involve these types of cell index were used for testing. The best correct classification rate over all these images is 90.5%, achieved by the use of the network of 18 hidden nodes.

For comparison, a minimum distance classifier (MDC) [12] was also implemented. The same numbers of the training and testing feature patterns were used, as with the MFNN-based classifier. However, MDC is only able to achieve a correct classification rate of 80.5%.

5 Conclusion

This paper has presented a successful application of statistical image modelling techniques and artificial neural networks, for classifying cells in human resistance arteries. The method adopted for estimating the fractal dimension of a given cell image, which reflects the spatial information embedded in that image, is described. A specific multilayer feedforward neural network is employed to identify the underlying cell indices of estimated fractal feature patterns and hence those of the original cell images. A large real image data set taken from human ischaemia patients has been used in this study. Simulation results demonstrated that using the fractal dimension method to generate the global characteristics of normal and abnormal cell images, differentiating features between cell categories can be found, in terms of the roughness degree of the image surface. Given these features and additional standard statistic ones, the MFNN classifier is able to provide an automatic way to identify the cell images among different cell categories, be they normal or abnormal.

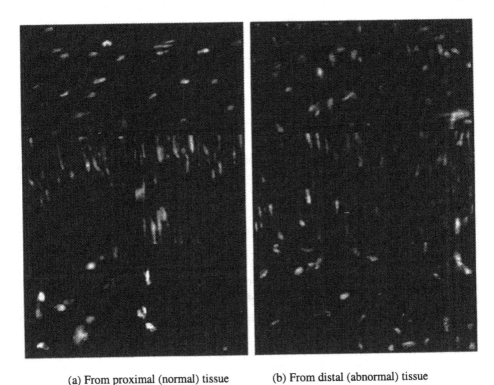

(a) From proximal (normal) tissue   (b) From distal (abnormal) tissue

Figure 1: Section images of cells, where the first, second and third rows respectively show adventitial, smooth muscle and endothelial cells

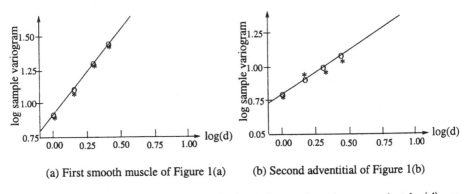

(a) First smooth muscle of Figure 1(a)   (b) Second adventitial of Figure 1(b)

Figure 2: Scatterplot of the logarithm of the isotropic sample variogram against $log(d)$. $*$: $(log(d_i), log(\hat{v}(d_i)))$ is obtained by (3); $\circ$: $(log(d_i), log(v(d_i)))$ is obtained by (2).

116

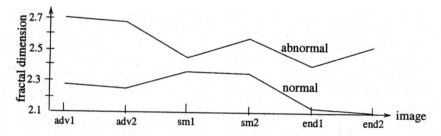

Figure 3: Fractal dimensions of the images shown in Figure 1.

References

[1] Mandelbrot BB. The fractal geometry of nature. San. Francisco: Freeman, 1982

[2] Keller J, Chen S, Crownover R. Texture description and segmentation through fractal geometry. Comp. Vision Graph, and Image Proc. 1989; 45:150-166

[3] Steve D, Hall P. Fractal analysis of surface roughness by using spatial data. J. R. Statist. Soc. B 1999; 61:1-27

[4] Wu CM, Chen YC. Texture features for classification of ultrasonic liver images. IEEE Trans. Medical Imaging 1992; 11:141-152

[5] Bishop CM. Neural networks for pattern recognition. Oxford University Press, Oxford,1995

[6] Jennings M, Graham J. A neural network approach to automatic chromosome classification. Phs. Med. Boil. 1993; 38:959-970

[7] Shang C, Brown KE. Principal feature based texture classification with neural networks. Pattern Recognition 1994; 27:675-687

[8] Rumelhant DE, Hinton GE, Williams RJ. Learning internal representations by error propagating. In: Parallel Distributed Processing. Rumelhant E, McClelland JL. (Eds.), MIT Press, 1986

[9] Mulvang MJ. Vascular remodelling of resistance vessels: can we define this? J. Cardiovascular Research 1999; 41:9-13

[10] Daly CJ, Gordon JF, McGrath JC. The use of fluorescent nuclear dyes for the study of blood vessel structure and function: novel applications of existing techniques. J. Cardiovascular Research 1992; 29:41-48

[11] Arribas SM, Gordon JF, Daly CJ, Dominiczak AF, McGrath JC. Confocal microscopic characterisation of a lesion in an acerebral vessel of the stroke-prone spontaneously hypertensive rat. Stroke 1996; 27:1118-1123

[12] Duda RO, Hart PE. Pattern classification and scene analysis. Jone Wiley and Sons, New York, 1993

Signal Processing in Medicine

Mutual Control Neural Networks for Sleep Arousal Detection

Theodoros Assimakopoulos

TKN, Tech. University Berlin,

Berlin, Germany

thass@ee.tu-berlin.de

Kyra Dingli

Sleep Laboratory, Royal Infirmary Edinburgh,

Edinburgh, UK

kdingli@srv1.med.ed.ac.uk

Neil J. Douglas

Sleep Laboratory, Royal Infirmary Edinburgh,

Edinburgh, UK

njdouglas@srv1.med.ed.ac.uk

Abstract

We demonstrate the use of artificial neural networks (ANN) to automatically detect arousal states during sleep. In this approach we model the attractor of the underlying process from time series and we show how the hidden control neural networks can be extended to model instationary behavior, by means of mutual control neural networks (MCNN). A verification of the model, based on polysomnographic recordings of 5 patients suffering from obstructive sleep apnea hypopnoea syndrome (OSAHS) is given.

1 Introduction

Most of the proposed artifical neural networks (ANN) models for sleep stage classification, follow the basic approach of low-grade of instationarity (e.g. [1],[2]). For the arousal detection task the grade of instationarity is not well known yet. For instance, EEG recordings in REM-sleep have a higher fractal dimension than in NREM-sleep, implying that in REM more non-linear modes are activated ([3]).

Considering the sleep process as a highly transient non-autonomous dynamical system, we model the attractor of the underlying process from time series and demonstrate how ANNs can be extended to handle instationary behavior, by means of mutual control neural networks (MCNN). The MCNN model deals with processes, which are highly transient rather than stationary, by (i) using modules, which share the same state space and (ii) introducing interaction between these modules. The modules are trained synchron, using time series (measurements) of different observables of the same system. The interaction between the modules is established through control variables.

In the following, we give first a short review concerning the state space reconstruction from time series followed by a descriprion of MCNN and its training algorithm. Thereafter, we investigate the neural model for arousal detection. Finally, conclusions and further work are stated in the last section.

2   Attractor Reconstruction and Neural Models

According to Takens' reconstruction theorem [4], a nonlinear model of a dynamical system could be derived directly from a systems' time series. Consider a sequence $\{x_t\}$ of t measurements generated through sampling on an observable of an m-dimensional dynamical system $f(\cdot)$ evolving on an attractor $\mathbb{A}$, with *fractal dimension* d_f. Under weak assumptions the fractal embedding theorem [5] ensures that, for $d_r > 2d_f$, the set of all *delayed coordinate vectors* $\mathbf{x}_{d_r,\tau} = \{t > t_0 : (x_t, x_{t-1}, \ldots, x_{t-(d_r-1)\tau})\}$ with arbitrary delay τ and arbitrary t_0, is an embedding of $\mathbb{A}$ in an n-dimensional *reconstruction* space. The minimal d_r is called *embedding dimension* d_e. The embedding preserves characteristic features of $\mathbb{A}$ and can be used to model the system. If the system is piecewiese stationary, due to slowly varying system parameters, it can be approximated as a sequence of stationary systems.

It is also well known, that multilayer feedforward neural networks are able to approximate arbitrarily well any continous real-valued function on a bounded subset of $\mathbb{R}^n$, [6],[7],[8]. In case of modeling of non-linear and slow time-varying systems with neural networks, introducing of a control signal, $q \in \mathbb{R}^n$, leads to the Hidden Control Neural Network (HCNN) that can approximate arbitrary closely any set $\{F_1, \ldots, F_N\}$ of continuous functions of the observable $\mathbf{x}$, ([9],[10],[11],[12]).The HCNN model provides a powerful tool for many time series prediction tasks, but it deals only with piecewise stationary systems. This model is acceptable for many problems, but not sufficient for the arousal detection task, because of the dynamical feature of its time series parameters[1].

3   The Mutual Control Neural Network

Our approach extends the benefits of the HCNN model. The MCNN model consists of a set $L = \{1, \ldots, L\}$ interacting modules, trained in parallel using traces of different observables. All modules in the model share the same state space S. Each module $l \in L$ experiences an influence[2] of the remaining $L \setminus \{l\}$ modules. Fig.1 depicts an example of the structure of MCNN with two modules. The input of each module consists of three parts: (i) the observable $\mathbf{x}^{(1)}$ (resp. $\mathbf{x}^{(2)}$), (ii) the hidden control parameter of the module, $q^{(1)}$ (resp. $q^{(2)}$, and (iii) the influence of the other modules in the model, e.g. module 1 "enforces"("corrects", or "disturbs") the module 2, through $q^{(1)}$, and vise-versa. The MCNN module $\lambda^{(l)}$, $l \in L$, can be described as:

$$\lambda^{(l)} = (\mathbf{w}^{(l)}, \mathbf{A}^{(l)}, \mathbf{B}^{(l)}, \mathbf{C}^{(l)}, \pi^{(l)}) \tag{1}$$

---

[1]In the arousal detection task, the slow time varying assumtion does not hold.
[2]This influence could be considered as "cross-talk" between different sources or "noise" induced from an unknown source, e.g. measurement errors.

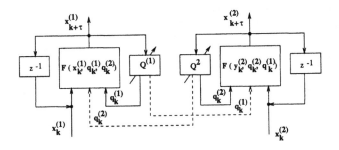

Figure 1: The MCNN structure, see text for explanation.

where:

$q_t^{(l)}$	State of module $l \in L$ at time t
$q_t^{(*)}$	"Enforcing" signal to module l at time t
$Q_t^{(l)}$	State sequence $\{q_1, q_2, \ldots, q_t\}$
$Q_t^{(*)}$	State sequence of the "enforcing" signal, defined similar to $Q_t^{(l)}$.
S_i	Values, the control signals $q_t^{(l)}$ take with $S_i \in S = \{S_1, \ldots, S_N\}$.
$\mathbf{x}_t^{(l)}$	Observable. $\mathbf{x}_t^{(l)}$ takes values out of a bounded set $\{u_1, u_2, \ldots, u_M\}$.
$X_t^{(l)}$	Sequence of observation $\{x_1^{(l)}, x_2^{(l)}, \ldots, x_t^{(l)}\}$.
$\mathbf{w}^{(l)}$	Weights of the neural network module l.
$\mathbf{A}^{(l)} = \{a_{ij}^{(l)}\}$	*transition* $N \times N$ *matrix, where* $$a_{ij}^{(l)} = P(q_{t+1}^{(l)} = S_j | q_t^{(l)} = S_j), 1 \le i, j \le N$$
$\mathbf{B}^{(l)} = \{b_{jk}^{(l)}\}$	*emission of observables, where* $$b_{jk}^{(l)} = P(\mathbf{x}_t^{(l)} = u_k | q_t^{(l)} = S_j), 1 \le j \le N, 1 \le k \le M$$
$\pi_{ij}^{(l)}$	initial probability that $q_1^{(l)} = S_i$ and $q_1^{(*)} = S_j, 1 \le i, j \le N$
$\mathbf{C}_t^{(l)}$	$\mathbf{C}_t^{(l)} = \{c_{ij}^{(l)}(k)\}$, where $$c_j^{(l)}(k) = P(\mathbf{x}_t^{(l)} = u_k, q_t^{(l)} = S_j | q_t^{(*)} = S_i), 1 \le i, j \le N, 1 \le k \le M$$

The state $\Omega_t = \cup_{l=1}^{L} q_t^l$ of the overall model results as the concatenation of all q_t^l and its interpretation depends on the chosen coding (about coding and data representation in neural networks see [13]).

4 Training of MCNN Modules

Training of each MCNN module is accomplished by minimising the prediction error. The error is a function of the hidden control input sequence Q_T and the "enforcing" signal[3] sequence Q_T^*:

$$E(\mathbf{w}, Q_T, Q_T^*) = \sum_{t=1}^{T} ||\mathbf{x}_t - F(\mathbf{x}_{t-\tau}, q_t, q_t^*)||^2 \tag{2}$$

---

[3] Superscript l omitted.

which is equivalent to

$$\min_{Q_T, Q_T^*} E(\mathbf{w}, Q_T, Q_T^*) \Leftrightarrow \max_{Q_T, Q_T^*} P(X_T, Q_T, Q_T^* | \mathbf{w}) \tag{3}$$

The Q_T^* is extern, thus considered as unique to the module, and we result in

$$\hat{\mathbf{w}} = \arg\min_{\mathbf{w}} \{\min_{Q_T} E(\mathbf{w}, Q_T, Q_T^*)\} \tag{4}$$

which implies the following training algorithm[4]:

1. Appropriate initialisation of $\mathbf{A}, \mathbf{B}, \mathbf{C}, \pi, \mathbf{w}$ for all modules $l \in \mathcal{L}$. In case of no a-priori knowledge of the input pattern distribution assume an uniform distribution for $\mathbf{B}$ and Gaussian for $\mathbf{C}$. $\mathbf{w}$ can be initialised with small random values drawn from uniform distribution.

2. For all $l \in \mathcal{L}$ repeat steps (a) and (b) until the termination condition is satisfied:

   (a) For the present values of Q_T, Q_T^* optimise $\mathbf{w}$:

   $$\hat{\mathbf{w}} = \arg\min_{\mathbf{W}} E(\mathbf{w}, Q_T, Q_T^*)\} \tag{5}$$

   using the backpropagation algorithm.

   (b) Determine the optimal sequence Q_T (using the Viterbi algorithm [16] to estimate $\hat{Q}_T$):

   $$\hat{Q}_T = \arg\min_{Q_T} E(\hat{\mathbf{w}}, Q_T, Q_T^*)\} \tag{6}$$

   In case of an ergodic model, we can perform this optimization as

   $$\hat{q}_t = \arg\min_{Q} ||\mathbf{x}_t - F_{\mathbf{w}}(\mathbf{x}_{(t-\tau)}, q_t, q_t^*)||^2 \tag{7}$$

5 Experiments

5.1 Data Aquisition, Parameter and Feature Estimation

Polysomnographic data of five patients were aquired as a part of a polysomnographic digitalised recording study, [17], consisting of: (i) **Electroencephalography**: two central C3A2, C4A1, two occipital channels O1A1, O2A2, two outer canthi electrographic channels (EOG) and a submental EMG. (ii) **Breathing**: consisting of detectors for the thoracic and abdominal movement, oronasal flow, oxygen saturation and snoring detection. (iii) **Movement**: anterior tibial EMG and infrared camera. (iv) **Electrocardiogram**, ECG. All patients' data were scored by a skilled clinician (sleep stages, respiratory events, arousals). The DC offset was subtracted from all data, followed by a subsequent scaling in $[0, 1]$. The EEG signals were filtered through a FIR low pass filter (cutoff frequency = 25Hz). Based on [3] and on our own estimates of the fractal dimensions of the EEG time series[5], we choose as embedding dimension

---

[4]See [14], [15] for methods to estimate $\pi, \mathbf{A}$ and [15] for a detail description of the training algorithm.
[5]EEG is regarded as the more *chaotic* signal of the polysomnography.

$d_e = 27$. From the available raw data we generate time series covering the sleep time between 1:00am and 3:00am. From the power spectra of the time series, calculated using 1sec steps and a Hanning window of 64 bins we generated two time series: a so-called max-spectrum-time series (msTS) using the frequency with the maximum power, and a spectral gravity-time series (gTS) using the gravity under the spectral curve. The gravity is defined as $g = \frac{\sum_{i=0}^{N=bins/2} Power\_at\_Freq(i)*Freq(i)}{\sum_{i=0}^{N=bins/2} Power\_at\_Freq(i)}$. The resulting vector sets were bisected in training and in validation sets, both of the same length. In a first experiment we combined the C3A2 channel with the left EOG (interpolated to the C3A2 sampling frequency). The MCNN model used consisted of two modules, each with 28 input units, one binary valued $q_t^{(l)}$ ($q_t^{(*)}$ resp.) state unit, one linear output unit, and 35 hidden units. The time delay τ was set to unity. The modules were trained simoultaneously

5.2 Results

First we evaluated the model for each patient separately using the raw signal time series. The same experiments performed with HCNN models resulted in poor arousal detections while a strong state oscillation in the transition paths was observed. Training with the spectra-time series or the gravity obtains an increasing recognition rate. In all cases the model relaxed down to two states, while the values of both q-units where equal, i.e. $(q^{(1)}, q^{(2)}) = (0,0)$ or $(q^{(1)}, q^{(2)}) = (1,1)$, resulting in the a sleep state sequence of (q_{sleep}, q_{awake}). Next table depicts the average arousal detection on the validation set (calculated over ten trials).

	HCNN	MCNN
raw	49.6% ± 7.5%	71.1% ± 11.2%
msTS	53.3% ± 8.4%	91.4% ± 7.3%
gTS	51.6% ± 18.9%	89.7% ± 9.9%

The two-sided χ^2 test showed that the MCNN outcomes are highly correlated with the visual scoring ($p < 0.001$) with an overlap of 97.8% (96.7% − 98.8% between patients). A slightly increased time spent in wake, as interpreted by the MCNN, is possibly due to the increased time spent in wake subsequent to apnoeas and hypopnoeas. Time spent in aroused condition after apnoeas/hypopnoeas (time window set at 5sec) increased about 16% (2-50% between patients), as detected by the MCNN.

6 Conclusion and Further Work

A new model for arousal detection is presented. The results indicate the superiority of our method in comparison to visual scoring in detecting arousals or changes in the state of sleep caused by apnoeas/hypopnoeas. The visual scoring, which is time consuming, expensive and the outcomes of which varry between sleep laboratories and technicians, is not of a high diagnostic validity in terms of arousal detection. Further validation of the method is required.

The high recognition rate, even using only the raw sampled data, suggests (i) that the model responds reliable to the underlying dynamics, and (ii) that a higher recognition rate is possible if an appropriate feature selection takes place. The first item is highly correlated with the issue of coding of the state parameters, especially in case of a MCNN with more than two modules. Both issues are parts of ongoing studies. In our current investigations we examine: (i) how the concatenation of the contol signals can be mapped onto $\mathbb{Z}$, (ii) what influence these mappings have on the learning algorithms, (iii) improvements of the model through better feature selection and (iv) topology and training algorithms of the modules.

References

[1] M. Sun, N.D. Ryan, R.E. Dahl, H.C. Hsin, S. Iyengar, and R.J. Sclabassi, "A neural network system for automatic classification of sleep stages", in Proc. 12th South. Biomed. Eng. Conf., pp. 54-64, 1991

[2] S.J. Roberts, M. Krkic, I. Rezek, J. Pardey, L. Tarassenko, J. Stradling and C. Jordan, "The use of Neural Networks in EEG Analysis", Proc. of IEE Colloquium on Sleep Analysis, Dec. 1995.

[3] C.L. Ehlers, J.W. Havstad, A. Garfinkel, and D.J. Kupfer, "Nonlinear analysis of EEG Sleep Stages", Neuropsychopharmacology, Vol. 5, Nr. 3,pp. 167-176, 1991.

[4] F. Takens, "Detecting strange attractors in turbulence", Vol. 898 of Lecture Notes in Mathematics, Dynamical Systems and Turbulence, pp. 366-381, Springer Verlag, 1981.

[5] T. Sauer, J.A. Yorke, and M. Casdagli, "Embedology", Journal of Statistical Physics, 65 (3/4):579-616, 1991.

[6] G. Cybenko, "Approximation by superpositions of a sigmoidal function", Mathematics in Control, Signals, and Systems, 2, pp. 303-314, 1989.

[7] J. Park and I.W. Sandberg, "Universal approximation using radial basis function networks", Neural Computation 3 (2), pp. 246-257, 1991.

[8] Y. Shin and J. Ghosh, "Ridge polynomial networks", IEEE Trans. Neural Networks, Vol. 6, pp. 610-622, 1995.

[9] T. Assimakopoulos, "Non-autonomous, dynamical system modeling using hidden control neural nets", Tech. Rep. 94-41, TU-Berlin, Berlin, Germany, 1994.

[10] E. Levin,"Hidden control neural architecture modelling of nonlinear time varying systems and its applications", IEEE Trans. on Neural Networks, 4(2), pp. 109-116, 1993.

[11] R.S. Huang, C.J. Kuo, L-L. Tsai, and O.T.C. Chen, "EEG pattern recognition - arousal states detection and classification", pp. 641-645

[12] S.K. Riis and A. Krogh, "Hidden neural networks: A framework for HMM/NN hybrids", ICASSP97, Vol. 4, pp. 3233-3236

[13] P.J.B. Hancock, "Data represenatation in neural networks: An empirical study", Connectionist Models Summer School, D.S. Touresky, G.E. Hinton, and T.J. Sejnowsky, Eds., pp. 11-20, San Mateo, 1988.

[14] L.R. Rabiner, "A tutorial on Hidden Markov Models and selected appliications in speech recognition", Proc. of IEEE, Vol. 77, No 2,pp. 257-286, Feb. 1989

[15] T. Assimakopoulos, "Mutual control neural networks: dynamical system modeling aspects and training issues", unpublished, TU-Berlin, Berlin, Germany, 1994.

[16] G.D. Forney, "The Viterbi algorithm", Proc. IEEE, Vol. 61, pp. 268-278, Mar. 1973.

[17] K. Dingli, S. Quispe-Bravo, I. Fietze, C. Witt, "Breathing and arousals: a retrospective analysis.", Eur. Respiratory Jour. 10, Suppl. 25, pp. 69s-70s, 1997.

Extraction of Sleep-Spindles from the Electroencephalogram (EEG)

Allan Kardec Barros

Bio-mimetic Control Research Center, RIKEN,
2271-130 Anagahora, Shimoshidami, Moriyama-ku, Nagoya 463, Japan

Roman Rosipal, Mark Girolami

Department of Computing and Information Systems, University of Paisley
Paisley, PA1 2BE, Scotland

Georg Dorffner

Dept. of Medical Cybernetics and Artificial Intelligence, University of Vienna
Freyung 6/2, A-1010 Vienna, Austria

and

Austrian Research Institute for Artificial Intelligence
Schottengasse 3, A-1010 Vienna, Austria

Noboru Ohnishi

Bio-mimetic Control Research Center, RIKEN,
2271-130 Anagahora, Shimoshidami, Moriyama-ku, Nagoya 463, Japan

Abstract

Independent component analysis (ICA) is a powerful tool for separating signals from their observed mixtures. This area of research has produced many varied algorithms and approaches to the solution of this problem. The majority of these methods adopt a truly *blind* approach and disregard available *a priori* information in order to extract the original sources or a specific desired signal. In this contribution we propose a fixed point algorithm which utilizes *a priori* information in finding a specified signal of interest from the sensor measurements. This technique is applied to the extraction and channel isolation of sleep spindles from a multi-channel electroencephalograph (EEG).

1 Introduction

Sleep spindles are particular EEG patterns which occur during the sleep cycle at 11.5 to 15 Hz, and are used as one of the features to classify the varying stages of sleep. The visual detection of sleep spindles is particularly difficult when multi-channel EEG recordings are used due to inter-channel degradation of the sleep-spindle signatures and other sources of unwanted signals which can be regarded simply as noise.

To aid the manual and possibly automatic detection of sleep spindle activity, it would be desirous to isolate the spindles in a single channel of EEG measurements. This could be possible by the application of a suitable form of the *independent component analysis* (ICA) transform for the extraction of an information bearing signal.

Preliminary work on sleep spindle isolation has been reported in Rosipal et al. [8]. ICA appeared recently as a promising technique for separating independent sources in biomedical signal processing [1, 6, 9]. ICA is based on the following principle. Assuming that the original (or source) signals have been linearly mixed in some otherwise unknown manner, and that these mixed signals are available as observations, ICA is a single neuron network which seeks to find in a *blind* manner a linear combination of the mixed signals which recovers the original source signals, possibly re-scaled and randomly arranged in the outputs.

However, extracting all the independent signals from, for example, a set of EEG measurements can be a lengthy and computationally demanding process. Therefore, in some cases, it may be important to extract only the subset of *desired* components however these may be defined. In the case under consideration, that of isolation of sleep spindles, we would seek to extract only the sleep spindles for further possible manipulation. To achieve this *desired signal* extraction we propose the use of a single neuron deflation algorithm along with the utilization of the *a priori* information available about the signal of interest. In fact, the deflation approach is important because we can extract only one independent component out of the measured signals, instead of separating all of them at once. However, as mentioned previously, the independent components are extracted in a non-deterministic manner, thus any one of the signals could potentially appear as the first extracted source. In solving this problem, we use the *a priori* information available about the desired signal as the input to a standard Wiener filter to initialize the signal extraction algorithm.

In the literature, Rosipal et al. [8] were the first to suggest that the application of general ICA algorithms could minimize the channel overlap in the transformed EEG, thus possibly isolating the sleep-spindle patterns. Our objective here is to enhance this isolation and possible subsequent classification by applying an ICA technique which will enforce the extraction of the sleep spindles alone. This then ensures that the electroencephalographer (or some automated classifier) would have only the sleep spindles at hand, after processing the measured EEG with the proposed new ICA algorithm.

2 Second Order Statistics

Let us first define the problem in a framework that will be used throughout this paper. Consider n *statistically mutually independent* signals $\mathbf{s}(k) = [s_1(k), s_2(k), \ldots, s_n(k)]^T$ arriving at n electrodes (or another type of receiver). Each electrode receives an unknown linear combination of the signals. For simplicity, we will drop the iteration index k and only use when necessary, then we have $\mathbf{s} = [s_1, s_2, \ldots, s_n]^T$. The mixed signals $\mathbf{v}$ are thus given by $\mathbf{v} = \mathbf{A}\mathbf{s}$, where $\mathbf{A}$ is an $n \times n$ invertible matrix[1]. We assume that we can observe only the mixture $\mathbf{v}$. Moreover, without loss of generality, let $\mathbf{x} = \mathbf{M}\mathbf{v}$, so that $E[\mathbf{x}\mathbf{x}^T] = \mathbf{I}$. In practice, this operation will speed up the convergence of the algorithm to be presented later, in addition to producing a simpler form of the updating technique.

Our purpose is to find from the mixed observation vector $\mathbf{x}$ one given component

---

[1]This is a simplified model of the more general linear mixing model where there may be modulation of the source and observation dimensions and sensor noise also may be included.

s_i of the source signal $\mathbf{s}$, using some *a priori* information included in a signal d, correlated with s_i, i.e., $E[ds_i] \neq 0$. In carrying this out, we use all the components of the input vector $\mathbf{x}$ so that we have $u = \mathbf{w}^T\mathbf{x}$, where $\mathbf{w}$ is a weight vector to be estimated by the adaptive algorithm, d is the *reference signal* and the error is given by $\varepsilon = d - u$. The weights are updated by the minimization of the mean squared error (MSE) given by $E[\varepsilon^2]$.

One can find easily that the optimum weight, which minimizes the MSE is $\mathbf{w}_* = \mathbf{R}^{-1}\mathbf{P}$, where $\mathbf{R}$ and $\mathbf{P}$ are defined as the correlation matrix of $\mathbf{x}$ and the cross correlation matrix of $\mathbf{x}$ and d respectively. Due to the spatial whitening of the observation signals $E[\mathbf{x}\mathbf{x}^T] = \mathbf{I}$ the Wiener solution optimum weight vector is then given by

$$\mathbf{w}_* = \mathbf{R}^{-1}\mathbf{P} = E[d\mathbf{x}]. \tag{1}$$

Barros et al. [1] have shown that in the presence of a reference signal, second order statistics are enough to provide separation of the source signals.

3 ICA for Biomedical Signal Processing

The final objective of ICA is also to separate signals, under the assumption that they are mutually independent, in other words, the joint probability density of the source signals is the product of the marginal densities of the individual sources, $p(\mathbf{s}) = \prod_{i=1}^{M} p(s_i)$. The difference with the method in the previous section is that most of the algorithms in the literature were proposed for separating the signals in a *blind* manner. Within this framework, ICA algorithms find a linear combination of the elements of $\mathbf{x}$ which gives the most independent components as the output. Usually this output is found using a matrix $\hat{\mathbf{W}}$, so that the elements of $\mathbf{z} = \hat{\mathbf{W}}\mathbf{x}$ are approximately mutually independent.

To find the matrix $\hat{\mathbf{W}}$, many different algorithms have been proposed in the literature. However, we are interested in the ability to extract or remove some given signal from a number of measured signals. Therefore, instead of finding an $n \times n$ matrix $\hat{\mathbf{W}}$ and from the output $\mathbf{z} = \hat{\mathbf{W}}\mathbf{x}$ pick up the desired signal (as in [7]), we are interested in finding only one component z_i of $\mathbf{z}$. Therefore, this component is given by $z_i = \hat{\mathbf{w}}_i^T\mathbf{x}$, where $\hat{\mathbf{w}}_i$ is one of the rows of $\hat{\mathbf{W}}$.

Dropping the index i for the purposes of exposition clarity, in the estimation of $\hat{\mathbf{w}}$, we combine the following updating rules:

$$\hat{\mathbf{w}}_{k+1} = E[\mathbf{x}(\hat{\mathbf{w}}_k^T\mathbf{x})^3] - 3\hat{\mathbf{w}}_k, \tag{2}$$

$$\hat{\mathbf{w}}_{k+1} = \mathbf{R}_{\mathbf{x}\tilde{\mathbf{z}}}\mathbf{R}_{\tilde{\mathbf{z}}z}, \tag{3}$$

where $\mathbf{R}_{\mathbf{x}\tilde{\mathbf{z}}} = E[\mathbf{x}\tilde{\mathbf{z}}^T]$ and, $\tilde{\mathbf{z}}(k) = [z(k-1) \cdots z(k-L)]^T$, and L is a delay. In the above equations, (2) is based on finding the extrema of the transformed output kurtosis [5]. As it may fail or show slow convergence in the case of low normalized values of kurtosis, we use (3) to update the algorithm [3]. It has been found that the combination of the two adaptive steps is a very efficient solution for the problem of blind source separation.

3.1 Proposed Algorithm

From the reasoning in the previous section, we propose the following fixed-point algorithm:

- Perform a principal component analysis (PCA) decomposition on the sample covariance matrix of the observation vectors $\mathbf{v}$ and project $\mathbf{v}$ onto the matrix of normalized eigenvectors $\mathbf{M}$ such that, $\mathbf{x} = \mathbf{Mv}$ and $E[\mathbf{xx}^T] = \mathbf{I}$.

- Take the initial vector $\hat{\mathbf{w}}_0 = E[d\mathbf{x}]/\|E[d\mathbf{x}]\|$. Iteration number $k = 1$.

- Update $\hat{\mathbf{w}}$ by, $\hat{\mathbf{w}}_{k+1} = E[\mathbf{x}(\hat{\mathbf{w}}_k^T\mathbf{x})^3] - 3\hat{\mathbf{w}}_k$. If the normalized value of the sample kurtosis of z is tending to zero, shift the updating to the kurtosis invariant algorithm

$$\hat{\mathbf{w}}_{k+1} = \mathbf{R}_{\mathbf{x}\hat{z}}\mathbf{R}_{\hat{z}z}, \tag{4}$$

- Divide $\hat{\mathbf{w}}_k$ by its norm and update $k = k + 1$.

- Test if $\|\hat{\mathbf{w}}_k - \hat{\mathbf{w}}_0\| < \zeta$, otherwise, change the current weight to the Wiener one, added to a small random deviation. This step is important to guarantee that the solution is *spatially close to* the Wiener one.

- Repeat the last three above steps until $\|\hat{\mathbf{w}}_{k+1}^T\hat{\mathbf{w}}_k\|$ approaches 1 (up to a small error)[2].

3.2 Signal Enhancement/Elimination

After obtaining the output $z = \mathbf{w}\hat{\mathbf{M}}\mathbf{v}$ using the algorithm described above, one may be interested either in keeping this signal, or in removing it from the sensors for subsequent manipulation or analysis. In order to accomplish this last option, we can simply use the Wiener filter as proposed above. This can be carried out by estimating the signal z using the previous ICA method, and computing its contribution to each element of $\mathbf{x}$. Thus, one can either have a vector of the contribution of z to each sensor as $\mathbf{y} = \mathbf{b}z$, or the sensors with z eliminated from it given by $\tilde{\mathbf{y}} = \mathbf{v} - \mathbf{b}z$. From (1), we find that the elements of $\mathbf{b}$ are estimated by

$$\mathbf{b} = E[\mathbf{v}z]. \tag{5}$$

4 Results

A seven minutes recording of 18 channels of EEG (Fp1, F8, F4, Fz, F3, F7, T4, C4, Cz, C3, T3, T6, P4, Pz, P3, T5, O2, O1) was used as the input data for the following experiments to demonstrate the utility of the proposed approach to biomedical signal processing. Electrodes were placed according to the international 10-20 system. The data were digitized with a sampling rate of 102.4 Hz.

---

[2]The MATLAB code for this algorithm is available upon request. Or, if the reader is interested in a version without a reference input, refer to the site in [2].

The measured signals were then filtered by a butterworth band-pass filter between 10 and 20 Hz. The signals were passed forward and backward through the filter to avoid phase distortion.

Channels HEO and Cz of the EEG were chosen as the reference inputs to extract the sleep spindles. The results are shown in Fig.1 for a window of 10 seconds it is clear that the sleep spindles have been extracted and therefore isolated within this single output channel. Notice that the outputs are already scaled in relation to the references using (5).

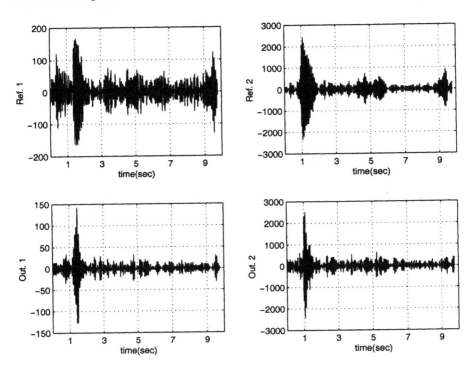

Figure 1: Example of two sleep spindles extracted from the raw data.

5 Conclusions

In this paper we have introduced an algorithm which will prove useful in biomedical signal processing where a specific underlying signal requires to be extracted from the possibly noisy multi-channel recordings. This algorithm is a modified version of the that originally developed in [1]. It is clear that this method is suitable for the extraction of independent components from the measured EEG. Experimental evidence of the algorithm's ability to extract pre-specified signals has been given using multi-channel EEG. The algorithm worked efficiently in extracting sleep spindles which were distributed throughout the measurement channels.

Acknowledgments

The authors thanks Prof. Zeitlhofer for providing EEG data set which was recorded at the Clinic of Neurology at the University of Vienna. This work was done as part of the Biomed-2 project BMH4-CT97-2040 SIESTA, sponsored by the EU Commission. RR is funded by a research grant for the project "Objective Measures of Depth of Anaesthesia"; University of Paisley and Glasgow Western Infirmary NHS trust, and is partially supported by Slovak Grant Agency for Science (grants No. 2/5088/98 and No. 98/5305/468). The Austrian Research Institute for Artificial Intelligence is supported by the Austrian Federal Ministry of Science and Transport.

References

[1] Barros AK, Vigário R, Jousmaki V, Ohnishi N. Extraction of Event-Related Signals from Multi-channel Bioelectrical Measurements *submitted to a journal*.

[2] Barros AK, Cichocki A. RICA - Reliable and robust program for Independent Component Analysis. Technical Report and MATLAB program of RIKEN. Web page: http://www.bmc.riken.jp/sensor/allan/RICA or http://go.to/RICA.

[3] Cichocki A, Barros AK. Robust Batch Algorithm for Sequential Blind Extraction of Noisy Biomedical Signals. In: Proceedings of ISSPA 1999, pp 363–366.

[4] Delfosse N, Loubaton P. Adaptive blind separation of independent sources: a deflation approach. Signal Processing 1995; 45:59–83.

[5] Hyvärinen A, Oja E. A fast fixed-point algorithm for independent component analysis. Neural Computation 1997; 9:1483–1492.

[6] Jung T-P, Makeig S, Westerfield M, Townsend J, Courchesne E, Sejnowski T. Independent component analysis of single-trial event related potentials. In: Proceedings ICA 1999, pp 173–179.

[7] Makeig S, Jung T-P, Ghahremani T, Bell AJ, Sejnowski T. Blind separation of event-related brain responses into independent components. In: Proceedings Natl Acad Sci USA 1997, 94:10979–10984.

[8] Rosipal R, Dorffner G, Trenker E. Can ICA improve sleep-spindles detection? Neural Networks World 1998; 5:539–547.

[9] Vigário R, Jousmaki V, Hamalainen M, Hari R, Oja E. Independent component analysis for identification of artifacts in magnetoencephalographic recordings. In: Advances in Neural Information Processing Systems. MIT press, 1997.

Analyzing Brain Tumor related EEG Signals with ICA Algorithms

M. Habl, Ch. Bauer, Ch. Ziegaus, E. W. Lang

Institute of Biophysics

University of Regensburg, 93040 Regensburg, Germany

F. Schulmeyer

Department of Neurosurgery

University Hospital, 93042 Regensburg, Germany

Abstract

Scalp EEG has been used as a clinical tool for the diagnosis and treatment of brain diseases. A generalized ICA algorithm modified by a kernel-based density estimation procedure is studied to separate EEG signals from tumor patients into spatially independent source signals. The algorithm allows artifactual signals to be removed from the EEG and isolates brain related signals into single ICA components. Their back-projection onto the scalp sensors provides topographic relations useful for a meaningful interpretation by the experienced physician.

1   Introduction

Scalp electroencephalography (EEG) is used routinely as a clinical tool for the diagnosis and treatment of brain diseases. Early identification and characterization of various brain tumors with non-invasive, low cost methods like EEG recordings would be especially valuable in this respect. Recently Makeigh et al. [1] demonstrated the ability of ICA algorithms to separate EEG signals collected with multiple electrodes into spatially stationary and temporally independent signals. The latter can be considered the signals generated by independent neural sources at different brain locations plus artifactual signals. Once the independent time courses of different brain and artifactual sources are extracted from the data, "corrected" EEG signals can be derived. It is only then that tumor related EEG signals can be clearly identified.

Since in scalp EEG the number of independent neural sources is generally unknown we analyze in this study EEG recordings from normal and tumor patients with a modified version of overcomplete ICA [2] and investigate the possibility to reliably isolate artifactual and tumor specific EEG signals into different independent ICA components.

2   Overcomplete ICA

Independent Component Analysis (ICA) was proposed to solve the *blind source separation* problem, to recover independent source signals, $\vec{s} = \{s_1(t), \ldots, s_M(t)\}^{\mathrm{T}}$ after they are linearly mixed by an unknown matrix $\mathbf{A}$. Nothing is generally known about the sources or the mixing process except that there are L different recorded mixtures, $\vec{x} = \{x_1(t), \ldots, x_L(t)\}^T = \mathbf{A}\vec{s}$. The task is to recover a possibly scaled and permuted version $\vec{u}(t)$ of the original sources.

Lewicki and Sejnowski [2] recently proposed an overcomplete ICA algorithm which might provide a method for identification when there are more sources than mixtures. We further generalize this algorithm by incorporating a kernel-based density estimation of the unknown prior densities $P(s_m)$ of the source signals. This modification provides a more flexible and adaptive approach and avoids any *ad hoc* assumptions concerning the source model.

2.1 The Mixing Model

Each component in the data vector $\vec{x} = (x_1, \ldots, x_L)^T$ represents the time course of the potential at a scalp electrode. The linear mixing model is described by

$$\vec{x} = \mathbf{A}\vec{s} + \vec{\varepsilon} \tag{1}$$

where $\mathbf{A}$ is an $L \times M$-matrix and $\vec{\varepsilon} = \vec{\varepsilon}(t)$ is an additive Gaussian noise. The number of sensors L may be less than (overcomplete), equal to (square) or greater than (undercomplete) the number of sources M. The latter are assumed to be mutually independent, hence their joint probability density can be factorized: $P(\vec{s}) = \prod_{m=1}^{M} P(s_m)$. The components of $\vec{s} = (s_1, \ldots, s_M)^T$ may correspond to independent neural activity patterns or to artifactual signals like eye movement and blinks, cardiac signals, muscle noise etc.

The *log likelihood* of a Gaussian additive noise is given by

$$\log P(\vec{\varepsilon}) = \log P(\vec{x} \mid \mathbf{A}, \vec{s}) \propto -\frac{1}{2\sigma_N^2}(\vec{x} - \mathbf{A}\vec{s})^2 \tag{2}$$

with σ_N^2 the noise variance.

2.2 Inferring the sources

Due to the additive noise and the possibly non-square mixing matrix $\mathbf{A}$ a probabilistic approach to estimating the basis coefficients (= sources) is most appropriate. The most probable decomposition $\vec{s}$ of the signal is found by maximizing the posterior distribution of $\vec{s}$

$$\hat{\vec{s}} = \max_{\vec{s}} P(\vec{s} \mid \vec{x}, \mathbf{A}) = \max_{\vec{s}} P(\vec{x} \mid \mathbf{A}, \vec{s}) P(\vec{s}). \tag{3}$$

A suitable approach for optimizing $\vec{s}$ in the case of finite noise ($\vec{\varepsilon} > 0$) and a non-Gaussian prior probability $P(\vec{s})$ is to use *conjugate gradient descent* on the negative *log posterior* $-\log P(\vec{s} \mid \vec{x}, \mathbf{A})$.

2.3 Learning the basis vectors

Basis vectors can be learnt by maximizing the probability of the data given the model

$$P(\vec{x}_1, \ldots, \vec{x}_K \mid \mathbf{A}) = \prod_{k=1}^{K} P(\vec{x}_k \mid \mathbf{A}) \tag{4}$$

where temporal independence of the data vectors is assumed. Computation of the likelihood requires marginalizing over all possible sources

$$P(\vec{x}_k \mid \mathbf{A}) = \int P(\vec{s}) P(\vec{x}_k \mid \mathbf{A}, \vec{s}) \, d\vec{s} \tag{5}$$

which formulates the problem as one of density estimation. Performing gradient ascent on the objective function $\log P(\vec{x}_1, \ldots, \vec{x}_K \mid \mathbf{A})$ and using a saddle point approximation of (5) around the posterior mode $\hat{\vec{s}}$ yields the learning rule [2]

$$\Delta \mathbf{A} \propto \mathbf{A}\mathbf{A}^\mathsf{T} \frac{\partial}{\partial \mathbf{A}} \log P(\vec{x} \mid \mathbf{A}) \approx -\mathbf{A}(\Phi(\hat{\vec{s}}) \, \hat{\vec{s}}^\mathsf{T} + \mathbf{I}) \tag{6}$$

where $\Phi(\hat{s}_m) = \partial \log P(\hat{s}_m)/\partial \hat{s}_m$ is called the *score function* and $\mathbf{I}$ is the identity matrix.

2.4   Estimating the source density

The proper choice of the prior probability of the basis coefficients $P(s_m)$ is crucial to any success of the decomposition. It is often assumed that the basis coefficients have a sparse distribution with the prior probability being modelled by super-Gaussian Laplace density $P(s_m) \propto \exp(-\lambda|s_m|)$. If the assumed prior density does not fit the unknown source density reasonably well, i.e. if the latter is sub-Gaussian, for example, a reliable extraction of some independent source signals is no longer possible. Instead of making any *ad hoc* assumptions, an appropriate prior can be estimated from the data during the training process. We use a kernel-based density estimation with either Gaussian or Laplacian kernels [3]. The kernel estimator is given as

$$\hat{P}(s_m) = \frac{1}{h_m n} \sum_{i=1}^{n} K\left(\frac{s_m - \hat{s}_m^{(i)}}{h_m}\right) \tag{7}$$

where h_m represents the *bandwidth* of each kernel $K(\ldots)$ and may be computed as $h_m = 1.06 \, \sigma_m n^{-1/5}$ where σ_m^2 is the variance of the basis coefficient $s_m(t)$.

Starting with an initial Laplacian prior a new estimate of the marginal density $P(s_m)$ can be obtained from the last n inferred sources $\hat{s}_m^{(i)}$ ($i = 1, \ldots, n$). We used about $n = 200$ kernels for the approximation to be used in equations (3) and (6).

3   Experimental Results

EEG data sets were recorded from 21 scalp electrodes placed according to the International 10-20 System with a sampling rate of 167 Hz. Additional signals from the ECG or EOG placements were not used for this analysis. ICA decomposition was performed on 40–50 sec EEG epochs from the data sets. Starting with a randomly initialized matrix $\mathbf{A}$, the 21-dimensional data vectors $\vec{x}$ were presented in cyclic order over 200000 iterations using (6) at a learning rate fixed to 0.00045. The calculations presented below refer to the square case $L = M$, as no better results could be obtained with the overcomplete case $L < M$. In the former $\hat{\vec{s}} = \mathbf{A}^{-1}\vec{x}$ could be used for quickly inferring the most probable sources (3) under the assumption of zero noise ($\sigma_N = 0$). Due to the time consuming density estimation learning took 20 min of CPU time on a SUN Ultra 10 workstation.

3.1 Results

Figure 1 shows a 2 sec portion of the recorded EEG time series and figure 2 its respective ICA component activations. The EEG was recorded from a patient with an identified cancerous Meningeom in the left frontal brain region. Three artifactual and two tumor related ICA components may be identified from figure 2. Component no. 4 reveals small periodic muscle spiking sampled at the left scalp area mainly (figure 4). Eye blink artifacts in the EEG data are isolated to ICA component no. 7. Any cardiac contamination in the EEG data is concentrated in ICA component no. 20 and shows up mostly in the sensor signals close to both ears. ICA components no. 12 and 17 show δ/ϑ-waves with a characteristic frequency of 3.5–4.5 Hz. These oscillations are slightly phase shifted with respect to each other. δ-waves are not observed with normal objects and result from swollen brain regions induced by the spreading malignant tumor. Back-projection to the scalp electrodes indicates, that at least ICA component no. 17 stems from the identified cancerous Meningeom. ICA component no. 12 originates from the central scalp region and is not yet characterized clearly.

By calculating with an overcomplete basis set ($M = 26$) we obtained the same five signals (figure 3) as described above for the square case ($M = 21$). But right now the eye blink artifact exists identically in six different channels. So using this algorithm for EEG data with more dimensions than the number of sensors has no advantage.

3.2 A Method of Identification of Artifactual and Tumor Related ICA Components

A major problem is the interpretation of isolated ICA components. Only few of them may be identified via their characteristic frequencies or signal shapes as artifactual or tumor related signals unequivocally. Others cannot be identified by simple means. A way to automate this identification will be presented next.

After performing the ICA analysis 15 times with randomly initialized matrices $\mathbf{A}$ different matrices $\mathbf{A}_i$ ($i = 1,\ldots,15$) resulted with only a subset of the basis vectors (columns of $\mathbf{A}$) being reproducibly obtained in almost all $\mathbf{A}_i$. One way to select these constant basis vectors is to calculate for all combinations $i < j$ the matrix $\mathbf{A}_i^T \cdot \mathbf{A}_j$ (with all columns of the matrix $\mathbf{A}_i$ being normalized), whose elements provide a similarity measure of the corresponding basis vectors.

However, better results were obtained when the ICA components $\vec{u}^i(t) = \mathbf{A}_i^{-1}\vec{x}(t)$ were compared directly via their related correlation coefficients

$$C_{mm'}^{ij} = \frac{|\sum_t u_m^i(t) u_{m'}^j(t)|}{\sqrt{\sum_t (u_m^i(t))^2 \sum_t (u_{m'}^j(t))^2}} \tag{8}$$

Two signals $u_m^i(t)$ and $u_{m'}^j(t)$ may be considered similar if $C_{mm'}^{ij} > 0.9$ holds. After calculating $C_{mm'}^{ij}$ for all combinations $i < j$ ($i,j = 1,\ldots,15$) and $m,m' = 1,\ldots,M$ a suitable search algorithm classifies all similar signals $u_m^i(t)$ into different classes. With the example given above 5 different classes were obtained containing 13–15 ICA components each. All others of the 15×21 signals have found no or just by accident only

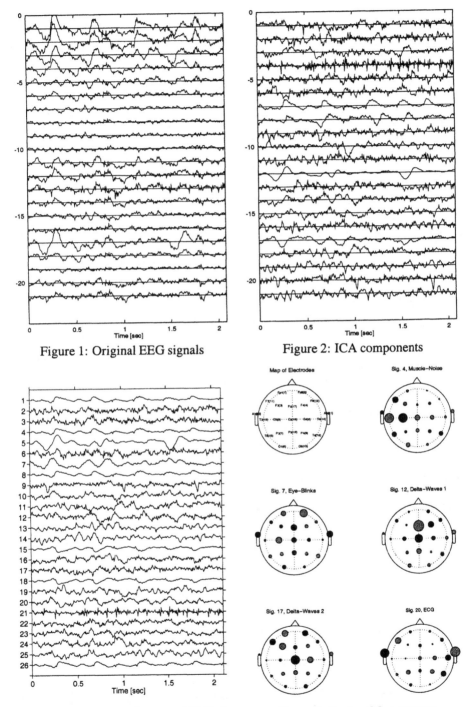

Figure 1: Original EEG signals

Figure 2: ICA components

Figure 3: Overcomplete case (26 dim)

Figure 4: Scalp of five sources

one similar partner. The reason why only 5 ICA components could be isolated reproducibly can be found in the densities of the non-reproducible components which were almost Gaussian. Hence these near Gaussian components could not be decomposed by the algorithm.

The upper limit of the entropy of a signal with unit variance is $\frac{1}{2}(1 + \log(2\pi)) \approx 1.42$ which is exactly obtained with a normally distributed signal. The entropy

$$E(u(t)) = -\int du\, p(u) \log(p(u)) \approx -\frac{1}{\sum_t 1} \sum_t \log(p(u)) \tag{9}$$

is thus a suitable measure to judge the results obtained with the ICA analysis. The farther $E(u_m^i(t))$ deviates from 1.42 the stronger is the probability for being an independent source signal. The corresponding calculation of the entropy of the 21 signals presented in Figure 2 yielded the following result:

m	1	2	3	4	5	6	7	8	9	10	11
$E(u_m)$	1.42	1.44	1.43	**1.34**	1.43	1.42	**1.22**	1.42	1.43	1.40	1.43
m	**12**	13	14	15	16	**17**	18	19	**20**	21	
$E(u_m)$	**1.25**	1.43	1.43	1.42	1.43	**1.31**	1.42	1.42	**1.33**	1.42	

4 Discussion and Conclusions

We have presented a flexible ICA algorithm which does not rely on *a priori* assumptions about the unknown source distributions. Though starting from an overcomplete ansatz the best results have been obtained with the square case, $L = M$. Comparing results of the ICA analysis starting with randomly initialized mixing matrices or by calculating the entropy of the isolated components within a single ICA analysis it was possible to identify consistently independent source signals, whether neural or artifactual. Most of the components isolated were almost normally distributed, hence cannot be considered independent source signals. Yet we have shown that tumor related EEG signals can be isolated into single independent ICA components. These signals were not observed in corresponding EEG traces of normal patients (not shown because of space limitations). Their back-projection onto the scalp electrodes allows a topographic assignment and thus a meaningful interpretation by an experienced physician.

References

[1] S. Makeigh, A. J. Bell, T.-P. Jung, T. J. Sejnowski *Independent Component Analysis of Electroencephalographic Data* Advances in Neural Information Processing Systems 8, D. Touretzky, M. Mozer and M. Hasselmo (Eds.), MIT Press, Cambridge MA, 145-151, 1996

[2] M. S. Lewicki, T. J. Sejnowski *Learning overcomplete representations*, Neural Computation, in press, 1999

[3] B. W. Silverman, *Density Estimation for Statistics an Data Analysis* Chapman and Hall, London, 1986

Isolating Seizure Activity in the EEG with Independent Component Analysis

Christopher James

Neural Computing Research Group, Aston University
Birmingham, United Kingdom
jamescj@aston.ac.uk

David Lowe

Neural Computing Research Group, Aston University
Birmingham, United Kingdom
lowed@aston.ac.uk

Abstract

We present a methodology for isolating the underlying seizure activity from the scalp EEG recorded during seizure onset in a number of patients. We use the method of Independent Component Analysis (ICA) to isolate both the temporal and spatial aspects of the seizure activity. Seizure related activity in the independent components is identified through visually analysing the spatio-temporal components along with the spectrogram of the temporal components. Subjectively, in four seizure EEGs analysed we can identify what appear to be the relevant seizure components. Furthermore, artifactual components are isolated from the seizure activity. This means that the scalp EEG can either be 'remapped' using only the identified seizure components, or further analysis on the seizure can be undertaken on the spatio-temporal components directly. Although subjective, the preliminary results indicate that ICA can be of benefit on pre-processing the epileptiform EEG for further analysis.

1 Introduction

The EEG is a recording of the electrical activity of the brain as measured at the scalp and is used in the diagnosis of many brain disorders, most particularly in the diagnosis of epilepsy. The ictal EEG includes the abnormal electrical activity generated during an epileptic seizure. Analysing seizures plays an important part in contributing towards the overall analysis of the type of epilepsy and is of particular interest in the analysis of which regions of the brain are involved with the seizure activity.

The EEG recorded in the long-term monitoring units at epilepsy centres, contains a lot of artifacts both from intra-cerebral and extra-cerebral sources. In analysing ictal EEG it is particularly useful to analyse the seizure onset in order to find which regions of the brain are contributing to the seizure activity at its earliest stages.

We propose a technique based upon Independent Component Analysis (ICA), a relatively new technique used to separate statistically independent components from a mixture of data [1,2]. ICA is finding its way into the analysis of EEG, however almost all of the work is related to extracting artifacts from routine EEG [3] or analysing averaged evoked responses. ICA does not require any data averaging nor is any assumption required for the generator model in the head, such as the constraint to use current dipoles, for example. We want to evaluate the potential of using ICA as a means of isolating seizure activity, particularly at the seizure onset, with the goals of removing artifactual/unrelated background EEG to facilitate subsequent analysis of the seizure – such as localisation of the seizure origin. We have already evaluated ICA in isolating Epileptiform Discharges (EDs) in the ongoing interictal EEG (i.e, without first averaging the EEG) [4,5]. Our previous studies, on synthetic data and arrays of real ED data, indicated that ICA can manage to isolate EDs (synthetic or otherwise) from the background EEG and even manages to separate sources which are close both in terms of their spatial origin and also in terms of the type of activity generated. In this paper we explore ICA on longer seizure segments. As far as we know, ICA has not been previously applied to the analysis of seizures and EDs in the ictal and interictal EEG.

2 Methods

2.1 Independent Component Analysis

ICA decomposes mixed input data into a set of independent components without any information about the distribution of the sources, i.e. blind separation of sources. Whereas Principal Component Analysis (PCA) uses second order spatio-temporal correlation information for data decomposition, ICA uses higher-order statistics. Several different implementations of ICA can be found in the literature [1,2,6,7,8] we settled for using the Fast ICA algorithm [8,9] because of its ease of implementation and speed of operation.

The assumptions we make in the analysis of scalp EEG are in keeping with the underlying assumptions of implementing ICA. We can assume that the scalp EEG is a linear summation of the electrical activity from various brain regions [10] and that the potential field distribution is spatially fixed and that the electrical strength only is changing within these regions [11]. Finally we assume that any epileptic activity is independent of any ongoing background EEG – this holds true at least for the seizure onset, early on in the evolution of a seizure.

2.2 Ictal EEG

The ictal scalp EEG data was obtained from four epileptic patients of the Montreal Neurological Institute and Hospital. The EEGs were digitally recorded at a rate of 200 samples/sec with a low-pass filter of 65 Hz and with 12 bits resolution. Twenty-five electrodes were used for recording the scalp EEG, placed according to the modified 10-20 electrode placement system [12]. The data was recorded with a reference at position FCz and the data matrices in each case were then mean corrected in the columns (i.e., an average referential montage was assumed). The recordings consisted of 20 s of seizure EEG with the seizure onset (as determined visually from the scalp EEG) occurring approximately mid-way into each segment. Fast ICA was then performed on the data matrices for further analysis.

2.3 Selecting relevant seizure components

The major disadvantage with using ICA on data such as this is that the resulting components are not rank ordered such is the case for PCA, for example. For this study, we opted to make our choice of the relevant seizure components visually by (a) looking for rhythmic structure to the temporal components occurring in the right time frame, (b) looking at the corresponding topographic maps given by the columns of the mixing matrix and looking for dipolar sources in the region of brain identified from the scalp recording and (c) constructing a spectrogram of each temporal component and looking for a *change* in cerebral activity which coincides with the seizure onset (as visually determined from the scalp EEG). This is a highly subjective approach, but it was considered as acceptable as a proof of principle study into the effectiveness of ICA in ictal EEG analysis.

3 Results

From the four seizures examined in this way, ICA managed to extract components which were deemed seizure components (at least under the criterion set out in the previous section). Notably, in each case, there was more than one component of 'clear' rhythmic activity centred around the region of brain suspected as being the focus from scalp EEG studies. Furthermore, the ability of ICA to extract sources which were clearly extra-cerebral in nature such as ocular artifacts, muscle contamination, ECG and electrode artifacts was proven in each case.

Figures 1a, 1b and 1c show the spatial and temporal components of (a) ocular artifact – IC6, (b) muscle contamination – IC7 and (c) seizure component – IC16, for patient 1. The scalp EEG for this patient indicated an epileptic focus centred around right fronto-polar region, which is reflected by the spatial component shown in Figure 1c (IC16).

Figure 2a shows the seizure EEG for patient 2 which has a left temporal lobe onset approximately mid-way through the recording which spreads rapidly. Figure 2b shows the resulting independent components and the spatial components and spectrograms for IC's 1, 6, 15 and 20. Both the temporal and spatial components indicate that IC's 1 and 6 are artifact and that no relevant cerebral activity seems to be isolated in these components. IC's 15 and 20 both show seizure activity (a) due to the rhythmic temporal components and (b) due to the strong focal nature of the topographic maps centred around the appropriate region of the brain. The spectrogram shows the evolution of cerebral activity at about mid-way through the data, corroborating the previous clues. Furthermore, IC20 seems to be representative of the seizure portion where it has more of a basal left temporal focus and IC15 appears to describe the seizure component with a more parietal focus.

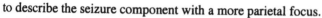

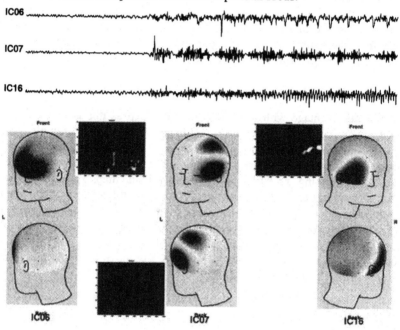

Figure 1. Three independent components from Patient #1 depicting (a) ocular artifacts – IC6, (b) muscle artifact – IC7, and (c) seizure component – IC16.

4 Conclusions

This preliminary study has shown that seizures can be isolated from a mixture of epileptic and background activities through ICA. More investigation is warranted to examine the relevance of there being more that one independent component related to seizure activity; are the multiple components a product of the ICA algorithm or due to separate cerebral activity contributing to the seizure development? In spatio-temporal source modelling, physiological noise, i.e., the background activity, mis-leads the computational process [13]. We suggest that applying source modelling to the field distribution of the seizure components isolated by ICA might be better than applying the source analysis to the original EEG data which is heavily, and un-avoidably, contaminated.

In ICA one of the drawbacks could be the appropriate choice of the relevant components, in this preliminary study we use the subjective method of choosing the seizure component(s) by looking at both the temporal and spatial components and at the spectrogram, given our prior knowledge that the seizure onset occurs roughly mid-way through the data. A method allowing a more objective choice of the rel-evant seizure components is the focus of our next study.

We conclude that ICA could be an invaluable tool in the analysis of the epileptiform EEG and we plan to pursue this study with more real epileptiform data and to perform source analysis on the extracted components allowing us to further corroborate/contrast our findings with spatio-temporal source analysis methods commonly in use today.

We gratefully acknowledge funding from EPSRC grant #GR/L94673 through which this work has been made possible, and the Montreal Neurological Institute and Hospital for the EEG data.

References

[1] Comon P. Independent component analysis, a new concept? Signal Processing, 1994; 36: 287-314

[2] Jutten C, Herault J. Blind separation of sources part I: An adaptive algorithm based on neuromimetic architecture. Signal Processing, 1991;24:1-10

[3] Vigário RN. Extraction of ocular artefacts from EEG using independent component analysis. Elctroenceph. clin. Neurophysiol., 1997;103:395-404

[4] James CJ, Kobayashi K, Lowe D. Isolating Epileptiform Discharges in the Unaveraged EEG using Independent Component Analysis. Medical Applications of Signal Processecing Colloquium, IEE, London, 6th October, 1999

[5] Kobayashi K, James CJ, Nakahori T, Akiyama T, Gotman J. Isolation of epileptiform discharges from unaveraged EEG by independent component analysis. Electroenceph. clin. Neurophysiol. 1999;110:1755-1763

[6] Bell AJ, Sejnowski T.J. An information-maximization approach to blind separation and blind deconvolution. Neural Computation, 1995;7:1129-1159

[7] Makeig S, Jung TP, Bell AJ, Ghahremani D, Sejnowski TJ. Blind separation of auditory event-related brain responses into independent components. Proc. Natl. Acad. Sci. USA, 1997;94:10979-10984

[8] Hyvärinen A, Oja E. A fast fixed-point algorithm for independent component analysis. Neural Computation, 1997;9:1483-1492

[9] The FastICA MATLAB Package. Available at http://www.cis.hut.fi/projects/ica/fastica/, 1998

[10] Nunez PL, Katznelson RD. Electric Fields of the Brain: The Neurophysics of EEG. New York: Oxford University Press, 1981

[11] Scherg M, Von Cramon D. Two bilateral sources of the late AEP as identified by a spatio-temporal dipole model. Electroenceph. clin. Neurophysiol., 1985;62:32-44

[12] Lagerlund TD, Sharbrough FW, Jack CR, Erickson BJ, Strelow DC, Cicora KM, Busacker NE. Determination of 10-20 system electrode locations using magnetic resonance image scanning with markers. Electroenceph. clin. Neurophysiol., 1993;86:7-14

[13] Achim A, Richer F, Saint-Hilaire JM. Methodological considerations for the evaluation of spatio-temporal source models. Electroenceph. clin. Neurophysiol., 1991;79:227-240

142

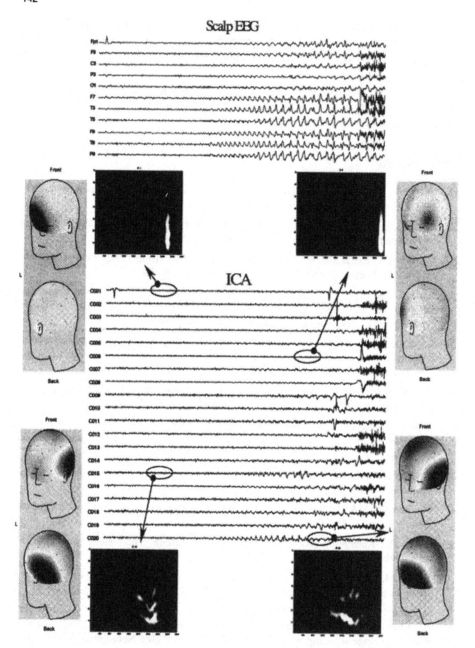

Figure 2. Part of the scalp EEG and independent components of a 20 second segment of seizure EEG from Patient #2. The top part of the figure shows the raw EEG (LH side channels only) and the bottom the ICs. Components 1, 6 15 & 20 have been highlighted along with their topographic maps and their respective spectrograms.

Seizure Detection with the Self-Organising Feature Map

Christopher James

Neural Computing Research Group, Aston University
Birmingham, United Kingdom
jamescj@aston.ac.uk

Katsuhiro Kobayashi

Department of Child Neurology, Okayama University Medical School,
Okayama, Japan

Jean Gotman

Department of Neurology and Neurosurgery,
Montreal Neurological Institute and Hospital, McGill University,
Montreal, Canada

Abstract

We have developed a system to detect the presence of seizures in the multi-channel scalp EEG. At the heart of the system is the Self-Organising Feature Map (SOFM) that has been trained on normal and epileptiform EEG segments of 12 patients (64 seizures). Following preliminary spatial analysis, auto-regressive (AR) parameters are extracted from variable width segments which have been delineated through the use of a non-linear energy operator. The AR parameters are used as feature vectors for the SOFM training process. Following initial training, probability values are automatically assigned to the 'prototype' seizure segments based on the consensus of 3 EEGers. The use of a self-organising network retains objectivity in calculating the prototype seizure segments.

Preliminary results (using the training set only) are given here. With a detection threshold of d_{th}=0.49 the Sensitivity and Selectivity were both measured at 75% with a corresponding false detection rate of 0.5 / hour. These preliminary results indicate that the system shows promise for use as a generic seizure detection system - i.e., a non-patient specific seizure detection system.

1 Introduction

For long term epilepsy monitoring, increasing numbers of channels are being used in the recording of digital EEG. The automated detection of seizures becomes imper-ative in order to be able to manage properly the large amount of data generated during a recording session. The ultimate aim of any automated seizure detection system is to have a 100% sensitivity and selectivity. In practice this is difficult to achieve but a reasonable balance of both measures is still considered acceptable.

Of the automatic detection systems in the literature e.g., [1,2,3,4], most are based on single channel analysis of scalp EEG [1] and in most cases are based on the frequency analysis of the EEG. A template method was proposed by [3], although this is not a general purpose method as it can only detect seizures similar to a template seizure from the same patient. ANN approaches have also been considered in the literature [4]. The use of the self-organizing feature map (SOFM) in clustering EEG data is not new, the SOFM was used for sleep staging by [5], for evoked response clustering by [6] and for spike detection by [7].

We present a proof-of-principle system that detects seizures with appreciable performance whilst maximising on the use of spatial and temporal cues in the EEG, as well as taking into account the different morphologies seen in seizure EEGs

2   The Multi-stage System

The system we propose is a hybrid system based on multiple stages, at the heart of which is the Self-Organising Feature Map (SOFM) ANN (Figure 1). The method brings into play the spatial and temporal features of the data as an integral part of the detection process. The multiple stages of the system are described next:

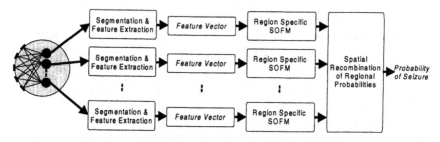

Figure 1: A block diagram of the multi-stage seizure detection system comprising: (1) Remontaging scalp EEG to regional brain activity through a multiple current dipole model. (2) Variable width segmentation of each channel using a non-linear energy operator. (3) Feature extraction from the segments in the form of a 25 parameter Auto-Regressive model. (4) Self-organised clustering using the SOFM. (5) Spatial recombination of the regional prob-abilities output by each SOFM.

2.1   Remontaging the Scalp EEG

Scalp EEG recordings in epilepsy can contain 30 or 40 EEG channels. Rather than analysing each channel independently, we want to integrate the information meaningfully, so the scalp EEG is first remontaged to a fixed set of current dipoles placed in regions of anatomical / neurophysiological interest of a 3-shell spherical model of the head. A Weighted Linear Spatial Decomposition solution is used to calculate each regional activity (R-EEG), this is based on previous work [8]. This has the effect of reducing the number of channels of scalp EEG to process, and decorrelates the EEG through a set of spatial filters.

2.2 Segmentation

The R-EEG is segmented into segments of variable length, each segment being approximately stationary. A method based on the Non-linear Energy Operator (NEO), a function that is proportional to amplitude and frequency of the signal, is used to this end [9,10]. The segmentation process results in segments of R-EEG with lengths varying from 2 to 10 or 20 seconds.

2.3 Feature Extraction

A feature vector is extracted from each (variable length) segment in order to be presented to the ANN for training and testing purposes. We model each segment as an Auto-Regressive (AR) process of fixed model order (p=25) [11] and use the parameters of the model as the feature vector. It is possible that the EEG can be modelled by AR models of orders down to 2, but model order varies considerably depending on the segment being modelled. A variable model order would be ideally suited to cover all eventualities, however as we require a fixed model order we opted for a higher order system so as to be able to model the more complex waveforms (based on [11]).

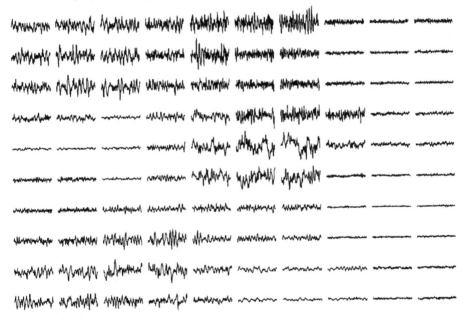

Figure 2: The inverse filtered AR parameters at each node of a 10x10 SOFM trained with AR parameters extracted from the R-EEG segments in the brain region denoted LO (or Left Occipital).

Because we parameterise the EEG segments through the AR parameters, it becomes possible to reconstruct the representative waveforms at each node of the ANN by passing white noise through the inverse AR filter allowing for a visualisation of the segment types represented by the trained SOFM (as explained in the next section).

2.4 The Self-Organising Feature Map

The SOFM ANN is described best in [12] and is at the heart of the detection system. During training, through a self-organising process, the SOFM becomes "tuned" to the specific characteristics underlying the data used in its training. Following training the SOFM results in an N-dimensional topological representation of the most prominent (statistically) features in the training data set. The information is held in the form of weight-vectors which are of an identical order as the input vectors. In our implementation the SOFM assigns a probability value to incoming EEG segments based on the similarity of the input wave morphology to those held in the trained weight vectors of the SOFM. As the feature vectors are made up of AR parameters we replace the usual Euclidean distance as a similarity measure with the Itakura distance measure [13], which reflects the goodness-of-fit of the two sets of AR parameters, rather than the distance between the parameters themselves. Each node of the SOFM is characterised by a feature vector which is representative of the underlying EEG epochs of the training data (see Figure 2). A sub-set of the training set is chosen and labelled as seizure/non-seizure based on a consensus of 3 EEGers and is used to calibrate the trained SOFM and assign a probability value to each node according to the frequency with which each node 'wins'.

The whole training process is repeated for each channel of R-EEG such that a different, region specific, SOFM is trained for each channel. This is important as different brain regions will exhibit characteristic artifacts that tend to be region specific.

2.5 Spatial Recombination

The variable length segments of the R-EEG, which have been assigned a probability value, are first aligned into concurrent 2 s epochs. Then channels are grouped regionally and a simple linear summation is performed followed by smoothing over a 6 s floating window. A seizure is detected if the probability in any brain region exceeds a variable threshold at least twice over 10 consecutive seconds of EEG.

3 Subjects

Epileptiform EEGs of 12 patients undergoing long term monitoring at the Montreal Neurological Institute and Hospital were used to form a limited training set. This consisted of 64 seizures and sampled segments of normal EEG covering all times of day and night giving a total of 30 hours of multi-channel EEG. The seizure EEG segments were labelled (based on a consensus of 3 EEGers) and used to auto-matically assign probabilities to the trained SOFMs. Sixteen 2D SOFMs, consisting of 10x10 lattices were trained, one for each R-EEG region.

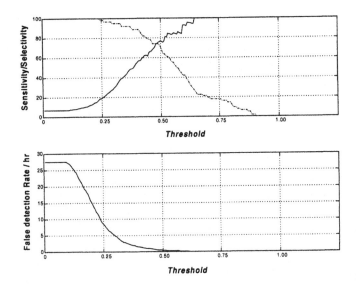

Figure 3: The performance of the system (tested using the training set as input) across all threshold values. The sensitivity is shown by the dashed line and the selectivity by the solid line. At cross-over (d_{th}=0.49) both the sensitivity and selectivity are at 75% and the false detection rate is of 0.5/hour.

4 Results

We present here the preliminary results obtained by testing the system with the training set data. (Novel test data is being collected and will form part a future analysis of the system). By varying the detection threshold from 0.0 to 1.0 and presenting the test set, it is possible to measure the Sensitivity, Selectivity and False Detection rate of the system at each threshold level. The resulting values of Sensitivity, Selectivity and False Detection rate are depicted in Figure 3.

At cross-over (d_{th}=0.49) the sensitivity and selectivity are both 75% with a False Detection rate of 0.5/hour. Alternatively, with a False Detection rate of 1/hour the sensitivity is 85% and the selectivity 62%. In comparison [4] report a sensitivity of 90% with a mean FD rate of 1.29±1.32/ hour (over 162 records or 3681.9 hours). On the same data set applying the method of [2] resulted in a sensitivity of 74% and a mean FD rate of 3.02±2.78/ hour.

5 Conclusions

We present a multi-stage seizure detection system that is based around the SOFM. We contend that breaking down the detection problem into manageable stages, along with maximal use of spatial and temporal cues, results in a system with optimised performance in comparison with other automated systems. In particular, the use of the SOFM ANN, a self-organised system, means that the network retains as much objectivity as possible when calculating the prototype seizure segments.

Although testing with the training set introduces bias in the results at this stage, preliminary results indicate that the system shows promise for use as a generic - i.e., non patient specific - seizure detection system. Further testing with novel seizure data is our next goal in the development of the system. This work is funded by Canadian MRC Grant #10-189 and in part by UK EPSRC Grant #GR/L94673.

References

[1] Murro AM, King DW, Smith JR, et al. Computerized seizure detection of complex partial seziures. Electroenceph. clin. Neurophysiol. 1991;79:330-333

[2] Gotman J. Automatic seizure detection: improvements and evaluation. Electroenceph. clin. Neurophysiol. 1990;76:317-324

[3] Qu H, Gotman J. A patient-specific algorithm for the detection of seizure onset in long-term EEG monitoring Possible use as a warning device. IEEE Trans. Biomed. Eng. 1997;44:115-122

[4] Gabor AJ. Seizure detection using a self-organising neural network: validation and comparison with other detection strategies. Electroenceph. clin. Neurophysiol. 1999;107:27-32

[5] Roberts S, Tarassenko L. New method of automated sleep quantification. Med. & Biol. Eng. & Comput., 1992;30(5):509-517

[6] Peltoranta, M. and Pfurtscheller, G. Neural-network based classification of non-averaged event-related EEG responses Med. & Biol. Eng. & Comput., 1994;32:189-196

[7] James CJ, Jones RD, Bones PJ, Carroll GJ. Detection of epileptiform discharges in the EEG by a hybrid system comprising mimetic, self-organized artificial neural network, and fuzzy logic stages. Clin. Neurophysiol. 1999;110(12):2049-2063

[8] James CJ, Kobayashi K, Gotman J. Using Weighted Linear Spatial Decomposition to investigate Brain Activity through a set of fixed Current Dipoles. *To appear:* Clin. Neurophysiol. 2000

[9] Kaiser JF. On a simple algorithm to calculate 'energy' of a signal. In: Proc. IEEE Intl. Conf. Acoust., Speech, and Signal Processing. 1990; April

[10] Agarwal R, Gotman J. Adaptive segmentation of electroencephalographic data using a nonlinear energy operator. In: ISCAS '99: Proc. IEEE Intl. Symposium on Circuits and Systems. 1999;4:199-202

[11] Lopes da Silva F, Djik A, Smits H. Detection of nonstationarities in EEGs using the autoregressive model - an application to EEGs of epileptics. In: CEAN: Computerised EEG analysis. Gustav Fischer Verlag, 1974, pp 180-199

[12] Kohonen T. Self-organizing maps. Springer-Verlag, Berlin, 1995

[13] Kong X, Lou XS, Thakor NV. Detection of EEG changes via a generalized Itakura distance. In: Proceedings of the 19th annual international conference of the IEEE Engineering in Medicine and Biology Society, 1997;19:1540-1542

Graphical Analysis of Respiration in Postoperative Patients Using Self Organising Maps

M.Steuer[1], P.Caleb[1], P.K.Sharpe[1], G.B.Drummond[2], A.M.S.Black[3]

[1] Intelligent Computer Systems Centre, University of the West of England, Bristol, UK

[2] Dept. of Anaesthetics, University of Edinburgh, Edinburgh, UK

[3] Dept. of Anaesthesia, Bristol Royal Infirmary Bristol, UK

Abstract

Previous work has shown that the relationships between respiratory pressures and dimensions can be used to investigate how analgesia, airway obstruction and hypoxia are related. These relationships can be more clearly visualised by plotting the different pairs of signals against each other to create a graphical representation of the respiratory mechanisms. This technique of visual classification has been automated using Self Organising Maps. Respiratory signal can be classified into categories on a breath by breath basis without having to explicitly indicate the start of inspiration. Having achieved this we have established a system for extending the research from a medical perspective by making it viable to conduct wider studies on a larger patient base.

1   Introduction

After major surgery patients may experience severe pain which is only controllable by pushing morphine-like analgesics to their limits of systemic or regional effectiveness. The consequence of this practice is a significant risk of disrupting the control of the upper airway, or breathing, and precipitating life-threatening hypoxia. An ideal monitoring system would not only detect respiratory distress, but also anticipate them from indicators present in the breathing patterns. Nimmo and Drummond [3] used a method involving visual analysis of respiratory pressures and dimensions to investigate how analgesia, airway obstruction and hypoxia are related. This involved recording four physiological signals, oesophageal and gastric pressures and chest and abdomen dimensions, and studying their phase relationships. Degrees of asynchrony of ribcage and abdominal movement were found to be consistent with degrees of airway obstruction [3].

In this work Self Organising Maps (SOMs) [1] were used to learn the graphical representations of the different categories of the breathing mechanism. The main feature of SOMs is that they use unsupervised learning to group data into categories. The system adapts in order to recognise groups of similar cases from within its set of training examples without being told in advance what the different categories might be. In doing so, it develops a set of prototypical vectors or centres in the data set to form regions or clusters. These regions can be identified and calibrated by projecting known data examples on to the trained map.

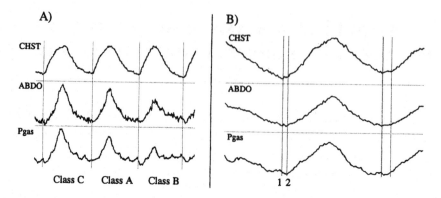

Figure 1: A) Segmented and classified part of the respiratory signal. B) Ambiguity resulting from explicitly defining the starting point of the breath in some cases.

2   Background

The data set used for this study was the same as that used by Nimmo and Drummond [3]. Data from eight out of the ten original patients was used. These included patients aged 40 or more, who had undergone major abdominal surgery involving an incision at least partly above the umbilicus. The exclusion criteria were patients with severe cardiac, respiratory or renal disease or an abnormal body weight. The signals were measured during the first night after surgery under observation at the Edinburgh Royal Infirmary. The average duration of the recording of each patient was about 5.5 hours. The following signals were used in our study: 1) nasal gas flow (NASAL), 2) chest dimension (CHST) 3) abdominal dimension (ABDO) 4) oesophageal pressure (Poes) and 5) gastric pressure (Pgas).

Chest and abdominal movement signals were measured using inductance bands around the chest and abdomen. Oesophageal and gastric pressure were measured using a modified nasogastric tube with an integral oesophageal balloon and a gastric balloon attached to the tube tip.

The category of each breath was determined according to the algorithm described by Nimmo and Drummond [3].

Pgas	ABDO	Class
normal	normal	A
paradox	normal	B
paradox	paradox	C

The starting point of inspiration was detected from the CHST signal. The ABDO signal normally rises simultaneously with the CHST signal, but sometimes, it first decreases slightly and then starts to rise after a delay. This is called an ABDO paradox. The same may happen to the signal Pgas. By analogy this case is called a Pgas paradox. The breaths were thus classified as shown in the table.

An example of a short sequence with the different classified types is shown in Figure 1-A. Class C changes to A and later to B. Single breaths are separated by vertical lines, which also indicate the beginning of inspiration.

When the rules summarised in the table above were implemented in software, the main problem encountered was the determination of the starting point of inspiration. As shown in Figure 1-B, even a small shift in the starting point can result in a completely different classification. As can be seen, it is very hard to say exactly where inspiration begins. For example, if the start of the first breath was determined as point 1, the breath would be classified as C. If it was at point 2, the classification would be A. The resulting classification is not very reliable in terms of accuracy but is adequate for giving an indication of the balance of classes present in each patient. This information is useful when constructing training and test sets.

2.1 Data Preparation

The raw signal contains a large amount of disturbance caused either by the patient (movements, coughing, talking and so on) or by problems with the technical equipment (slipping of the inductance bands, or general outages). All these sections of signal had to be removed, as the breathing signals were distorted or completely damaged by the artifacts. To mark the disturbed sections, a technique based on Wavelet analysis [2] was used.

The analysis of the signal was done on a breath by breath basis so the next step involved the segmentation of the signals into separate breaths. To achieve this, we used the signal CHST as an indicator of the start of the inspiration.

A point was marked as the starting point of a breath if it lay in a trough with predefined lengths of descending and rising sections. The exact detection of the starting point was not easy because of the noise contained in the signal. Due to the overlap of the signal of interest and noise bandwidths, noise filtering was not used as it would result in loss of detail information contained in the signal. This has the additional benefit that the presence of random noise helps neural networks to develop better generalisation capabilities, so noise in the signal was not considered harmful.

The selected signal which was to be used for testing the networks was then classified by a clinician. This provided a correct and reliable classification against which the results of the neural networks were evaluated.

Due to the practical problems with maintaining calibration of some of the measuring equipment through the entire recording period, drifts in the absolute values of the signal were noted. As a result, all the signals were normalised to a fixed height and width to enable standard visualisation. The shape of the signals is more significant than the actual value of the samples in this situation.

2.2 Experiments and Visualisation of the Results

The correct selection of inputs is a very important precondition for the network to solve a particular problem. Our choice was to use the graphs of the pairs of signals, shown in Figure 2, which were converted to vectors by digitising the graphs to a 19 x 19 pixel grid. These will be referred to as *plot vectors*. The resolution was found to be the optimum size for capturing all the major features of the graphs.

The resulting *plot vectors*, of which there were 15, were investigated separately, so in all, there were 15 self organising maps (SOMs) created per patient, one map

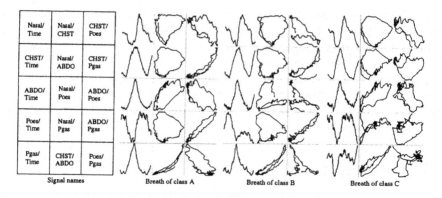

Figure 2: An example of the graphs of all the signal combinations used for the classification. The items in the table on the left correspond to the graphs on the right. The breaths shown in the graphs on the right demonstrate examples from all three classes.

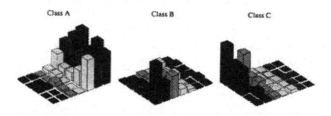

Figure 3: Test data projected onto a 7x5 node SOM network trained using ABDO/Pgas plot vector. The diagrams show the number of hits on each node for each class.

per *plot vector*. This also allowed us to investigate the importance of each signal individually.

3   Analysis of the Results

Several configurations of the SOM were tested to find the best. Maps of 7 by 5 neurons were found to give the best separation The trained maps were evaluated by projecting the test data classified by the clinician. One of the results is shown in Figure 3.

Analysis of the results showed several important facts.

- The self organising maps are capable of extracting features necessary to separate between the different classes of breaths.

- It is possible to reduce the number of signals while still maintaining satisfactory classification. This was experimentally tested by using each of the different *plot vectors* separately. In most cases, the SOM separated one class from the other two classes. So for the complete separation of all three classes at least two maps had to be used. E.g. one separated class A from all the others, the second separated class C from all the others and the remaining breaths were separated as belonging to the class B. None of the *plot vectors* alone seemed to contain sufficient information for good separation of all the three classes of breaths, so one map was not enough to perform this task. There was just one *plot vector* (ABDO/Pgas as shown in Figure 3) which approached this ideal of classification of all three classes at once, but the separation of the three classes was not as good as in the case of using at least two together.

- Some *plot vectors* did not result in any separation at all. The results are summarised in the following table. The code shows which class was separated from the others.

graph	separation	quality
Nasal/Time	-	
CHST/Time	-	
ABDO/Time	C	high
Poes/Time	-	
Pgas/Time	A	medium
Nasal/CHST	-	
Nasal/ABDO	C	high
Nasal/Poes	-	
Nasal/Pgas	A	high
CHST/ABDO	C	high
CHST/Poes	-	
CHST/Pgas	A	medium
ABDO/Poes	C	high
ABDO/Pgas	A,B,C	low
Poes/Pgas	A	medium

- These results were very stable across many variants of the SOM, different sizes and training coefficients.

- Experiments were also conducted using slightly different variants of the graphs. The difference was in the length of the signal used. The data included: 1) Inspiration only, 2) Inspiration and expiration, 3) Inspiration and expiration, including small overlaps with neighbouring breaths. The starting point of the inspiration was taken from the previous process of segmenting the signals into separate breaths. The results obtained using the first two data sets were similar, but the third data set, where the starting point of inspiration was not explicitly stated, gave the best results.

4 Conclusions

The main result of this work was to show that it is possible to automate the recognition of single breaths as classified by Nimmo and Drummond [3].

Although the algorithm for the classification can be coded in software, using self organising maps provides a more suitable approach in terms of possibilities of enhanced expert interpretation of results through visualisation. Additionally, experiments have shown that the exact boundaries of the breaths need not be explicitly specified as the self organising maps seem capable of extracting the relevant information more reliably.

The process of classification using self organising maps is based on recognition of the shape of the breathing signals, so the absolute values of the signals are not needed at all. This is also an important benefit because it is not easy to keep some of the components of the measuring system calibrated during the whole course of the overnight measurement.

On evaluating the importance of the signals in contributing to achieving a high level of classification, our experiments showed that at least one of the invasively measured signals, namely Pgas, is necessary for correct classification and the information carried by this signal could not be extracted from the other, more easily measurable, signals.

On the other hand some signals did not seem to be necessary at all (e.g. Poes). Also, the signal CHST, which was supposed to be used for the exact determination of the start of inspiration, is not necessarily needed.

Currently, alternative methods of data representation and feature extraction are being explored along with comparisons being made with other classification techniques. We will also be investigating repeating these studies using further derived signals based on the combination of the present signals with a view to enhancing the level of the information available to the networks.

Having achieved the ability to automate the identification of single breaths, it will be possible to investigate the nature of the signals on a larger scale. The next stage will involve the correlation of the statistics concerning the frequency of occurrence and relationship between the sequence of occurrences of the different classes of the breaths, to the condition of the patients breathing.

References

[1] T. Kohonen, J. Hynninen, J. Kangas, and J. Laaksonen. *SOM_PAK: The Self-Organizing Map Program Package.* Helsinki University of Technology, Laboratory of Computer and Information Science; Report A31, 1996. Available at http://www.cis.hut.fi/nnrc.

[2] M. Misiti, Y. Misiti, G. Oppenheim, and J. Poggi. *Matlab Wavelet Toolbox User's Guide.* The MathWorks, Inc., 1996.

[3] A. F. Nimmo and G. B. Drummond. Respiratory mechanics after abdominal surgery measured with continuous analysis of pressure, flow and volume signals. *British Journal of Anaesthesia,* (77):317–326, 1996.

[4] A. Ultsch. *Self-organizing Neural Networks for Visualisation and Classifikation.* Department of Computer Science, University of Dortmund, P.O. Box 500, D-4600 Dortmund 50, Germany. Report.

Clinical Diagnosis

and

Medical Decision Support

Neural Network Predictions of Outcome from Posteroventral Pallidotomy

Jeffrey E. Arle

Department of Neurosurgery, The Lahey Clinic
Burlington, MA, USA
arle1neuro@pol.net

Ron Alterman

Department of Neurosurgery, The INN at Beth Israel Hospital
New York, NY, USA

Abstract

Neural networks have recently been utilized to make predictions in multivariate systems where simple univariate correlations often miss key predictive relationships. Toward this end, we have implemented an application of neural network analysis to the problem of patient selection and outcome prediction for posteroventral pallidotomy (PVP) in patients with Parkinson's Disease (PD). Forty six (46) cases were used to develop a variety of neural network structures that were able to learn and predict 6 month outcomes with a high degree of accuracy. Twelve variables were used as inputs to the nets. Networks could predict 93% of Unified Parkinson's Disease Rating Scale (UPDRS) outcomes (choosing between less than or greater than 50% improvement). Networks were further validated by using only random sets of input data and testing their predictions on known cases. Additionally, we used real-data created nets and tested them on random-data test 'cases'. Nets trained on random data were unable to learn at all (understandably). The networks created with real data, moreover, were no better than chance (50%) at predicting random outcomes. These results strongly suggest that the networks created were specific to the pallidotomy data.

1   Introduction

Several groups have documented outcomes in series of posteroventral pallidotomies (PVP) performed on patients with Parkinson's Disease (PD) [e.g.1,2]. All have shown the great potential benefit for these patients with each of the classic symptoms including dyskinesia, bradykinesia, on-off phenomena, rigidity, gait and voice disturbances, and tremor. The most responsive symptom, however, seems to be dyskinesia. and secondarily rigidity and bradykinesia. We have sought to refine patient selection further for PVP by taking into account multiple variables to predict outcome. We used a non-linear approach, employing neural networks to analyze the interaction of input variables in PD and particular aspects of the PVP itself to predict outcome with this procedure.

2   Methods

2.1   Patient Selection

All PVP's performed at our institution between 12/91 and 8/96, a total of 154 patients, were reviewed for preoperative, demographic, intra-, and postoperative follow-up data. Forty six (46) cases were found to have a uniform database, including 6 month follow-up United Parkinson's Disease Rating Scale (UPDRS) scores and preoperative UPDRS scores. All patients had initially responded to L-Dopa therapy, and at the time of surgery were predominantly hampered by dyskinesias, bradykinesia, rigidity, and/or on-off fluctuations.

2.2   Surgical Technique

The details of the surgical procedure for PVP in this study have been described elsewhere [3]. The procedure is performed under local anesthesia. Single cell recording is obtained to delineate the floor of the Globus Pallidus interna (Gpi) and the transition from Globus Pallidus externa (Gpe) to GPi for optimum placement and a follow-up MRI is performed within 1 week of surgery.

2.3   Patient Evaluation

All patients were assessed using the UPDRS, which were administered 12 hours off and then on medications, for pre-operative baselines and follow-up.

2.4   Neural Network Development

Networks were developed using commercially available software (NeuralWorks – Neuralware Inc., Pittsburgh, PA), allowing a wide range of network structures, parameter manipulations, and learning algorithms. The data were coded and placed in file formats in such a manner as to enhance training and allow for cross-validation of test sets of the real data. Typical techniques for development and refinement were as described in prior studies [4,5]. Briefly, back-propagation algorithms in small networks with one or two hidden layers were analyzed. The decision level for output layer nodes was set to 0.975, and variations of learning rules, momentum terms, activation functions, and network sizes were adjusted until a 'best' network for a given set of inputs was found. All of a network's predictions were obtained by cross-validation using a randomly held-aside set of the original data. In this manner, the network being trained was tested on data from the original set of patients that it had never seen before in training. These data were cases randomly removed from the original data set for the purposes of testing the network once trained. Final networks were 'bootstrapped' by performing this cross-validation 5 times with the results averaged in order to determine the network's overall predictive ability. This second layer of validation helped eliminate variability in a network's potential performance on a single random set of the test data.

The number of 'cases' in the training set of data was increased by randomly altering individual cases with a small amount of Gaussian noise signal (10% of the range) so as to create 'new' cases. All of these new artificial *training* cases were based on original data, but increased the overall size of the training set to enhance network learning and prevent overfitting. This method has been used effectively in other neural network studies and is often essential to eliminate entrenchment of the network in the training set and frequent overtraining which often occurs with clinical data. All *test* sets used in cross-validation, however, were composed only of actual data cases from the series.

Additionally, an equivalently-sized set of training 'cases' was created using entirely random values in the same range as the original training sets (rather than the slightly noise-altered 'new' cases used for the original training). This 'data' was used to create another neural network, which failed to learn the randomized 'cases'. In reverse, the original 'best' network was also tested using the randomized test 'cases' but, as hoped, was only correct 50% of the time.

2.5 Input Variables Evaluated

Variables evaluated in this study included age, sex, laterality of the lesion made, years with a diagnosis of PD, preoperative UPDRS score, whether the patient's primary symptomatology was dyskinesia or other symptoms, which extremities were primarily affected (upper, lower, or both), total span of the GP (GPi + Gpe, taken in mm from the microelectrode recording raw data) along the lesioning trajectory, the impedance measurement at the intended target site, the number of trajectories taken during entire procedure, and the number of lesions made.

3   Results

Patients had been diagnosed with PD for an average of 13.1 years (range 3-45 years) before coming to surgery. The average age was 60.5 years (range 37-81 years) and included 27 male and 19 female patients. Statistical tests on individual variables showed that the two outcome groups were similar. Only the preoperative UPDRS scores were found to be statistically different between the two groups (p=0.038).

Figure 1 shows the final network structure for the 'best' network found for predicting outcome between the two groups (>50% UPDRS improvement, or <50% UPDRS improvement). In general, relatively small networks with 2 hidden layers and both feedforward and feedback connections worked best.

When length in the GP and the number of trajectories were left out, networks could predict 93% of UPDRS outcomes (choosing between less than or greater than 50% improvement). Interestingly, adding these data (length in GP and the number of trajectories) and training a new 'best' network with this additional information lowered the predictive ability to 79%. Other combinations of variables yielded 'best' networks with predictive estimates between 60% and 85% accuracy.

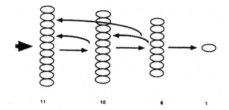

Figure 1: Schematic of the 'best' network obtained to predict outcome after PVP. The eleven input variables were connected through two 'hidden' layers of 10 and 8 nodes respectively in feedforward and feedback paths. The outcome node was considered 1 if its value was >0.975 and 0 otherwise.

As a further check of the specificity of the trained networks, some networks were trained with a completely random set of values within the same ranges as the original data, and then tested on a set aside group of original cases. They were no better than 50% accurate in predicting outcomes. In addition, the 'best' network was tested against completely random set of cases, again made from random values within the same range as real data. It was no better than 50% accurate as well in predicting outcome.

Table 1: Statistical analysis of particular individual variables. Student's t-tests were performed to determine whether there are significant relationships between certain variables and outcome. X=mean, SD=standard deviation. None of the variables was significantly different in the two outcome groups, except for pre-operative UPDRS, which obtained a p value of 0.038. This suggests that having a lower level of function (higher baseline UPDRS) prior to surgery plays an independent role in a patient receiving greater benefit from the surgery.

Outcome

	>50% Improvement	<50% Improvement	p value
Number	n = 28	n = 18	
Age	x = 60.5 years SD = 7.8 years	x = 60.4 years SD = 11.0 years	0.974
Years with PD	x = 13.0 years SD = 6.3 years	x = 13.3 years SD = 10.2 years	0.909
Pre-Op UPDRS	x = 45.8 SD = 17.7	x = 34.1 SD = 18.6	0.038*
Target Impedence	x = 525.9 SD = 95.9	x = 481.9 SD = 112.0	0.130
# Lesions	x = 4.75 SD = 1.24	x = 4.167 SD = 0.92	0.090
# Trajectories	x = 1.57 SD = 0.64	x = 1.94 SD = 1.16	0.167

4 Conclusions

Our results are encouraging, in that over 90% of the outcomes could have been predicted using the neural network. While there are possible weaknesses in using neural networks to analyze multivariate systems, they generally perform better than the best multiple regression analyses and can better account for differentially weighted factors that may be non-linearly related. In this study, we also had the networks tested on random data outcomes to appreciate their specific generalization to pallidotomy cases as a third level of validation of the network testing. They were, as hoped, unable to predict random outcomes any better than chance (50%).

References

[1] Hariz M. Correlation between clinical outcome and size and site of the lesion in computerized tomography guided thalamotomy and pallidotomy. Stereotact Funct Neurosurg 1990;54/55:172-185.

[2] Bakay RAE, DeLong MR, Vitek JL. Posteroventral pallidotomy for Parkinson's disease. J Neurosurg 1992;77:487-488.

[3] Dogali M, Sterio D, Fazzini E, Kolodny E, Eidelberg D, Beric A. Effects of posteroventral pallidotomy on Parkinson's disease. In: Battistin L, Scarlato G, Caraceni T, and Ruggieri S, eds. Advances in Neurology 1996;69:585-590.

[4] Arle JE, Morriss C, Wang ZJ, Zimmerman RA, Phillips PG, Sutton LN. Prediction of posterior fossa tumor type in children by means of magnetic resonance image properties, spectroscopy, and neural networks. J Neurosurg 1997;86(5):55-61.

[5] Arle JE, Devinsky O, Perrine K, and Doyle W, Neural Network Analysis of Preoperative Variables and Outcome from Epilepsy Surgery, J Neurosurg, 1999; 90(6):998-1004.

Survival Analysis: A Neural-Bayesian Approach

Bart J. Bakker, Bert Kappen and Tom Heskes*
Foundation for Neural Networks, Geert Grooteplein 21
6525 EZ Nijmegen, The Netherlands

Abstract

In this article we show that traditional Cox survival analysis can be improved upon when written in terms of a multi-layered perceptron and analyzed in the context of the Bayesian evidence framework. The obtained posterior distribution of network parameters is approximated both by Hybrid Markov Chain Monte Carlo sampling and by variational methods. We discuss the merits of both approaches. We argue that the neural-Bayesian approach circumvents the shortcomings of the original Cox analysis, and therefore yields better predictive results. As a bonus, we apply the Bayesian posterior (the probability distribution of the network parameters given the data) to estimate p-values on the inputs.

1   Introduction

The goal of survival analysis (in medical terms) is to estimate the chances of a patient's survival as a function of time, given the available medical information. A well-known way to conduct such an analysis is Cox's proportional hazards method [1]. In this method the hazard function $h(t;x)$, which estimates the probability density of death occurring at time t, is a product of two independent parts. The first part is the proportional hazard, $h(x) = \exp(w^T x)$, which depends on patient information x only, the second part is a time-dependent baseline hazard $h_0(t)$.

The proportional hazards method can be implemented in the form of a multi-layered perceptron (MLP) with exponential transfer functions. Introducing sensible priors on the parameters, we adopt a Bayesian approach. The resulting posterior can be approximated either by drawing samples, using for example Hybrid Markov Chain Monte Carlo (HMCMC) sampling, or a form of "ensemble learning". This last term was coined by Hinton and van Camp [2] and has been applied to MLP's in [3].

In practice, medical experts do not work with probability distributions over model parameters directly: they rather use the concept of p-values. The Bayesian posterior distribution will therefore be used to "analytically" calculate p-values for the inputs to the network (i.e. patient information). The proposed methods are tested on a medical

*This research was supported by the Technology Foundation STW, applied science division of NWO and the technology programme of the Ministry of Economic Affairs. We would like to thank David Barber for stimulating comments and discussions.

database of 929 ovarian cancer patients, of whom (next to their medical information) the time of death or censoring (extraction from the research group for reasons other than ovarian cancer) has been recorded. More information about this database can be found in [4].

2 Cox Survival Analysis

Given the hazard function $h(t;x) = \exp(w^T x)h_0(t)$ the survivor function $F(t;x)$, indicating the probability to survive time t, and the function $f(t;x)$, indicating the probability density of dieing at time t can be formulated as

$$F(t;x) = \exp\left[-\int_0^t dt'\, h(t';x)\right]; f(t;x) = -\frac{\partial F(t;x)}{\partial t} = h(t;x)F(t;x).$$

The likelihood function $P(D|w)$, expressing the probability to observe the data in database D given the model parameters w, then immediately follows as

$$P(D|w) = \prod_{v \in \text{uncensored}} f(t^v;x^v) \prod_{\mu \in \text{censored}} F(t^\mu;x^\mu).$$

The first product is over the patients of whom the time of death is known. An element in the second product specifies the estimated probability of censored patient μ to be alive at time t^μ, the time patient μ was taken out of the study.

An advantage of classical Cox analysis is that the optimal parameters of the proportional and the time-dependent hazard can be found sequentially. The optimal choice for the parameters w (w^{ML}) depends only on the ordering of the times of death of the patients (see [1]); all other time-dependent information can be modelled in the function $h_0(t)$. Disadvantages of this approach are the hazards tendency to become highly non-smooth, and the danger of strongly over-fitting the data.

3 Neural Interpretation

A discrete version of the main elements of Cox analysis (hazard and survivor function) can easily be represented by an MLP with exponential transfer functions, as can be seen in Figure 1 (concentrate for the moment only on the solid lines). The input x consists of the elements of patient information. The weights w in the first layer (input to hidden) correspond to the parameters w in the proportional hazard function $\exp(w^T x)$ which, as can easily be seen, is the output of the hidden unit. Equating the weights v in the second layer (hidden to output) to minus the integral upto time t_i of the base-line hazard (Fig. 1), the output $F_i(x)$ of the network at neuron i equals the probability for a patient with characteristics x to survive upto time t_i. The likelihood function now becomes

$$P(D|w) = \prod_{v \in \text{uncensored}} f_{i(v)}(x^v) \prod_{\mu \in \text{censored}} F_{i(\mu)-1}(x^\mu), \tag{1}$$

where $i(\mu)$ is the first discrete point in time after death or censoring of patient μ.

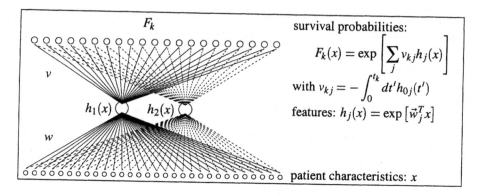

F_k

survival probabilities:

$$F_k(x) = \exp\left[\sum_j v_{kj} h_j(x)\right]$$

with $v_{kj} = -\int_0^{t_k} dt' h_{0j}(t')$

features: $h_j(x) = \exp\left[\vec{w}_j^T x\right]$

patient characteristics: x

Figure 1: Neural interpretation of survival analysis.

Given this network structure, it is easy to extend the model beyond proportional hazards. By adding more hidden units to the second layer (dashed lines), we may be able to model complex input-output relations, which cannot be found in the proportional hazard approach. In this article, we will focus on the network with one hidden unit.

4 Bayesian Inference

The solution proposed in this paper is a Bayesian approach. Instead of merely searching for the maximum likelihood solution w^{ML}, we seek to construct a probability distribution (hence the term "ensemble learning") over all possible values of the parameters in our network. This distribution will not only depend on the data in our database, but also on prior knowledge about the nature of the problem. Using Bayes' formula, the prior and the data likelihood can be transformed into a posterior distribution. The first prior

$$P(v|\gamma) \propto \exp\left[-\frac{\gamma}{2}\sum_{ij} f(|i-j|)[v_i - v_j]^2\right] \propto \exp\left[-\frac{\gamma}{2}v^T \Gamma v\right],$$

where $\Gamma_{ij} = -f(|i-j|), \Gamma_{ii} = \sum_{j\neq i} f(|i-j|)$ and we choose $f(x) = e^{-x^2}$, prevents the hazard from becoming too sharp as a function of time. It introduces a preference for survivor functions which decay exponentially. The second prior

$$P(w|\lambda) \propto \exp\left[-\frac{\lambda}{2}w^T \Lambda w\right], \quad \text{where} \quad \Lambda = \frac{1}{\#\ patients}\sum_\mu x^\mu x^{\mu T},$$

prevents large activities of hidden units, i.e., prefers small weights. Incorporation of the covariance matrix Λ makes this preference independent of a (linear) scaling of the inputs x.

The posterior distribution of the parameters w and v and hyperparameters λ and γ given the data follows from Bayes' formula:

$$P(w, v, \lambda, \gamma | D) = \frac{P(D|w, v)P(w|\lambda)P(v|\gamma)P(\lambda)P(\gamma)}{P(D)},$$

with $P(D|w, v)$ the likelihood as in (1) and $P(D)$ an irrelevant normalizing constant. We choose Gamma distributions $P(\lambda)$ and $P(\gamma)$ for the hyperparameters λ and γ [1]. The posterior $P(w, v|D)$ in principle follows by integrating out these hyperparameters.

5 Approximation of the Posterior

Since analytic calculation of $P(w, v|D)$ is impossible, the posterior can either be approximated by drawing samples or by fitting an analytical distribution. The sampling procedure we use here is HMCMC, a method which employs Markov Chain Monte Carlo sampling to draw samples from the network parameters w and v, while keeping the hyperparameters constant, and uses Gibbs sampling to estimate the distribution of the hyperparameters for fixed weights. This yields a large collection of samples w, v, λ, γ.

Another approximation of $P(w, v|D)$ can be made by fitting an analytical distribution. Following Barber and Bishop [3] we approximate the joint posterior distribution of weights and hyperparameters $P(w, v, \lambda, \gamma | D)$ by a factorized distribution of the form

$$P^*(w, v, \lambda, \gamma) = Q(w, v)R(\lambda)S(\gamma),$$

with $Q(w, v) = \mathcal{N}(\hat{w}, \hat{v}, C)$ [2], a Gaussian distribution, and $R(\lambda)$ and $S(\gamma)$ for the moment unspecified. As a distance measure between $P(w, v, \lambda, \gamma | D)$ and $P^*(w, v, \lambda, \gamma)$ we use the Kullback-Leibler (KL) divergence. This error measure depends both on the choice of parameters $\{\hat{w}, \hat{v}, C\}$ and on the distributions $R(\lambda)$ and $S(\gamma)$.

Let us first suppose that $R(\lambda)$ and $S(\gamma)$ are given. It can be shown [3] that the KL divergence now depends only on the parameters $\{\hat{w}, \hat{v}, C\}$, which can be found by minimizing the remaining error function, for example using conjugate gradient methods (all integrals in the KL divergence can be calculated analytically). Now suppose that we know $Q(w, v)$ and would like to optimize for $R(\lambda)$. It is easy to show that, with a Gaussian prior $P(w|\lambda)$ and a Gamma distribution for $P(\lambda)$, the optimal $R(\lambda)$ is also Gamma distributed (see e.g. [3] for details). The procedure for $S(\gamma)$ is completely equivalent.

The approximate posterior $P^*(w, v, \lambda, \gamma)$ can now be found by sequentially finding $\{\hat{w}, \hat{v}, C\}$ with R and S fixed, and estimating R and S with the Gaussian parameters fixed. Since both steps decrease the value of the total divergence, this iterative procedure will converge to an (at least locally) optimal distribution $P^*(w, v, \lambda, \gamma)$.

To compare the predictive qualities of the maximum likelihood solution of the Cox model and both approximations of the Bayesian posterior, we studied fifteen random divisions of our database into a training set and a validation set. The results

---

[1] In the Gamma distribution, $\Gamma(\lambda|\sigma, \tau) \propto \lambda^{\sigma-1}\exp(-\tau\lambda)$, the ratio $\frac{\sigma}{\tau}$ signifies the value for λ we deem most likely *a priori*. The ratio $\frac{\tau^2}{\sigma}$ signifies the strength of this believe.

[2] $\mathcal{N}(\hat{w}, \hat{v}, C) \propto \exp\left[-\frac{1}{2}(y - \hat{y})^T C^{-1}(y - \hat{y})\right]$, where $y = \{w, v\}$ and $\hat{y} = \{\hat{w}, \hat{v}\}$.

Table 1: *p*-values (in percentages) of the remaining inputs after pruning, for both the sampling method (p_s) and the variational method (p_v).

Input	p_s	p_v	Input	p_s	p_v
# tumours after surgery	0.00	0.00	Patient's length	3.04	1.84
Patients performance	0.00	0.00	Creatinine Clearance	2.29	2.59
FIGO stage	–	0.03	CALV1 CAR1/AUC	–	3.96
Leucocytes	0.00	0.24	Dose Intensity		
Carboplatin	–	0.94	Adrimmycine	–	4.36

clearly showed that the Bayesian approach yields significantly better predictions than standard Cox (Student's *t*-tests yielded *p*-values smaller than 1×10^{-7} for both the sampling and the variational approach). There is no significant difference between both approximations to the Bayesian posterior ($p = 0.39$).

6 Application: *p*-Values

For medical scientists, the calculation of *p*-values is a well-known tool in statistical analysis. In general, a *p*-value expresses the probability of the observed data (or data showing a more extreme departure from a certain null-hypothesis) when this null-hypothesis would be true. Consider for example the network described in section 3: if in the "true" model one of the inputs, say input k, would actually have no effect on the output, the weight w_k should be zero. Now, the *p*-value of this input can be defined as the probability that a value for weight w_k, at least as extreme as the expectation value $\hat{w}_k$ we obtained, can be found "by accident". In the variational approach we can calculate *p*-values simply by integrating $Q(\hat{w}_0, \hat{v})$ over the area where w_k is larger than $\hat{w}_k$, where $\hat{w}_0$ is $\hat{w}$ with $\hat{w}_k$ set to zero. Since this is not possible in the HMCMC approximation, in this case we have to define and calculate "pseudo *p*-values" (see [5] for more details).

Calculation of *p*-values for the 42 inputs of the complete model showed that most of the inputs were irrelevant ($p_k \approx 1$), which either means that the major part of the inputs has no effect on the patients' prognosis, or that the loss of any single "irrelevant" input can be compensated for by the remaining inputs. We therefore reduced our network step by step, each time removing the least relevant input (largest *p*-value) and refitting the distribution $P^*(w, v, \lambda, \gamma)$ to the remaining data. This "backward elimination" procedure was iterated until only relevant ($p_k < 0.05$) inputs were left. The resulting *p*-values can be seen in Table 1. It can be seen that the HMCMC method is more critical than the variational method: although all parameters which are relevant within the sampling approach are also relevant in the variational approach, within the latter approach four more inputs are left after pruning.

7 Discussion

In this article we have shown that standard Cox survival analysis can be significantly improved upon by using Bayesian neural network analysis. By choosing smart priors, we eliminated the disadvantages normally connected to Cox's method and obtained better survival predictions, at least on our database. Apart from prediction, the Bayesian treatment allowed us to easily calculate p-values on the input parameters, yielding a greater insight into the data.

Looking more closely at ways to approximate the Bayesian posterior, we showed that, given a similar amount of computation time, both methods yield equally good results. Of course, the sampling approach may well improve when more samples are drawn (in the limit of an infinite amount of samples, it will approximate the posterior arbitrarily close). However, this would not only require a lot more computation time (the improvement with the number of samples is very slow), it also yields an unmanageable bulk of data to be used in further calculations, instead of a neat Gaussian distribution as in the variational approach. The advantages of the latter already became clear in section 5, where p-values could readily be calculated within the variational approach, whereas using the HMCMC sampling method required the definition of pseudo p-values. Our final conclusion is that where both sampling and variational Bayesian techniques improve upon standard maximum likelihood Cox analysis, the variational approach is preferable: its "output", a Gaussian distribution over model parameters, is much easier to handle than a HMCMC sample with similar predictive capabilities.

References

[1] D. Cox and D. Oakes. *Analysis of Survival Data*. Chapman Hall, London, 1984.

[2] G. Hinton and D. van Camp. Keeping neural networks simple by minimizing the description length of the weights. In *Proceedings of the 6th Annual Workshop on Computational Learning Theory*, pages 5–13, New York, 1993.

[3] D. Barber and C. Bishop. Ensemble learning for multi-layer networks. In *NIPS 10*, pages 395–401, Cambridge, 1997. MIT Press.

[4] H. Kappen and J. Neijt. Neural network analysis to predict treatment outcome. *The Annals of Oncology*, 4:S31–S34, 1993.

[5] B. Bakker and T. Heskes. A neural-Bayesian approach to survival analysis. In *ICANN'99*, 1999.

Identifying Discriminant Features in the Histopathology Diagnosis of Inflammatory Bowel Disease Using a Novel Variant of the Growing Cell Structure Network Technique

Dr Simon S Cross

Department of Pathology, University of Sheffield Medical School
Sheffield, UK
s.s.cross@sheffield.ac.uk

Andrew J Walker & Dr Robert F Harrison

Department of Automatic Control & Systems Engineering, University of Sheffield
Sheffield, UK
r.f.harrison@sheffield.ac.uk

Abstract

We describe a novel variant of a growing cell structure network system which can be used as a statistical classifier and a data visualisation tool. We applied the technique to the histopathology diagnosis of chronic idiopathic inflammatory bowel disease and the division of this into Crohn's disease and ulcerative colitis. The system gave areas under receiver operating characteristic curves that were not significantly different from logistic regression. It produced visualisation of the data which enabled an intuitive interpretation of the importance of individual input variables to the prediction which agreed with expert published opinion.

1 Background

Symptoms of bowel disease, such as abdominal pain or diarrhoea, can be produced by a number of different conditions from those without a definite intrinsic bowel pathology, such as irritable bowel syndrome, to intrinsic inflammation of the bowel, chronic idiopathic inflammatory bowel disease (CIIBD). CIIBD can be divided into two distinct patterns, Crohn's disease (CD) and ulcerative colitis (UC), and the treatments for these can vary. The diagnosis in these bowel diseases is made by a multifactorial process including clinical features, blood tests and radiology; but histopathology examination of endoscopic colorectal biopsies is usually held to be the gold standard. However studies have shown that whilst quite accurate histopathology does not provide the correct diagnosis in all cases.[1] In this study we use a novel variant of the growing cell structure (GCS) network technique to investigate which histopathology features are important in the diagnosis of CIIBD with a view to developing a decision support system.

2 Methods

2.1 Study population

The total study population was 809 large bowel endoscopic biopsies reported in the Department of Histopathology, Royal Hallamshire Hospital, Sheffield between 1990 and 1995 (inclusive). The diagnosis was confirmed by the finding of typical endoscopy appearances seen on video photographs in the clinical notes, subsequent bowel resection, pattern of disease on radiological investigation or microbiological culture results. In cases without confirmation by subsequent resection specimens this final diagnostic outcome was made with review of the patient's case notes.[1] The biopsies were a mixed population of single distal biopsies and colonoscopic series from initial presentation and follow-up of disease. By verified outcome the study population contained 165 biopsies from normal subjects, 473 from cases of ulcerative colitis and 171 of Crohn's disease.

2.2 The observed features

The biopsies were examined (blind to all clinical details) by a single experienced observer (SSC) using a computer interface which implements the British Society of Gastroenterology (BSG) Guidelines for the Initial Biopsy Diagnosis of Suspected Chronic Idiopathic Inflammatory Bowel Disease[2] with digitised images representing examples of each histopathological feature[3]. Some of the features are dichotomous variables, e.g. the presence or absence of mucosal granulomas, whilst others are ordinal categories, e.g. mucin depletion classified into none, mild, moderate or severe. The observer recorded each feature by choosing the digitised image that was most similar to the image seen down the light microscope and using a mouse to click on this, which recorded the observation in the computerised database. Observation was spread over a period of 9 months with no more than 30 biopsies observed in a single day.

2.3 Partitioning of the dataset

The dataset was analysed in two partitions. The whole dataset was used with a dichotomous outcome of normal or CIIBD (with Crohn's disease and ulcerative colitis cases combined as a single category). The order was randomised and the first 540 cases were used as the training set and the other 269 cases as the test set. 370 cases from the whole dataset were selected with a dichotomous outcome of Crohn's disease or ulcerative colitis. These cases were defined by their outcome and the presence of active inflammation (as indicated by polymorphs in the lamina propria). The order of these was randomised and the first 247 cases were used as the training set and the other 123 cases as the test set. The input data in both partitions were normalised to the interval (0,1). This corresponds only to a shift in origin for the binary features and can be justified for the categorical features on the grounds that they are ordinal and that increasing value corresponds to an increasing effect.

2.4 Growing cell structure system analysis

Our novel variant of the growing cell structure (GCS) system performs as a conventional statistical classifier but also produces visualisation of each input variable in relation to the known outcome. The details of our GCS system have been described elsewhere[4] but the process is basically divided into two steps. The input variables for each case in the training set are used in an unsupervised clustering process which can be visualised as a two dimensional projection of a network of nodes. The network starts with three nodes and nodes are added iteratively until the best performance is achieved (as determined by the area under the ROC curve for the test set). The frequency of each class at each node is then converted to a posterior probability using a Parzen windows implementation of Bayes' theorem.[5] These probabilities are visualised as a colour contour overlay over the network grid and similar overlays are produced for each input variable. Visual comparison of the input feature overlays with the outcome overlays indicates which input features have an association with a particular outcome, and which input features correlate with each other.[4] The GCS system was used on both partitions of the dataset with all input variables included and just the BSG selected variables. The sensitivities and specificities were calculated for the points on the ROC curves that gave the greatest overall accuracy.

3 Results

The GCS system produces pictures which are best viewed in colour but which are reproduced in monochrome here (figures 1&2).

Table 1. Areas under the receiver operating characteristic curves for logistic regression and the GCS system for the test set of the normal versus CIIBD dataset.

	Logistic regression		GCS system	
	All variables	BSG selected	All variables	BSG selected
Area under ROC curve	0.8436	0.8775	0.8393	0.8532
SE of area under ROC curve	0.0245	0.0211	0.0230	0.0231
Sensitivity for CIIBD	78%	80%	76%	77%
Specificity	89%	89%	86%	89%

Table 2. Areas under the receiver operating characteristic curves for logistic regression and the GCS system for the test set of the Crohn's disease versus ulcerative colitis dataset.

	Logistic regression		GCS system	
	All variables	BSG selected	All variables	BSG selected
Area under ROC curve	0.7564	0.7771	0.7202	0.7334
SE of area under ROC curve	0.0464	0.0442	0.0466	0.0465
Sensitivity for UC	84%	67%	66%	74%
Specificity	62%	79%	66%	62%

There was no significant difference between the area under the ROC curves for either method on the same input variables (at a 5% level).

4   Discussion

As statistical classifiers the GCS systems for both decisions give a similar level of performance as logistic regression and these levels of performance are better than some published studies of human performance[1] so the systems have potential as decision support systems. However the advantage of the GCS systems over logistic regression is the intuitive manner in which the relevance of each input variable to the decision can be discerned and the correlation between different input variables can be identified. Although a similar assessment can be obtained by using variants of logistic regression, such as a principled Bayesian method with correlation analysis, the results are not so easily assimilated by non-statisticians and this is important in the medical domain.

A relatively simple 6 node network produced the greatest area under the ROC curve for the distinction between normality and CIIBD (figure 1). The colour overlay for the posterior probability of CIIBD shows highest values in the bottom left corner of the network (seen as a dark area in the monochrome reproductions). Looking at the colour overlays for the input variables it can be seen that there are corresponding dark areas on the overlays for mucosal surface, crypt architecture, transmucosal increase in cellularity, cryptitis - extent, cryptitis - polymorphs, crypt abscesses - extent, crypt abscesses - polymorphs, lamina propria polymorphs, epithelial changes, and mucin depletion. This suggests that all these features are of value in discriminating normality from CIIBD and that they are all highly correlated. Other features, such as patchy increased lamina propria cellularity, have high values in a small area of the high value region for CIIBD suggesting that they are present in a subset of cases of CIIBD. These features are all identified in the British Society for Gastroenterology's guidelines for the initial biopsy diagnosis of CIIBD, these guidelines being based on an evidence-based review of the world wide published literature.[2]

The division between CD and UC appears more complex with a 12 node network required to produce optimal separation. The highest values of posterior probability for CD are in the bottom left hand side of the network and there is close correlation between that area and high value areas for patchy increase in lamina propria cellularity, lamina propria granulomas and submucosal granulomas which agrees with the features identified as discriminant in the gastrointestinal literature.[2]

This study shows how this novel GCS system provides clear visualisation of multidimensional input data and good statistical classification. It provides a standalone classification system but is also a technique which is complementary to conventional statistical techniques, such as logistic regression, in the identification of discriminatory features.

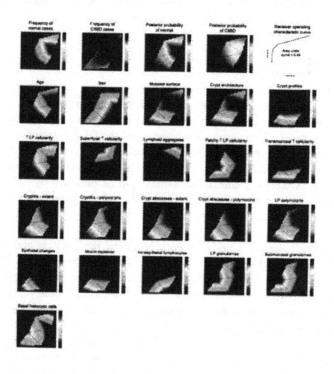

Figure 1. The maps of frequencies and posterior probabilities for the division into normal and CIIBD together with the maps of each different input feature and a ROC curve.

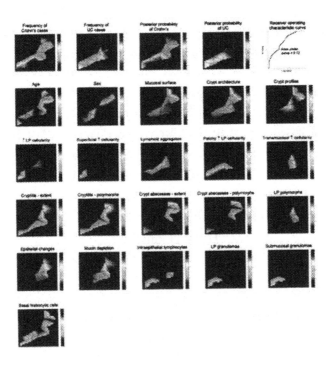

Figure 2. The maps of frequencies and posterior probabilities for the division into Crohn's disease and ulcerative colitis together with the maps of each different input feature and a ROC curve.

5   References

[1]   Dube AK, Cross SS, Lobo AJ. Audit of the histopathological diagnosis of non-neoplastic colorectal biopsies: achievable standards for the diagnosis of inflammatory bowel disease. *J.Clin.Pathol* 1998;51:378-81.

[2]   Jenkins D, Balsitis M, Gallivan S, Dixon MF, Gilmour HM, Shepherd NA, Theodossi A, Williams GT. Guidelines for the initial biopsy diagnosis of suspected chronic idiopathic inflammatory bowel disease. The British Society of Gastroenterology Initiative. *J.Clin.Pathol.* 1997;50:93-105.

[3]   Cross SS, Dube AK, Underwood JCE, et al. A visual image-based data entry system for reporting non-neoplastic endoscopic colorectal biopsies. *J.Pathol.* 1997;181:17A

[4]   Walker AJ, Cross SS, Harrison RF. Visualising Biomedical Datasets Using Growing Cell Structure Networks: A Novel Diagnostic Classification Technique. *Lancet* 1999;354: 1518-1521.

[5]   Parzen E. On estimation of a probability density function and mode. *Annals of Mathematical Statistics* 1962;33:1065-76.

Classifying Pigmented Skin Lesions with Machine Learning Methods

Stephan Dreiseitl

Dept. of Software Engineering for Medicine
Polytechnic University of Upper Austria, Hagenberg, Austria
email: Stephan.Dreiseitl@fhs-hagenberg.ac.at

Harald Kittler

Department of Dermatology
University of Vienna Medical School, Vienna, Austria

Harald Ganster

Institute for Computer Graphics and Vision
Technical University Graz, Austria

Michael Binder

Decision Systems Group, Brigham and Women's Hospital
Boston, MA, USA

Abstract

We use a data set of 1619 pigmented skin lesions images from three categories (common nevi, dysplastic nevi, and melanoma) to investigate the performance of four machine learning methods on the problem of classifying lesion images. The methods used were k-nearest neighbors, logistic regression, artificial neural networks, and support vector machines. The data sets were used to train the algorithms on the following tasks: to distinguish common nevi from dysplastic nevi and melanoma, and to distinguish melanoma from common nevi and dysplastic nevi. Receiver operating characteristic curves were used to summarize the performance of the models.

Three of the methods (logistic regression, artificial neural networks, and support vector machines) achieved very good results (area under curve values about 0.96) on the problem of distinguishing melanoma from common and dysplastic nevi, and good results (area under curve values about 0.82) for the problem of distinguishing common nevi from dysplastic nevi and melanoma. The performance of k-nearest neighbors models was about 3 percentage points worse than that of the other methods. These results show that for classification problems in the domain of pigmented skin lesions, excellent results can be achieved with several algorithms.

1 Introduction

The classification of pigmented skin lesions is a hard task even for expert human dermatologists [1]. With rising incidence rates of melanoma worldwide, it is becoming increasingly important to correctly diagnose PSLs. Machine learning can be used to support dermatologists in the diagnostic process. The purpose of this paper is to investigate which of a number of machine learning algorithms is best suited for the tasks of distinguishing between different types of lesions. The algorithms considered in this paper are those best suited for mostly real-valued variables: k-nearest neighbors, logistic regression, artificial neural networks, and support vector machines. The benchmark problems for comparing these four algorithms were the task of distinuishing common nevi from dysplastic nevi and melanoma, and the task of distinguishing melanoma from common and dysplastic nevi.

2 Material

For the experiments in this paper, we used a data set from the pigmented lesion unit of the Department of Dermatology, University of Vienna Medical School. The data set contained 1290 common nevi, 224 dysplastic nevi, and 105 melanoma for a total of 1619 lesions. The images of the lesions were obtained using a digital epiluminescence microscopy (ELM) system (MoleMax II, Derma-Instruments, Austria). The diagnosis of the lesions was established by histopathology (melanoma and dysplastic nevi) and one-year follow-up examinations (common nevi), respectively.

From the images, 107 morphometric features were extracted. In addition to the features obtained from the image, we used six clinical data items that were recorded for each lesion as variables in the data set. The data set was split into training and test sets for the machine learning algorithms. The training sets contained 600 common nevi, 144 dysplastic nevi, and 65 melanoma, and the test sets contained 690 common nevi, 80 dysplastic nevi, and 40 melanoma. To determine the influence of different data set splits on the methods, a total of 100 different splits were used in each method. The data set splits were the same for each of the methods used. To ensure that no method was given an advantage due to the scaling of the data set, each variable was transformed to be zero-mean and unit variance over the whole data set.

3 Methods

The machine learning algorithms used for this paper were k-nearest neighbors, logistic regression, artificial neural networks, and support vector machines. With the exception support vector machines, these methods have been applied extensively to medical problems and need no further explanations. The fundamental principle behind support vector machines is summarized briefly.

Support vector machines are a machine learning paradigm that is based on statistical learning theory [2, 3]. The main algorithm computes separating hyperplanes that maximize the margin between two data sets. The problem can be formulated as a Lagrangian maximization problem that can be solved by quadratic optimization methods. In the Lagrangian formulation, the data points enter the problem statement only in the form of dot products. This observation is the key that allows the computation of nonlinear separating surfaces: First, a function Φ maps the data points nonlinearly to a higher-dimensional space. Then, by computing the dot products and separating hyperplanes in this space, the algorithm effectively calculates nonlinear boundaries in the original data space. The trick is to combine nonlinear transformation and dot product computation in one step: By using *kernel functions*, one need not even know the transformation Φ, nor the dimensionality of the space that Φ maps to. This is because a kernel function K is a function that can be written as $K(x,y) = \Phi(x) \cdot \Phi(y)$ for some transformation Φ and any two data vectors x, y. To check whether a function K is a kernel function, one need only check whether K satisfies *Mercer's condition* [2]. Several functions, such as polynomials and the radial basis function (RBF) kernel $K(x,y) = e^{-\|x-y\|^2/2\sigma^2}$ satisfy this condition.

To evaluate the results of the different methods, we used the theory of receiver operating characteristic (ROC) curves [4]. The area under the ROC curve (AUC) is an indicator of the discriminatory power of a test. It is equivalent to the probability that the test can correctly rank two cases, one from each of the classes, on its ordinal scale. A test with very poor discriminatory power has an AUC value of $1/2$, and a perfect test has an AUC value of 1.

4   Experiments

Each of the four algorithms was run on each of the 100 different splits of the data set into training and test data. In the following, we refer to the problem of distinguishing common nevi from dysplastic nevi and melanoma as *task 1*, and to the problem of distinguishing melanoma from common and dysplastic nevi as *task 2*. We summarized the performance of each method on each task by averaging the AUC values for each of the 100 different training/test set splits. Along with the results of the individual methods, we briefly list details and parameter settings that were used.

k-nearest neighbors   This method is popular for the fact that there is only adjustable parameter k in the algorithm, once a distance metric is fixed. For our experiments, we used the standard Euclidean distance. Therefore, each variable, having been rescaled to be zero-mean and unit variance, contributed equally to the result. We used values of $k = 10, 20, \ldots, 100$, and found the algorithm to be rather insensitive to the choice of k. For task 1, all settings of k resulted in AUC values between 0.7710 ($\sigma = 0.0200$) for $k = 10$ and 0.7943 ($\sigma = 0.0219$) for $k = 50$. For task 2, the AUC values ranged between 0.8983 ($\sigma = 0.0277$) for $k = 10$ and 0.9332 ($\sigma = 0.0234$).

Logistic regression   This method is especially popular in the medical domain because the model parameters can be interpreted as changes in log odds. It is therefore possible to determine which of the input variables contribute significantly to the output, and which ones can be eliminated. This is often advantageous in data sets with a high number of variables, as is the case here. As with k-nearest neighbors, there is only one parameter to adjust—the significance level for inclusion of variables in the model. We used the logistic regression implementation of the SAS system (SAS Institute, Cary, NC, USA), with stepwise variable selection. The significance level for addition and removal of variables was set to 0.05. Only the eight most significant variables were included in the model.

For task 1, logistic regression achieved an AUC value of 0.8288 ($\sigma = 0.0168$), and for task 2, an AUC value of 0.9677 ($\sigma = 0.0175$).

Artificial neural networks   A very popular method for its ability to construct models in a learning-like manner, neural networks have been applied extensively in different medical domains. Earlier algorithms relied extensively on a number of model parameters (number of hidden nodes, learning rate, momentum etc.), and optimal solutions could only be achieved after long searches through this parameter space. Nowadays, more advanced algorithms and techniques for preventing overtraining have mostly eliminated the need for parameter tuning. In our experiments, we used the NETLAB scripts (Neural Computing Research Group, Aston University, UK) for networks with 20 hidden neurons trained with the conjugate gradient learning algorithm. No further parameters were needed. Overtraining was avoided by splitting the training set into a "real" training and a holdout set; training was terminated when the error on the holdout set started to increase.

The results of the neural networks were an AUC value of 0.8263 ($\sigma = 0.0177$) for task 1, and an AUC value of 0.9680 ($\sigma = 0.0122$) for task 2.

Support vector machines   This paradigm has recently been the focus of much research interest for its sound theoretical foundations and the possibility to give a bound on the generalisation error of methods based only on the performance on a training set and the capacity (VC-dimension) of the model. Since this is still a "new" algorithm, there have been very few applications on medical problems.

There is a wide variety of possible SVM models, because there are many possible kernel functions. It is generally not possible to know a priori which the kernel function will work best on which problem. The two most popular kernel functions are polynomials and Gaussian RBF kernels. Both these kernel functions have adjustable parameters: For polynomials, this is the degree of the polynomial, for RBF kernels, the width of the function. We used the SVM-Light implementation (Dept. of Computer Science, University of Dortmund, Germany) with the following parameter settings: polynomial degrees 1 to 3, RBF inverse variances γ as 10^{-2} to 10^{-6}, and default settings for other parameters. The results for the various parameter settings are summarized in Table 1. There are no entries for RBF kernels with parameters 10^{-3} and 10^{-5}, as these results lie between those for parameters 10^{-2} and 10^{-4} and 10^{-4} and 10^{-6}, respectively. Also shown in this table are the minimum and maximum AUC values over all 100 data set splits, as well as the average maximum sensitivity and specificity as determined by the upper-leftmost point on the ROC curve.

5   Discussion

Three of the four methods (logistic regression, artificial neural networks, and support vector machines) obtained excellent results on task 2, and good results on task 1. The performance of k-nearest neighbors was only about 3 percentage points worse. A synopsis of the best results of all methods, together with minimum and maximum AUC and average maximum sensitivity and specificity values is given in Table 2. Although it is not possible to statistically compare the results in this table due to the overlap of training and test sets in the different runs, it is obvious that the three best methods perform on the same level. It is somewhat surprising that the nonlinear methods of neural

Table 1: Performance comparision of SVM models on tasks 1 and 2.

	polynomial kernel			Gaussian RBF kernel		
	$d=1$	$d=2$	$d=3$	$\gamma=10^{-6}$	$\gamma=10^{-4}$	$\gamma=10^{-2}$
task 1						
avg. AUC	0.8131	0.7379	0.7377	0.8189	0.8305	0.7863
std. dev.	0.0185	0.0223	0.0334	0.0184	0.0149	0.0183
min. AUC	0.7574	0.6702	0.5000	0.7699	0.7939	0.7488
max. AUC	0.8508	0.7909	0.7814	0.8659	0.8623	0.8257
avg. sens.	0.7571	0.7043	0.7209	0.7362	0.7820	0.7459
avg. spec.	0.7183	0.6598	0.6402	0.7443	0.7360	0.7164
task 2						
avg. AUC	0.9184	0.8544	0.9051	0.9644	0.9700	0.9471
std. dev.	0.0268	0.0390	0.0346	0.0132	0.0132	0.0178
min. AUC	0.8276	0.7593	0.7863	0.9337	0.9282	0.9100
max. AUC	0.9709	0.9551	0.9849	0.9892	0.9936	0.9852
avg. sens.	0.8448	0.7820	0.8370	0.9114	0.9205	0.8903
avg. spec.	0.8845	0.8855	0.8950	0.9141	0.9497	0.9124

networks and support vector machines do not perform better than the linear method of logistic regression. This could be due to the fact that the variable selection incorporated in logistic regression improves performance. While there is some dependency on correct choices of parameter values in the SVM models (see Table 1), the other methods needed no parameter tuning to obtain comparable results. Nevertheless, SVM models offer a promising new development, since they also allow to identify which cases lie on the decision boundary (the *support vectors*). SVM algorithms could thus be used to reduce large data sets to the cases that define a decision boundary.

6 Conclusion

We analysed the performance of four machine learning algorithms on the problem of classifying pigmented skin lesions. The results are encouraging, as logistic regression, artificial neural networks and support vector machines are all able to distinguish between melanoma and other lesions with high accuracy. The next problem now is how to best use such algorithms in clinical environments to increase the quality of care. One possible application is in an intelligent tutoring system for dermatologists. Such a system could make "educated guesses" at the correct diagnosis of an unknown lesion and guide the user to the correct diagnosis by presenting images of similar lesions from a data base. Previous research has shown that advanced techniques in melanoma detection, such as ELM, increase the quality of diagnoses only after sufficient training [5]. Further research is needed to determine how decision support tools could be used in the training process.

Table 2: Performance comparision of all four methods on tasks 1 and 2.

	k-NN	log. regr.	ANN	SVM poly. $d = 1$	Gauss. $\gamma = 10^{-4}$
task 1					
avg. AUC	0.7943	0.8288	0.8263	0.8131	0.8305
std. dev.	0.0219	0.0168	0.0177	0.0185	0.0149
min. AUC	0.7482	0.7899	0.7866	0.7574	0.7939
max. AUC	0.8404	0.8726	0.8730	0.8508	0.8623
avg. sens.	0.7292	0.7686	0.7714	0.7571	0.7820
avg. spec.	0.7287	0.7395	0.7417	0.7183	0.7360
task 2					
avg. AUC	0.9332	0.9677	0.9680	0.9184	0.9700
std. dev.	0.0234	0.0175	0.0122	0.0268	0.0132
min. AUC	0.8469	0.8948	0.9371	0.8276	0.9282
max. AUC	0.9775	0.9949	0.9929	0.9709	0.9936
avg. sens.	0.8493	0.9240	0.9143	0.8448	0.9205
avg. spec.	0.9044	0.9405	0.9397	0.8845	0.9497

7 Acknowledgements

The authors wish to thank Staal Vinterbo and Holger Billhardt for their contribution to running the experiments. Funding was provided in part by Austrian Science foundation grants FWF-P11735-MED (HK,HG,MB) and FWF-J1661-INF (SD), EU-TVP-grant: IN 301044 I (HK,HG,MB) and a personal grant from the Max Kade foundation (MB).

References

[1] C.M. Grin, A.W. Kopf, B. Welkovich, R.S. Bart, and M.J. Levenstein. Accuracy in the clinical diagnosis of malignant melanoma. *Arch of Dermat*, 126:763–766, 1990.

[2] V.N. Vapnik. *The Nature of Statistical Learning Theory*. Springer Verlag, 1995.

[3] V.N. Vapnik. *Statistical Learning Theory*. John Wiley & Sons, 1998.

[4] J.A. Hanley and B.J. McNeil. The meaning and use of the area under the receiver operating characteristic (ROC) curve. *Radiol*, 143:29–36, 1982.

[5] M. Binder, M. Schwarz, and A. Winkler *et al*. Epiluminescence microscopy. A useful tool for the diagnosis of pigmented skin lesions for formally trained dermatologists. *Arch of Dermat*, 131:286–291, 1995.

An Assessment System of Dementia of Alzheimer Type Using Artificial Neural Networks

Shin Hibino[1], Taizo Hanai[1], Erika Nagata[1], Michitaka Matsubara[2]

Kazutoshi Fukagawa[2], Tatsuaki Shirataki[2], Hiroyuki Honda[1]

Takeshi Kobayashi[1]*

*Corresponding author

Email address: takeshi@nubio.nagoya-u.ac.jp

[1]Department of Biotechnology, Graduate School of Engineering, Nagoya University

Nagoya, Japan

[2]Nagoya City Rehabilitation Center

Nagoya, Japan

Abstract

An assessment system of dementia of Alzheimer type (DAT) from electro-encephalogram (EEG) was investigated. The system consisted of two artificial neural networks (ANN) models; a model for distinction of DAT patients from non-DAT patients and an estimation model of severity of the DAT. First, EEG data of the DAT patients and the non-DAT patients were collected using 15 electrodes on the scalp. Then, power spectrum of each data was calculated by the fast Fourier transform. The power spectrum was divided into 9 frequency bands, and relative power values were calculated. The regions with 4.0-6.0, 6.0-8.0 and 8.0-13.0 Hz were used to input. The severity of DAT was assessed by Hasegawa's dementia rating scale (HDS-R). The relative power values and the averaged absolute power value were inputted into each ANN model. Using the acquired ANN model, DAT patients were distinguished from non-DAT patients completely. The average error of the ANN model for HDS-R score was 2.64 points out of 30. These models were found useful in order to distinguish DAT patients and quantify the severity of DAT from EEG.

1   Introduction

Coming of aging society, there are 580 million older people aged over 60 in the world now, and estimated that it will increase almost 1970 million in 2050. With the rapid increase of the population of older people, the number of dementia patients, approximately 29 million now, will rise up to 80 million in a few decades, so that interests in dementia become greater than ever before. Particularly, dementia of Alzheimer type (DAT), a typical case of dementia, has been studied very much from the viewpoint of clinical pathology and molecular biology. According to the DSM-III-R [1], diagnosis of dementia can be done with some intelligence or mental tests such as Wechsler adult intelligence test (WAIS-R) [2], Mini-Mental state

(MMS) [3], and in Japan, Hasegawa's dementia rating scale-revised edition (HDS-R) [4]. If patients do not intend to take the tests, it is difficult to carry out the tests, and more objective estimation methods of DAT should be established. Recently, methods of computerizing and digitizing for electroencephalography (EEG) have been progressing. On the other hand, analysis of EEG data is difficult because of the complexity of EEG. Knowledge information processing methods such as artificial neural networks (ANN) are useful to analyze those complicated data. For example, Pritchard et al. [5] reported classification of Alzheimer's disease patients from normal subjects with non-linear analysis in around 90% accuracy. And Riquelme et al. [6] reported classification of dementia from normal, depression, and anxiety subjects over 90% accuracy. We also tried to evaluate severity of cerebral disease quantitatively, and estimated aphasia patients faculty of speech from their EEG data using ANN in the previous paper [7]. It is generally known for DAT patients that degree of DAT is related to their EEG abnormalities. When the degree of DAT is not still serious, frequency and amplitude of α-waves become lower than normal people, and θ-waves appear. When DAT get worse, δ-waves appear and at last EEG becomes flat. Thus, in the present study, we tried to distinguish DAT from normal subjects using ANN. Estimation of severity of the disease is also demonstrated.

2 Materials and Methods

2.1 Subjects

Subjects were outpatients of Nagoya City Rehabilitation Center. Thirty-eight DAT patients (male: 13, female: 25, aged 51-90 and average 72.8) were selected. They were diagnosed as DAT with NINCDS-ADRDA [8], MRI, and PET. Seventy-five data were obtained from 38 patients. As controls, 27 data from 24 normal people (male: 9, female: 15, aged 48-81 and average 69.7) were also used. All 102 data were used to analyze. All subjects were right handed.

2.2 EEG recording and signal processing

EEG was recorded from the 15 electrodes (F_3, F_4, F_Z, C_3, C_4, C_Z, T_3, T_4, T_5, T_6, P_3, P_4, P_Z, O_1, O_2) on the scalp, in agreement with the international 10-20 system with bimastoid average reference, in a darkened soundproof shield room (temperature and humidity were constant) under eyes closed and resting state. EEG was amplified at the time constant 0.1 sec, and low pass filtered at 60 Hz. Then, artifact-free 5.12-second epoch was selected from EEG, and the signal was A/D converted with sampling frequency of 100 Hz. Under the same condition, we repeated the sampling five times, and calculated the absolute power spectrum with the fast Fourier transform. The absolute power spectra collected five times were averaged, and separated into nine frequency bands (δ: 3.0-4.0 Hz, θ_1: 4.0-6.0 Hz, θ_2: 6.0-8.0 Hz, α_1: 8.0-9.0 Hz, α_2: 9.0-11.0 Hz, α_3: 11.0-13.0 Hz, β_1: 13.0-15.0 Hz, β_2: 15.0-20.0 Hz, β_3: 20.0-25.0Hz). Then the relative powers of each frequency band were computed and those of $\alpha_1 \sim \alpha_3$ were averaged (averaged α). Since EEG amplitude of DAT patients became lower with the progress of DAT, average value of all channels' absolute power was also calculated. However, in this study, δ and β were not used for analysis, and θ_1, θ_2, averaged α were inputted into ANN, since influences of artifacts by eye movements and blinks were not completely negligible in δ, and relations of brain function decline and changes of β are not reported. As electroencephalograph and frequency analyzer, EEG-neurofax 4400 (Nihon Koden Co., Tokyo, Japan) and Pathfinder MEGA (Nicolet Co., Wisconsin, USA) were used.

Table 1. Contents of HDS-R and allotment of points.

No.	Question	Points
1	How old are you?	1
2	What is the date today?	4
3	Where is here?	2
4	Memorize 3 words and answer them	3
5	Take 7 from 100 one after another and answer them	2
6	Answer backward the given numbers	2
7	Answer the words memorized in No.4	6
8	Memorize the names of 5 goods and answer them	5
9	Say the name of vegetables as much as possible	5
	The sum	Out of 30

2.3 Assessment of DAT

Degree of intelligence disturbance of DAT patients was assessed by HDS-R. Though MMS is used widely in the world, we used HDS-R in this study, since the correlation coefficient between HDS-R and MMS is very high (0.94) [4], HDS-R is written in Japanese, and it is easy to examine HDS-R. Table 1 shows the contents of HDS-R. Compared with MMS, HDS-R has no questions about accomplishment of the command, creation of the text, and drawing of the figure. It was reported that HDS-R could distinguish DAT patients from non-DAT patients appropriately, when the threshold made 21 points out of 30 [4].

2.4 Construction of ANN model

Structure of ANN used in this study was composed of three layers. ANN inputs were relative powers of the three frequency bands and an averaged absolute power of all channels, and the output was DAT diagnostic result or score of HDS-R. Learning was done the by the back propagation algorithm with an inertia term (α). The slope of sigmoid function (β) was set to one for calculating fast. Parameters were set to give the best results, learning coefficient: 0.05, inertia term: 0.8, maximum number of learning: 30000. Learning was stopped when the estimation error (average value of the mean square error for learning and evaluation data) became minimum to avoid overfitting. Two thirds of all data were used to learn ANN, and others were used to evaluate of the universality of the ANN. To optimize the network structure, the number of hidden units was changed from n-5 to n+10 (n: number of input units) and fixed at the number which gave the least estimation error. All input values and teaching values of ANN were normalized linearly at the range from 0.1 to 0.9. In this study, we constructed two models using ANN; first was a model for discrimination between DAT and non-DAT, second was a model for estimation HDS-R score of DAT patients. To classify DAT patients by DAT discrimination model, all data were separated in non-DAT group and DAT group. Non-DAT group included normal people and DAT patients scoring over 21 points on HDS-R, and DAT group contained DAT patients scoring less than 20 points. There were 40 data in the non-DAT group and 62 data in the DAT group. Teaching values were set to 0.9 in non-DAT group and 0.1 in the DAT group. DAT discrimination model and estimation model used the same data set. Sixty-eight data were used for learning. Remaining 34 data were used for evaluation. Input variables of ANN were composed of the relative power of the three frequency bands from 15 channels, and the averaged absolute power of all channels (46 variables). Input channels and frequency bands were selected by parameter increasing method (PIM). PIM increased input variables of ANN model one after another. When the estimation error became minimum, it was known that those inputs variables were in the optimum combination.

3   Results

3.1   Discrimination of DAT patients with ANN

First, we built an ANN model to discriminate DAT from non-DAT group. When we set the threshold on 0.5 for both learning and evaluation data, all 62 data in the DAT group and 40 data in the non-DAT group could be classified in the correct group. The model could distinguish between DAT group and non-DAT group completely. As shown in Figure 1 (a), the input variables selected by PIM for the ANN model were $(O_1: \theta_1)$, $(O_2: \theta_2)$, $(P_3: \theta_2)$, $(P_Z: \theta_2)$, $(P_4: \theta_2)$, $(T_4: \theta_2)$, $(T_4: \alpha)$, $(T_5: \theta_1)$, and $(F_4: \alpha)$. Thirteen hidden units were used.

3.2   Estimation of HDS-R score with ANN

Second, we tried to construct an ANN model to estimate HDS-R score from EEG power spectrum. Figure 2 shows the results of HDS-R score estimation by the ANN model. The horizontal axis means a patient's actual score of HDS-R, and the vertical axis indicates the estimated score with the model. If the data are plotted close to the diagonal line, precision of estimation is considered high. The average estimation error of HDS-R score for learning and estimation data was 2.64 out of 30 points. For this HDS-R score estimation model, the input variables selected by PIM were $(P_Z: \theta_1)$, $(T_4: \theta_2)$, $(T_4: \alpha)$, $(T_5: \theta_1)$, $(C_3: \theta_1)$, $(C_3: \theta_2)$, $(C_Z: \alpha)$, $(C_4: \theta_2)$, $(F_3: \alpha)$, and $(F_Z: \alpha)$ (Figure 1 (b)). Eight hidden units were used. The selected channels for this ANN model were different from those of the DAT patients discrimination model. Only three channels $(P_Z, T_4, \text{and } T_5)$ were common in both models.

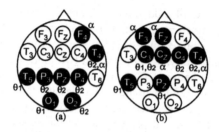

Figure 1: Channels and frequency bands selected for the models of (a) DAT classification and (b) estimation of HDS-R score (filled circles: selected, unfilled: not selected)

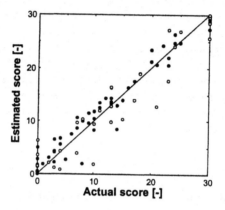

Figure 2: Estimation results of HDS-R score with ANN. Black circles represent scores for learning data, and white for evaluation data.

4 Discussion

For the DAT patients discrimination model, most selected channels were positioned in the occipital and the parietal lobes, and a frequency band θ_2 was selected many times. In the early phase of DAT, brain atrophy begins at the temporal lobe, so there are a few tendencies of abnormality and atrophy at the occipital lobe. Thus, if brain atrophy and abnormal tendencies of EEG were found in the occipital, ANN seems to judge the patient to be DAT. ANN selected the channels over the language center (P_3, P_Z, and T_5) and their contra lateral channel (P_4). The reason of selection is conceivable since HDS-R is a test with conversations. Two frequency bands were selected in T_4. T_4 corresponds to minor hemisphere, which attends sensorial functions such as image recognition at right-handed people. Therefore, it is interesting that T_4 was selected to assess the score of this language test.

For the HDS-R score estimation model, it is appropriate that the channels over the language center (T_5, P_Z, C_3, and F_3) were also selected. Since C_3, C_4, and C_Z are channels around the central sulcus, these regions are considered to connect with sensory motor functions such as utterance in the test. F_3 and F_Z locate on the frontal lobe that is related to integration of thought (especially, related to linguistic thought), and this channel may be used to assess patients' thought ability. Thus, this model could estimate subjects' language and integration ability with EEG spectrum.

For practical use, addition of data and analysis of input-output relations are necessary. Therefore, we are using fuzzy neural networks (FNN) to analyze the relationships between EEG and HDS-R score now. As for the assessment of DAT, HDS-R assesses patients' memory by conversation, and has no implements to assess functions of minor hemisphere. We are using other tests for DAT (ADAS and MMS), and get good results for estimation of ADAS score. These systems would be useful as diagnostic assistants and construction of home-care systems.

References

[1] American Psychiatric Association. Diagnostic and statistical manual of mental disorders: DSM-III-R, third edition revised. American Psychiatric Association, Washington D.C., 1987

[2] Wechsler D. Manual for the Wechsler Adult Intelligence Scale-Revised. Psychological Corporation, New York, 1981

[3] Folstein MF, Folstein SE, McHugh PR. Mini mental state; a practical method for grading the cognitive state for the clinician. Journal of Psychiatry Research 1975; 12:189-198

[4] Katou S, Hasegawa K, Shimogaki M. Construction of Hasegawa's dementia rating scale revised edition (HDS-R). Ronen Seishin Igaku Zassi 1991; 2:1339-1347

[5] Pritchard WS, Duke DW, Coburn KL, Moore NC, Tucker KA, Jann MW, Hostetler RM. EEG-based, neural-net predictive classification of Alzheimer's disease versus control subjects is augumented by non-linear EEG measures. Electroencephalography and clinical Neurophysiology 1994; 91:118-130

[6] Riquelme LA, Zanuto BS, Murer MG, Lombardo RJ. Classification of quantitative EEG data by an artificial neural network: A preliminary study. Neuropsychobiology 1996; 33: 106-112

[7] Hibino S, Hanai T, Matsubara M, Fukagawa K, Shirataki T, Honda H, Kobayashi T. Assessment of aphasia using artificial neural networks. Japanese Journal of Medical Electronics and Biological Engineering 1999; 37:140-145

[8] McKhann G, Drachman D, Folstein M, Katzman R, Price R, Stadlan EM. Clinical diagnosis of Alzheimer's disease. Neurology 1984; 34:939-944

A New Artificial Neural Network Method for the Interpretation of ECGs

Holger Holst, Lars Edenbrandt

Department of Clinical Physiology, Lund University

Lund, Sweden

holger.holst@klinfys.lu.se

lars.edenbrandt@klinfys.lu.se

Mattias Ohlsson

Department of Theoretical Physics, Lund University

Lund, Sweden

mattias@thep.lu.se

Hans Öhlin

Department of Cardiology, Lund University

Lund, Sweden

hans.ohlin@kard.lu.se

Abstract

Acute myocardial infarct could be a life-threatening disease. Early treatment could be life saving, but treatment of patients not suffering from infarct may cause serious complications. Therefore a rapid decision regarding diagnosis and treatment is of great importance. The physician at the emergency department has to rely on the electrocardiogram (ECG) for the diagnosis. A reliable computer-aided interpretation would be of great value. The purpose of this study was to develop a decision-support system for the diagnosis of acute myocardial infarct using a method that can estimate the error of an artificial neural network.

The material consisted of 3088 ECGs from both patients with and without acute myocardial infarction. The ECG material was randomly divided into three groups. The first group was used to train the neural network for the diagnosis of acute myocardial infarct. The second group was used to calculate the error of the network outputs and the third group was used to test the network performance and to obtain error estimates. The performance of the neural network, measured as the area under the receiver operating characteristic (ROC) curve, was 0.81. The 25% test ECGs with the lowest error estimates had an area under the ROC curve as high as 0.92, i.e. most of these ECGs were correctly classified. The results indicates that neural networks can be trained to diagnose acute myocardial infarct and to signal when the advice is given with high confidence or should be considered more carefully.

1 Introduction

Computerized electrocardiographs have proved to be highly accurate [1], but many physicians do not rely upon computer interpretations because computerized electrocardiographs sometimes make serious mistakes. Even though the computerized electrocardiographs of today are rule-based, many of their drawbacks apply to neural networks as well.

The output values of an artificial neural network could be interpreted as Bayesian probabilities. However, output values close to 0 or 1, i.e. with a very low or very high probability for a certain diagnosis, are not always correct, the reason being that the training set is not entirely representative of the test cases. Such a condition needs to be fulfilled in order to have a probabilistic interpretation of the output signal [2]. These mistakes are not common but it makes it difficult to rely on the advice when used in medical decision support systems. It is probably impossible to construct a method that could classify all ECGs with high confidence, but even if a small number of ECGs from patients with infarct could be classified with a very high confidence, these could receive treatment without further delay.

The purpose of this study was to develop a decision-support system for the diagnosis of acute myocardial infarct using a method [3] that can estimate the error of an artificial neural network.

2 Material and methods

2.1 Study population

The study was based on ECGs recorded on patients at the emergency department of the University Hospital in Lund, Sweden, from January 1990 to May 1997. ECGs with severe technical deficiencies and pacemaker ECGs were excluded. All ECGs recorded on patients who (i) were admitted to the coronary care unit after the recording and (ii) were discharged from the hospital with the diagnosis "acute myocardial infarction", comprised the infarct group (n=1544). The remaining ECGs were used as "control ECGs". The number of control ECGs was reduced to the same number as infarct ECGs, i.e. 1544 ECGs were randomly selected from the 10697 control ECGs. The total material of 3088 ECGs was randomly divided into the following three groups: training (D_{Tr}), validation (D_V) and test (D_{Te}) groups. The numbers of ECGs in the different groups are presented in Table 1.

Table 1: Study population

	Training	Validation	Test	Total
Infarct group	522	514	508	1544
Control group	522	514	508	1544
Total number of patients	1044	1028	1016	3088

2.2 Feature extraction

The ECGs were recorded using computerised electrocardiographs and thereafter stored digitally. Measurements of amplitudes and durations of the electrocardiographic complexes were performed by the recording software in the electrocardiographs. A total of 84 measurements from each ECG were analysed. In order to reduce redundant information and extract the most important features from the 84 measurements, the dimensionality was reduced from 84 to 15 dimensions by principal component analysis.

2.3 Training

The 1044 ECGs of the training set were used in two separate procedures. One was to train an artificial neural network and the other was to create a partitioning of the 15-dimensional data space in which any single data point, denoted x_i, represents an ECG.

2.3.1 Artificial neural network

A multi-layer perceptron architecture [4] was used, with 15 input, 10 hidden and one output node. The target value 1 represented acute myocardial infarct and the target value 0 represented a control ECG. During the training process, the connection weights were adjusted using the Langevin extension of the back-propagation updating algorithm [5]. The network weights were frozen after the training session. A 3-fold cross-validation procedure was used to determinate the optimal error for early stopping in order to avoid overtraining.

2.3.2 Partitioning of data space

The partitioning of the 15-dimensional data space was performed using hierarchical clustering [6]. The training ECGs were divided into groups, or clusters, such that ECGs with similar appearance were assigned to the same cluster. The training set was split into smaller and smaller clusters, but no cluster was allowed to contain less than 20 data points. A cluster is characterised by a reference point, the cluster centre, and each data point (training ECG) is assigned to the closest cluster centre. The result of the clustering procedure is embedded in the final positions of the cluster centres.

2.4 Validation

The validation set D_V was processed through the network and, for each data point, a validation error E_k was recorded. E_k is the absolute value of the difference between the network output and the corresponding target value for validation point x_k. Since the cluster assignments for each of the validation set points are known from above, one can compute the errors E_a corresponding to the different clusters y_a according to

$$E_a = \frac{\sum_{k \in a} E_k}{n_a} \tag{1}$$

where the sums runs over all data points in cluster a. n_a is the number of data points in cluster a. This is the key relation, where cluster assignment gives an error estimate. E_a is simply related to the confidence limit CL_a for cluster a through the definition

$$CL_a = \frac{t_{95}E_a}{\sqrt{n_a}} \tag{2}$$

in the case of 95% confidence level, where $t_{95} = 1.96$

2.5 Test

The next step is to define a probability P_{ia} that data point x_i belongs to cluster center y_a. P_{ia} obviously must fulfill the normalization condition $\sum_a P_{ia}=1$. P_{ia} is given by

$$P_{ia} = \frac{e^{-(x_i-y_a)^2/T}}{\sum_b e^{-(x_i-y_b)^2/T}} \tag{3}$$

The parameter T governs the degree of fuzziness for the probability . For a large T all clusters are equally probable for a given data point ($P_{ia} = 1/K$). On the other hand, in the limit $T \rightarrow 0$ an either-or situation is obtained ($P_{ia'} = 1$ and $P_{ia} = 0$, $a \neq a'$). In this study $T = 1$ is used. Finally, one can compute CL_l for a given data point l in the test set as follows:

$$CL_l = \sum_a P_{la}CL_a \tag{4}$$

where P_{la} has been computed using the cluster centers y_a and the formula given above. CL_l is called error estimate measure in the text.

3 Results

The network outputs and error estimates of the 1016 test ECGs are presented in Fig. 1. The performance of the neural network in the test set is presented as ROC curves in Fig. 2. The area under the curve was 0.81. The test ECGs with the lowest error estimates were also studied separately. An ROC curve was calculated from the 25% test ECGs (n=254) with lowest error estimates (Fig. 2). The area under this ROC curve was as high as 0.92, i.e. most of these ECGs were correctly classified. The area under the corresponding ROC curve for the remaining 75% (n=762) ECGs was 0.76. Thus the error estimates can be used to identify a subgroup of correctly classified test ECGs.

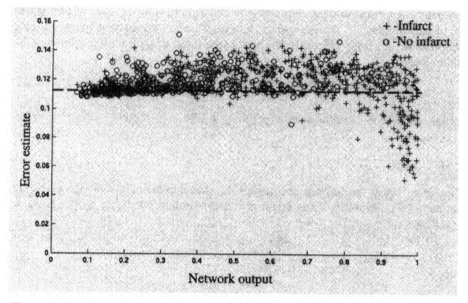

Figure 1: Network outputs and error estimates for the test group. The mean error estimate is indicated with a line.

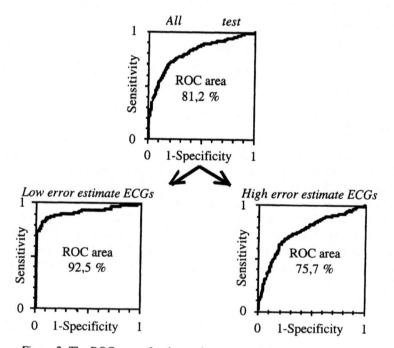

Figure 2: The ROC curve for the total test set and for the two subgroups.

4 Conclusions

The results of this study show that neural networks can be trained to diagnose myocardial infarction from the ECG with high accuracy. These findings are in accordance with previous studies where networks have shown to be even better than experienced physicians [7, 8]. In this study, it was also shown that an indication of neural network confidence by the use of the error estimate makes it possible to identify a subgroup of ECGs, which are more likely to be correctly interpreted. Of special interest was the fact that the neural network correctly and with high confidence detected many of the ECGs, recorded on patients with acute myocardial infarction. This could be of great clinical importance since this could help the physician to identify patients who will benefit from a rapid diagnostic decision and early treatment. Therefore we believe that this method could increase the possibility that artificial neural networks will be accepted as a reliable decision support system in clinical practice.

Acknowledgements

This study was supported by grants from the Swedish Medical Research Council (K99-14X-09893-08B), the Swedish Foundation for Strategic Research and the Swedish National Board for Industrial and Technical Development, Sweden.

References

[1] Willems JL, Abreu-Lima C, Arnaud P. The diagnostic performance of computer programs for interpretation of electrocardiograms. N Engl J Med 1991; 325:1767-1773

[2] Richard D, Lippman R. Neural networks estimate of Bayesian *a posteriori* probabilities. Neural Comput 1991; 3:461-483

[3] Holst H, Ohlsson M, Peterson C, Edenbrandt L. A confident decision support system for interpreting electrocardiograms. Clin phys 1999; 19, 5:410-418

[4] Rumelhart DE, McClelland JL. Parallel distrubuted processing, vol 1-2. MIT press, Cambridge, Massachusetts, 1996

[5] Rögnvaldsson T. On Langevin updating in multilayer perceptrons. Neural Comput 1994; 6:916-926

[6] Kashigan SK. Multivariate statistical analysis - A conceptual introduction. Radius Press, New York, 1991

[7] Hedén B, Ohlsson M, Rittner R, Pahlm O, Haisty WK, Petterson C, Edenbrandt L. Agreement between artificial neural networks and human expert for the electrocardiographic diagnosis of healed myocardial infarction. J Am Coll Cardiol 1996; 28:1012-1016

[8] Hedén B, Öhlin H, Rittner R, Edenbrandt L. Acute myocardial infarction detected in the 12-lead ECG by artificial neural networks. Circulation 1997; 96:1798-1802

Use of a Kohonen Neural Network to Characterize Respiratory Patients for Medical Intervention

Andrew A. Kramer, Ph.D.

Future Analytics, Inc
Middleburg, VA 20118 USA

Kramer@futureanalytics.com

Diane Lee, M.S.

MEDai, Inc
Orlando, FL USA

Randy C. Axelrod, M.D.

Sentara Health Services
Norfolk, VA USA

Abstract

Chronic Obstructive Pulmonary Disease (COPD) is one of the leading causes of respiratory hospitalisations in adults in the USA. Prognosis correlates highly with early diagnosis, however the disease may go unnoticed in its early stages. A database of 25,000 individuals with respiratory problems was received for further investigation. The reported rate of COPD in this population was 5.8%, which is fairly low. An unsupervised neural network using the Kohonen architecture was applied to the data in order to cluster patients into groups based on risk factors for COPD. The network consisted of five output neurons. After training characteristics of the groups were examined. Three of the groups consisted of patients with a high percent of risk factors for COPD. Patients in two of those groups were correctly diagnosed as having COPD, but patients in the third group were under-diagnosed for COPD. These patients should be re-examined by a pulmonologist for possible treatment of COPD. Thus Kohonen neural networks may be a useful tool for clustering patients into groups for differential medical intervention.

1 Introduction

Chronic obstructive pulmonary disease (COPD) is a disease category that includes emphysema and chronic bronchitis. These diseases are characterized by obstruction to air flow and frequently coexist.

Emphysema causes irreversible lung damage as the walls between the air sacs within the lungs lose their ability to stretch and recoil. Elasticity of the lung tissue is lost, causing air to be trapped in the air sacs and impairing the exchange of oxygen and carbon dioxide. As a result airflow is obstructed. Symptoms of emphysema include cough, shortness of breath and a limited exercise tolerance. Diagnosis is made by pulmonary function tests, along with the patient's history and physical examination. Chronic bronchitis is due to an inflammation and eventual scarring of the lining of the bronchial tubes. Symptoms of chronic bronchitis include chronic cough, increased mucus, frequent clearing of the throat and shortness of breath.

Mortality and morbidity due to COPD is high. An estimated 16 million Americans suffer from COPD, with an annual cost to the nation of approximately $32 billion[1]. The quality of life for a person suffering from COPD diminishes as the disease progresses. At the onset, there is minimal shortness of breath. People with COPD eventually may require supplemental oxygen and may have to rely on mechanical respiratory assistance.

The prognosis for COPD is enhanced considerably by early diagnosis and intervention. However this requires that the person undergo a lengthy physical examination. Usually patients seek care and definitive diagnosis as a result of one or more serious respiratory episodes, after significant tissue damage has already occurred. It would be desirable if a means could be developed of identifying individuals at high risk for developing COPD. Health data on individuals within a population may help identify the combination of characteristics that suggest an individual is likely to develop COPD.

Data on a subset of patients referred to a large health plan were obtained. The incidence of diagnosed COPD in this group of patients was 5.8%, much lower than reported for adults in the U.S. It is likely that these patients were under-diagnosed for COPD. An analysis was undertaken to determine if patients could be clustered into discrete groups based on their health data. The objective was to isolate one or more groups as candidates for increased medical intervention.

2 Materials and Methods

2.1 Description of Population

Data were obtained from a major health care organization in the U.S. on a portion of their subscriber base. The population consisted of 25,615 individuals who had a history of respiratory problems, including asthma. Information for each individual included demographic data, medical conditions and treatments, detailed pharmaceutical data, and health care costs. In all there were over 200 variables available for analysis.

2.2 Analytical Methods

2.2.1 Data Preprocessing

Exploratory analyses were conducted in order to reduce the size of the input space. Categorical variables with a frequency of less than 1.0% were removed, as were quantitative variables with sparse variation. A correlation matrix was created to detect instances of very high multicolinearity ($r > 0.9$), and one of the highly correlated variables was removed. The remaining variables were normalized to a range of $0 - 1$.

2.2.2 Neural Network Architecture and Processing

Since there was no outcome variable and the analysis was investigative in nature an unsupervised neural network design was chosen. Specifically, the self-organizing network based on the model initially formulated by Kohonen[2] and described in Simpson[3] was used. The input vector consisted of 68 variables of diverse types (see 2.1 above). There were five nodes in the output layer of the network. The starting neighbourhood size was four and allowed to decrease to zero. All weights were initially set to 0.5 with a learning rate of 0.4. The stopping criteria were a minimum of 400 epochs or a reduction in the learning rate to 0.001, whichever occurred first.

Network training took place using a random presentation of observations, with the distance metric being Euclidean. When training was stopped a data set consisting of the denormalized input vector and the winning output category for each observation was exported to SAS®, where all statistical analyses were conducted.

3 Results

Table 1 gives the frequencies that observations were placed in the five network categories for a single run. (These results were quite stable over additional runs, with group assignment correlations of 90%.) Each group consisted of between 20-25% of the total with the exception being Group 5 containing only 10%.

Table 1: Number of observations placed in each output category

	Group 1	Group2	Group 3	Group 4	Group 5	Total
N	6309	5500	5452	5740	2614	25615
%	24.6%	21.5%	21.3%	22.4%	10.2%	100.0%

Figure 1 shows the percent of respiratory disorders in each of the five groups. Clearly Groups 4 and 5 do not contain patients with COPD. Groups 1 and 2 have similar percentages of patients with chronic bronchitis and emphysema, but patients in Group 3 were less likely to be diagnosed with COPD.

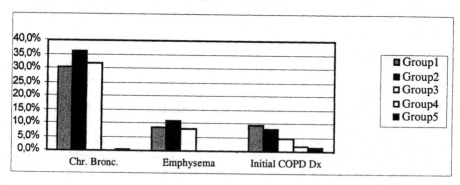

Figure 1: Percent of patients affected with respiratory disorders in the five output groups

Figure 2 gives the percentage of other chronic conditions for each group. It appears that patients in Groups 1 and 2 have a much higher incidence of other chronic conditions than patients in Group3. Perhaps the lower recorded diagnosis of COPD in Group 3 is due to those patients not being seen as often by physicians.

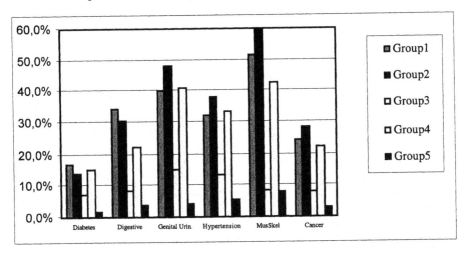

Figure 2: Percentage of patients with other chronic conditions in the five output groups

Figure 3 below shows that patients in Group3 have fewer medical conditions diagnosed and identified drug claims. This results in their costs due to respiratory diseases being a high % of their total medical costs. Group 1 is distinguished from Group 2 in that 100% of the former group had emergency room visits (not shown in Figure 3), as compared to 0% in the latter.

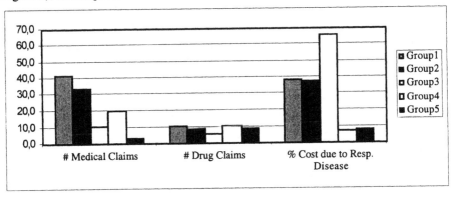

Figure 3: Mean number of other medical conditions, insurance claims, and % of total costs due to respiratory costs

The five groups of patients can be characterized as follows:

1. COPD, multiple medical problems, and at least 1 emergency room visit
2. COPD, multiple medical problems, and no emergency room visits
3. Respiratory disease usually unaccompanied by other conditions
4. No COPD, multiple medical problems
5. No COPD, usually no medical problems

4 Conclusion

Using a Kohonen network on medical data allowed for an informative grouping of patients. From this grouping it was apparent that diagnosis of COPD was highly correlated with a patient having respiratory symptoms accompanied by other medical conditions, and thus being evaluated more frequently by a physician. This was especially true for patients who had at least one emergency room visit. Such groupings allow for those organizations responsible for care delivery to approach these populations with distinct care management strategies. The findings here suggest that patients with existing respiratory disease unaccompanied by other chronic conditions should be evaluated more carefully for a diagnosis of COPD.

References

[1] American Lung Association Publication, "Lung Disease Data 1998-99"., New York, 1999.

[2] Kohonen T. The SOM Methodology. In: Deboeck G. and Kohonen T. (eds) Visual Explorations in Finance with Self-Organizing Maps. Springer-Verlag, Berlin, 1998, pp 159-167.

[3] Simpson, P. Artificial Neural Systems. New York, N.Y.: Pergamon Press, 1990.

Determination of Microalbuminuria and Increased Urine Albumin Excretion by Immunoturbidimetric Assay and Neural Networks

Dr. Bela Molnar

2nd Dept. of Medicine, Semmelweis Medical University
Budapest, Hungary
mb@bel2.sote.hu

Dr. Rainer Schaefer

Roche Diagnostics, Dept. of Biometry & Laboratory-IT
Tutzing, Germany
rainer.schaefer@roche.com

Abstract

A new technique is demonstrated for the determination of urine albumin concentration. A commercially available microalbuminuria test was combined with neural network analysis of reaction kinetic data. In total 102 patient urine samples were analyzed [27 diabetes patients, 21 with nephrosis or nephritis, 54 with hypertension]. Due to the prozone-effect in standard immunoturbidimetric assay technology we found 1 sample in the diabetes group, 6 in the nephrosis/nephritis group and 4 in the hypertension group, that yielded false negative results, i.e. misleading low instead of high urine protein concentrations. By means of albumin dilution series in a range of 0 to 40,000 mg/l a non-monotonous calibration curve (Heidelberger curve) was obtained. The measured kinetic data were split for training, testing and validation of a backpropagation neural net. It could be demostrated that such a net yields a correct correlation between measured signals and concentration, even for the difficult task of classification between very high and very low concentrations. Moreover also the false negatively assigned patient data were all classified correctly.

1 Introduction

Urine albumin is a potential marker of early organ damage in diabetic nephropathy, diabetes, and pre-eclampsia [1]. The amount of albumin in urine can be used as a monitoring marker for both the efficacy of the therapy and the progression of organ failure [2]. The early detection of renal involvement in diabetes or hypertension requires specific tests with high accuracy in the low concentration range (0 to 300 mg/l). In the severe and late form of these diseases, high concentration samples up to 40,000 mg/l have been reported [3]. The wide range of albumin concentration in urine (between 0 – 40,000 mg/l) still causes serious analytical problems despite of the many available assay principles, like the immuno-turbidimetry [4,5].

In routine-laboratory-testing the immuno-turbidimetry is a common technique: A sample containing an unknown amount of a human protein (=antigen) reacts with a fixed amount of antibody (against that protein). The kinetics of the antigen-antibody-complex formation (followed by precipitation) is recorded as the time series of the photometric absorbance-signals. In the range of antibody access the slope of the kinetics increases with the protein concentration. However, when the antigen/antibody-ratio is $\gg 1$, excessive protein can block the limited number of antibody binding sites and prevent immuno-complex formation. The slope of the kinetics then decreases with increasing protein concentrations. Thereby in the concentration (c) vs. slope (or ΔA) diagram, i.e. the "Heidelberger curve" (see below, Fig. 2, upper), one ΔA value corresponds both to a low concentration (common) and a very high concentration (rare, occasionally ignored).

Lately, the application of artificial intelligence techniques has led to the solution of several medical pattern recognition problems [6,7]. In addition to the rule-based expert systems, we can now evaluate signals/data with new techniques, developed after biological pattern recognition procedures. This method, named "artificial neural networks", simulates the essential functions of small-organized neuron groups [8]. They have been applied to medical decision support systems. e.g. in the fields of microscopic image analysis [9] and clinical chemistry [10].

In our study we demonstrate how false negative results from immuno-turbidimetric microalbuminuria tests can be avoided with the application of trained artificial neural networks, thereby presenting a solution to the "prozone effect" problem.

2 Methods

2.1 Urine samples and dilution series

Centrifuged urine samples from a routine clinical chemistry lab were grouped according to the patient's diagnosis. There were a total of 27 urine samples from diabetes patients (12 insulin dependent, 15 non-insulin dependent), 21 urine samples from patients with known chronic renal disease (8 from nephrosis syndrome, 13 from chronic glomerulonephritis) and 54 urine samples from hypertension patients.

In addition dilution series were prepared from human albumin lyophilizates (Roche, Germany). Each series had 50 concentrations equally distributed over the entire range from 0 to 40,000 mg/l in order to provide sufficient data/measuring points for training, testing and validation of the neural networks.

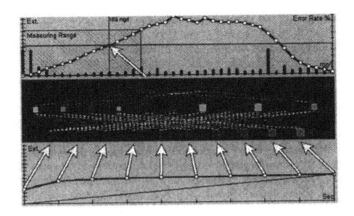

Figure 1: The principle of reaction kinetic analysis by neural networks. The reaction kinetics is recorded (x: time, y: absorption; bottom). A neural network is used for the evaluation of these signals, the number of neurons (10) in the input layer corresponds to the measuring points of the reaction kinetics (middle). The number of output neurons is equal to the classification criteria (2), an active "left" neuron assigns the kinetic data to the low concentration range, an active "right" neuron points to the high concentration range (x: concentration, upper).

2.2 Reagents, instrumentation and assay procedure

From each urine sample the albumin content (sAlb assay, Roche, Germany) and the microalbuminuria (MAU) was determined using the regular assay procedure. Prozone effect was declared if high sAlb and low MAU values were obtained.

The determinations were performed on the Keysys® clinical chemistry analyzer and on Roche/Hitachi clinical chemistry analyzers for reference. Reaction kinetics data were recorded online and stored on a PC using the OASE software (internal development by Roche Diagnostics).

2.3 Neural network analysis

The neural network analysis was performed by the NeuralWorks Professional II software (NeuralWorks, Pittsburgh, PA, USA). Preferentially back-propagation neural networks were used for training [11], in which the neurons were ordered in one input layer, one or two hidden layer(s), and one output layer.

The preferred neural network had 10 input neurons and 2 output neurons. In the hidden layer the number of neurons varied between 7 and 10. We used hyperbolic tangent transfer function and normalized cumulative delta learning rule. The learning rate was gradually changed between 0.1 and 0.6, the momentum term between 0.01 and 0.1. Changes of the connection weights were performed after the evaluation of the sequentially recalled samples. Training was continued until more than 95% of the training set samples were classified correctly. Overtraining was controlled by the iterative recall of a test set.

3 Results

In the routine MAU values of 108.3±117.21 mg/l were found for diabetes patients, 265.4±106 mg/l for the nephritis group, and 101.2±89.5 mg/l for hypertension patients (mean±sd). By 1 sample in the diabetes group, by 6 in the nephrosis/nephritis group and by 4 in the hypertension group, false low MAU results were detected.

The Heidelberger curve of the dilution series showed the maximum at 1200 mg/l. The measuring range of the standard test is between 0-450 mg/l. The right side of the Heidelberger curve "re-enters" the measuring range at a concentration of 4,500 mg/l. The microalbuminuria range (30 to 300 mg/l) on the left side of the Heidelberger curve corresponds approx. to the 4,500-15,000 mg/l range on the right side. Concentrations above 18,000 mg/l can yield negative results (x<30 mg/l) (Fig.2).

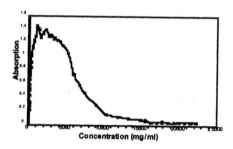

Figure 2: Heidelberger curve of the MAU assay. The maximum can be found at 1,200 mg/l, the Heidelberger curve "re-enters" the measuring range at 4,500 mg/l.

Typical shapes of the reaction kinetics are observed for the different absorbance ranges and concentrations. In the very low absorbance range (<0.05) the kinetics from the left side (very low albumin concentrations) are basically straight lines, the kinetics from the right side (very high concentrations) are similar with a slightly higher initial slope. In higher absorbance ranges kinetics from the left side increase continously during measurement, whereas kinetic data from high concentration samples show a decreasing slope, i.e. the immune-complex formation is already finished, when the last measuring points are recorded (Fig.3).

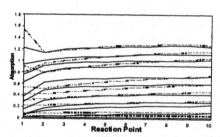

Figure 3: Kinetics of the measurements from the MAU Heidelberger curve. The dashed lines correspond to low concentration samples from the left side of the Heidelberger curve.

For this classification problem a standard back propagation neural network was used with a relatively high number (7-10) of hidden neurons. With typical parameter settings (learning rate 0.6, momentum term 0.1, random learning of samples) training results were poor, with frequent misclassifications in the low absorbance range of the Heidelberger curve. The reason was that the differences of the kinetic data were not very distinctive in this range. So either no convergence was observed or the network tended to get trapped in one of the many local minima. In the course of learning it was important to gradually reduce the rate of the weight changes by minimizing the learning rate and momentum term (0.01, and 0.001). This way the training became successful, all of the patient samples (102) were classified correctly.

Cross-validation of the trained network was done by the evaluation of reaction kinetics measured with another reagent lot and another Keysys® analyzer (Fig.4). Occasional errors were only found at the highest absorbance values, i.e. in a range without clinical relevance.

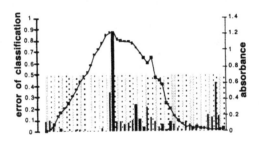

Figure 4: Cross-validation of the trained neural network. X: the concentration of the test samples. Y: (left, bars) result of the left-right-classification by the neural network, the error probability estimation yields 0, if the net is "sure" and the classification is correct, a value of 1 indicates the net is "sure" but the result is false, misclassification occurs above the error of 0.5. Y: (right, dots) the obtained absorption value (ΔA value).

4 Discussion

The follow-up studies of diabetes, hypertension and nephrosis syndrome patients require new and safe techniques for the broad concentration range urine albumin determination. Our measurements from routine samples show that today 3 to 5% of the samples sent for MAU determinations may be misclassified. By using a trained neural network all of the specimens with high urine albumin concentration can be classified correctly.

The method described here is convenient and easy to implement. As the training can be performed at the manufacturer's site, only a small module with few parameters must be added to the system's software. The program downloads the measured reaction kinetics data through the system's RS232-interface, performs a simple weight matrix multiplication and delivers its "left-right" decision.

The clinical results state that albumin in urine determinations require additional techniques for the evaluation of concentration range between 0-20 g/l. The back-propagation neural networks can be used in the prozone effect detection by the analysis of reaction kinetics. As a general conclusion, artificial neural networks are useful pattern recognition tools for the classification of photometric signals with the potential to safely extend the common measuring range of diagnostic tests.

References

[1]   Mogensen CE. Microalbuminuria, a marker for organ damage. Science Press London. 1993

[2]   Ward KM. Renal function [microalbuminuria] Review. Anal chem. 1995; 67: 383-91

[3]   Ballantyne FC, Gibbons J, O'Reilly D. Urine albumin should replace total protein for the assessment of glomerulal proteinuria. Ann Clin Biochem. 1993; 30:101-103

[4]   Newman DJ, Thakkar H, Medcalf EA et al. Use of urine albumin measurement as replacement for total protein. Clin nephrol. 1995; 43: 103-109

[5]   Kessler MA, Menitzer A, Petek W et al. Microalbuminuria and borderline-increased albumin excretion determined with a centrifuge analyser and the Albumin Blue 580 fluorescence assay. Clin Chem. 1997; 43: 996-1002

[6]   Bartels PH, Weber JE. Expert systems in histopathology. Introduction and overview. Anal Quant Cytol Histol. 1989; 11:1-7

[7]   Place JF, Truchaud A, Ozawa K. Use of artificial intelligence in analytical systems for the clinical laboratory. Clin Chim Acta. 1994; 231:5-34

[8]   McCullogh, WS. Pitts, W. Bull. Math. Biophys. 1943; 5:115-133

[9]   Molnar B, Szentirmay Z, Bodo M et al. Application of multivariate, fuzzy set and neural network analysis in quantitative cytological examinations. Anal Cell Pathol. 1993; 5: 161-175

[10] Astion ML, Wener MH, Thomas RG et al. Overtraining in neural networks that interpret clinical data. Clin Chem. 1993; 39: 1998-2004

[11] Rumelhart DE, McClelland JL. Parallel distributed processing: Explorations in the microstructure of cognition. Vol.I. Foundation, MIT Press 1987.

Artificial Neural Networks to Predict Postoperative Nausea and Vomiting

A. Nawroth

European Media Laboratory
D-69118 Heidelberg, Germany
Andreas.Nawroth@EML.Villa-Bosch.de

R. Malaka

European Media Laboratory
D-69118 Heidelberg, Germany
Rainer.Malaka@EML.Villa-Bosch.de

L.H.J. Eberhart

Department of Anesthesiology, University of Ulm
D-89070 Ulm, Germany
leopold.eberhart@medizin.uni-ulm.de

Abstract

The effect of postoperative nausea and vomiting (PONV) is not a dangerous effect but it causes discomfort. It is therefore an issue to reduce this effect as much as possible. The reasons for its occurrence, however, are not yet clear. Many factors seem to be candidates for facilitating or depressing this post-anaesthetic side effect ranging from individual to treatment factors. Even though some prediction of PONV can be made on statistical analyses, there is still a high level of uncertainty. In this paper, we use neural networks for predicting PONV and show how feed-forward networks can be designed as problem-specific predictors. The resulting networks outperform the statistical approaches and simple black box approaches using neural networks. The methods applied for designing the networks and encoding the data are furthermore generalizable for other medical applications.

1   Introduction

Postoperative nausea and vomiting (PONV) are the most frequent side effects after anesthesia. It occurs in about one third of all patients undergoing surgery with general anesthesia. Although PONV it is not a life-threatening complication, it is a major source for patients' postoperative discomfort and impairs well-being during recovery from anesthesia. It is, therefore, the major reason for patients' dissatisfaction with perioperative treatment [1].

There are several effective measures to prevent PONV However, routine efforts for prophylaxis are not indicated for medical (adverse effects of antiemetics), and economical reasons. Thus, from the anesthetist's point of view, tools to identify patients with an increased risk to suffer from PONV are very much desirable. There have been some attempts in the past to create risk scores to predict PONV using logistic regression analysis. However, the performance of these scores is not satisfying. Using ROC-curve analysis, it has been shown that less than 75% of the patients can be predicted correctly [2]. Certainly, the main reason for this disappointing result is that PONV is a complex and multi-factorial problem and that the pathophysiology is only poorly understood.

These results indicate that there might be advantages for using connectionist models. These can reveal some inherent properties in the data that cannot be expressed in simple rules and that more complex non-linear dependencies than those that can be discovered by simple statistical models may exist. We employed artificial neural networks and investigated several different approaches for coding the data. Particular interest has been in the question of preserving knowledge from experts in the data set in the definition of the encoding of the input data. We therefore did not use a sheer black-box approach but combined expert knowledge with the general-problem-solving abilities of feed-forward neural networks. The results of this work are very encouraging. At the current state, our neural networks are the best predictors for PONV and they may soon be tested in clinical praxis.

2   Neural Networks for PONV Prediction

The basis for our neural network for PONV prediction is a data set with 1444 patterns from patients who had anesthetic treatment. Each pattern consists of 67 input variables that can be continuous (like age) or binary (like sex). The most prominent variable is the sex of the patient. It does imply a set of variables that are only valid for female patients, e.g., menstrual cycle or meno-pause, and it separates the whole data set in two distinct groups of patients, namely male and female, that show very different sensibility for PONV. The latter will be discussed later on together with the results. Both reasons led us to the decision to handle both groups separately and to build two sex-specific neural PONV predictors. A first simple approach using a feed-forward network with six most commonly used variables [3] [4] [5] showed a learning success only for RPROP (resilient back-propagation, [6]) while other methods such as simple back-propagation did not converge to acceptable solutions (this refers to a 6-6-2-topology). The results are given in Table 1.

The 1444 patterns have been split into the set A and the test set B. In the case of logistic regression analysis we used all patterns of A to compute the coefficients. In the case of neuronal networks we split A in a training set and a validation set. The total score considers the different frequency of male and female patients (see below).

The difficulties in this naive approach lie in both the complexity of the problem and the dimensionality of the input space combined with a small set of examples and a sparse covering of the input space through examples. Since we have 67 input variables, it is questionable if this high number of input variables really helps the network in finding good solutions. On the other hand, using just the six most commonly used variables did not yet yield satisfactory results. Therefore, the dimensionality of the input space must be reduced in an intelligent way. But even if

Table 1: PONV prediction with statistical methods and simple neural networks. The table gives the classification rate (percentage of correct classification) for three methods and for the whole data set (total score), and the four partial data sets given by sex and classification output. The logistic regression and the neural network (6-6-2-topology) are trained using the same set of learning data. The results shown are those from the independent test data (not used for learning). For more details on the procedures applied, see text.

Classification rates:	logistic regression analysis	best net wrt. the validation set	best net wrt. the test set
total score	67.17%	67.29%	67.86%
female: no PONV	58.57%	45.71%	45.71%
female: PONV	57.00%	68.75%	68.75%
male: no PONV	96.67%	98.67%	98.67%
male: PONV	8.00%	2.00%	6.00%

we have a good model, some uncertainty still exists in the data themselves since the output classification, i.e., whether PONV occurs or not, is measured by a subjective response given by the patients in some individual four point rating from no to strong effects. Another problem lies in the sparseness of the input space. According to the statistical properties of the data some regions are not well covered by data which may lead to many local minima and to long training times.

Another problem occurs for the PONV classification of male patients. Here, only 25% percent of the patients showed PONV and thus a constant prediction of no PONV symptoms would be correct in 75% of the cases. This is basically the result of the predictors shown in Table 1 which leads to very good classifications for negative male PONV (greater than 96%) and bad rates for the positive case (2-8%). A good neural network will have to perform better than this trivial constant prediction.

3 Strategies for Problem-Adapted Learning of PONV Predictions

In the following, we show how to treat these problems in order to derive neural networks that can predict PONV symptoms. Even though these methods are specific to our selected application, i.e., PONV, the steps can easily be applied to other problems in an analogous way.

In order to reduce the dimensionality of the full input space, we had to throw off variables which are unnecessary for predicting PONV. The transformation of the input data space using a principal component transformation led to at least 2% worse results. Basically, this may be due to the fact that principal components are computed with no respect to the variables importance for causing PONV, but just on the basis of their distribution properties. We applied stepwise logistic regression analysis to determine the relevant input variables. This method considers the individual importance of each input variable and allows us to determine the most important ones. Best results have been achieved by using the 18 most important input variables for the female PONV predictor and 21 input variables for the male

Table 2: PONV prediction with statistical methods and more complex neural networks using the input variables determined by stepwise logistic regression analysis. The logistic regression and the neural network (female: 18-9-2-topology, male: 21-10-2-topology) are trained using the same set of learning data. See also Tab.1 and text for further details.

Classification rates:	logistic regression analysis	best net wrt. the validation set	best net wrt. the test set
total score	71.31%	68.47%	69.04%
female: no PONV	65.71%	55.71%	55.71%
female: PONV	65.00%	67.50%	67.50%
male: no PONV	96.00%	90.00%	90.67%
male: PONV	16.00%	24.00%	26.00%

PONV predictor. Using this new input variable set the score for both the logistic regression analysis and the neuronal network improved with significance.

A good example for the sparseness in the input space and the problems associated with it is the variable indicating the patient's age. For some statistical reasons, it can occur that all patients of one age suffer PONV, e.g., all 16 year old patients, but all 17 year olds do not encounter any problems. Such a strong dependence of PONV on the variable age, however, is very unlikely and almost certainly an artifact of the statistical properties of the data set. These statistical fluctuations indicate high irregularities that may lead to predictors that adapt perfectly to those artifacts given in the statistical sample of the training data. In this case the data should be accumulated in an appropriate way such that, for every range of ages the mean values smooth out the peaks. Continuous variable values should always be tested for their statistical variation, and if necessary, a large number of sparsely set values should be compressed into smaller numbers of range values that middle out extreme fluctuations. Classical statistical procedures can be applied and show that the information content in our transformed data set is not significantly reduced. Applying logistic regression analysis to the transformed data set reveal equal performance, or performance that is not worse than 1% from the original. Even better results could be achieved applying a local metric, i.e., the ranges for the parameters were selected such that each class contained approximately the same number of data.

4 Simulation Results

In order to compare the methods described above, we split the whole sample in two sub-samples. Sub-sample A contains about 75% of the data. In the case of logistic regression analysis it is used for determining the exponential coefficients. In the case of the neural predictor it was split into A_1 – the training set (contains about 50 % of the data) – and A_2 – the validation set (contains about 25% of the data). Subset B, the test set contains about 25% of the data and is only used for measuring the

Table 3: PONV prediction with statistical methods, more complex neural networks using the input variables determined by stepwise logistic regression analysis and transformed data applying z-Transformation and the method of local metric. The logistic regression and the neural network (female: 18-9-2-topology, male: 21-10-2-topology) are trained using the same set of learning data. See also Table 1 and text for further details.

Classification rates:	logistic regression analysis	best net wrt. the validation set	best net wrt. the test set
total score	71.31%	74.77%	76.48%
female: no PONV	65.71%	64.29%	64.29%
female: PONV	65.00%	75.00%	75.00%
male: no PONV	96.00%	96.00%	96.67%
male: PONV	16.00%	26.00%	36.00%

performance of the two predictors. All scores in the following tables refer to the performance of the predictors on this subset. As discussed above, we trained sex-dependent PONV predictors using the data for male and female patients, respectively. Thus, the data consists of four groups: female non-PONV (291 patients), female PONV (334), male non-PONV (613), and male PONV (206).

In order to get the best neuronal network we trained 25 networks and took the top five (according to their performance on the validation set). The score on the test set of the network that performed best on the validation set is shown in the second row. The third row shows the performance of the network out of the five, with the best results on the test set.

In Table 2, the classification results are given for the approach using stepwise logistic regression analysis for the selection of the 18 and 21 most important variables (the number of variables was determined using varying numbers and by choosing the best). It can be seen that the total scores are better for all methods. The logistic regression analysis is even better than the neural network approach (71.31% vs. 68.47% and 69.04%). A striking result is the prediction of the NN for male patients. In the simple network with only 6 input variables (Table 1), the prediction was basically a non-PONV classification for all patients. With the new input variable set, it makes positive predictions for the male PONV case as well.

These results indicate that more input variables yield better prediction results. But it is not just the number of input variables that increases the networks capabilities. Adding additional 10 or more random input variables to the six variables used in the first approach (Table 1) yielded only total scores of about 62%. Thus, even though they increased the information content of the input, they decreased the network performances. This may have several reasons. On the one hand, more variables imply more weights and it is more difficult for neuronal networks to find a good solution in the weight space (there are three times more weights in a 16-8-2 net than in a 6-6-2 net). On the other hand, larger networks are more likely to be disturbed by overtraining effects. It is therefore important to select the right set of input variables and the stepwise logistic regression analysis constitutes an excellent tool to do that.

Merging both approaches, selecting and transforming the data yields the results in table 3. The scores of the artificial neuronal networks improved from 67.29% to 74.77% and from 67.86% to 76.48% (Table 1) in comparison to the first approach. The increase in the predicting reliability is most striking in the case of male patients. The correct classification rate for positive male cases is still above 96% and for negative cases the scores rises from 2% to 26% and from 6% to 36% (cf. Table 1), respectively.

5 Conclusion

In this paper we used neural networks for predicting PONV symptoms. Previously used statistical methods had correct classification rates that were below 75%. Our neural network approach introduced here, outperformed these approaches and yielded rates of 76%. Moreover on our particular data set, the classical statistical approach using a logistic regression analysis only led to 71.3% correct classifications. Therefore, our resulting neural predictor is among the best known methods for predicting PONV.

These results are achieved using techniques that can also help to design predictors for other medical applications. Of paramount importance is the right choice of input variables and an intelligent data transformation (z-Transformation, compressing set values, applying local metric) that helps the neuronal network to make use of its powerful abilities. In this case the stepwise logistic regression analysis reveals itself as a powerful tool in determining the important variables. Moreover, these methods are generic in the sense that they can also be used together with non-neural methods and, as shown above, even enhance the capability of the logistic regression.

References

[1] Kapur P.A. 1991 *Editorial: The big "little problem"*. Anesth Analg 73: 243-245

[2] Eberhart L.H.J., J. Högel, W. Seeling, A.M. Staack, G. Geldner, M. Georgieff 1999 *Evaluation of three risk scores to predict postoperative nausea and vomiting*. Acta Anaesth Scand 43: *in press*

[3] Apfel C.C., C.A. Greim, I. Haubitz, C. Goepfert, P. Usadel, P. Sefrin, N.A. Roewer 1998 *Risk score to predict the probability of postoperative vomiting in adults*. Acta Anaestesiol Scand 42: 495-501

[4] Koivuranta M., E. Läärä, L. Snare, S. Alahuhta 1997 *A survey of postoperative nausea and vomiting*. Anaesthesia 52: 443-449

[5] Palazzo M., R. Evans 1993 *Logistic regression analysis of fixed patient factors for postoperative sickness: a model for risk assessment*. Br J Anaesth 70: 135-140

[6] Riedmiller M., H. Braun 1993 *A direct adaptive method for faster backpropagation learning: the RPROP algorithm*. In: Proceedings of the IEEE International Conference on Neural Networks (ICNN) 586-591

Acute Myocardial Infarction: Analysis of the ECG Using Artificial Neural Networks

Mattias Ohlsson

Department of Theoretical Physics, Lund University

Lund, Sweden

mattias@thep.lu.se

Holger Holst and Lars Edenbrandt

Department of Clinical Physiology, Lund University

Lund, Sweden

holger.holst@klinfys.lu.se, lars.edenbrandt@klinfys.lu.se

Abstract

This paper presents a neural network classifier for the diagnosis of acute myocardial infarction, using the 12-lead ECG. Features from the ECGs were extracted using principal component analysis, which allows for a small number of effective indicators. A total of 4724 pairs of ECGs, recorded at the emergency department, was used in this study. It was found (empirically) that a previous ECG, recorded on the same patient, has a small positive effect on the performance for the neural network classifier.

1 Introduction

For patients attending the emergency department with chest pain, an early diagnosis of acute myocardial infarction is important because of the benefits of immediate and correct treatment. Different diagnostic methods have been studied, but the 12-lead ECG together with patient history is still the best method for early diagnosis of acute myocardial infarction.

The ECG diagnosis can be a difficult problem and misdiagnosis do occur. The use of a computer-based ECG interpretation program is therefore valuable in order to detect infarction ECGs. Artificial neural networks (ANN) is one computer-based method that has shown to be even better than experienced physicians in the ECG diagnosis, regarding myocardial infarction [1, 2].

In clinical practice the physician include in their ECG analysis a comparison between the current ECG and a previous one of the same patient (if such one is available) as an aid in the decision making. The purpose of this study was to construct a neural network classifier for the diagnosis of acute myocardial infarction and to see if a previous ECG, recorded on the same patient, can increase classification performance.

2   Study Population

The study was based on patients who present to the emergency department of the University Hospital in Lund, Sweden during the period January 1990 - June 1997 and who had an ECG recorded and stored at that occasion. The 12-lead ECGs were recorded using computerized electrocardiographs (Siemens-Elema AB, Solna, Sweden). Each of these ECGs were analyzed as follows:

- (i) If the ECG was recorded on a patient who was admitted to the coronary care unit after the ECG recording, and was discharged with the diagnosis "acute myocardial infarction" and (ii) an earlier ECG recorded on the same patient was found in the ECG database of the hospital (not necessarily recorded at the emergency department).

then this pair of ECGs was defined as an *acute infarction* case. On the other hand:

- (i) If the ECG was recorded on a patient not suffering an "acute myocardial infarction" at that occasion and (ii) an earlier ECG recorded on the same patient was found in the ECG database of the hospital (not necessarily recorded at the emergency department).

then this pair of ECGs was defined as a *control* case. Only ECGs with severe technical deficiencies and pacemaker ECGs were excluded.

The acute infarction group consisted of 924 pairs of ECGs and the control group consisted of 3800 pairs of ECGs. A separate test group of 1000 pairs (200 infarctions and 800 controls) were randomly selected from the total set of 4724 ECGs.

3   Feature Extraction

3.1   12-Lead ECG

The digitized ECG consists of 12 leads where each lead represents one heart beat complex, approximately 1000 sample values long (1000 Hz). From each complex one extracts standardized measurements[1], that includes amplitudes, durations and areas. Figure 1 shows a generic ECG complex with P-, QRS- and T-waves. Common measurements are Q-, R-, S-amplitudes and sample values along the ST-segment. From a physiology point of view acute myocardial infarction cause changes in the ST-segment. We used the following measurements from each of the 12 leads:

1. QRS-duration (duration of the QRS wave).

2. QRS-area (area of the QRS wave).

3. Q-, R- and S-amplitudes.

4. ST-amplitudes (3 measurements taken from the start, 2/8 and 3/8 of the ST-segment).

---

[1]Computation of these measurements is performed by the recording software (Siemens Elema in our case).

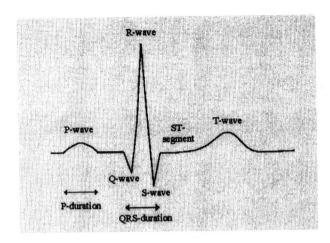

Figure 1: A generic ECG complex with a P- and QRS-wave followed by an ST-segment and an ending T-wave.

5. ST-slope (the slope at the beginning of the ST-segment).

6. T-max/min (maximum/minimum of the T-wave).

In total 122 measurements from each ECG were selected for further analysis.

For each patient there was two ECGs, the current and one previous ECG. Since the objective was to find out whether a previous ECG can help the classifier to determine acute myocardial infarction, we had more than one dataset. The first consists of the current ECGs, i.e. the ECGs recorded at the emergency department at the time of admittance to the coronary unit. The other datasets contained information about the previous ECG. Two such datasets were constructed, one that simply contained measurements from the previous ECG and one the contained measurements of the difference between the current and the previous ECG. The difference δ between each measurement α was taken as, $\delta_\alpha = \alpha_{current} - \alpha_{previous}$.

3.2 Principal Component Analysis

There is a high degree of correlation between measurements from the 12 leads. This correlation comes from the fact that 4 of the 12 leads are simple linear combinations of two other, e.g. lead II = lead I + lead III. There is also a natural correlation among some of the leads because they are physically close to each other. The 122 original measurements can therefore be reduced to a smaller set of more "effective" variables. We used principal component analysis (PCA) to achieve this reduction.

Prior to the PCA the 122 variables were grouped into the 6 groups listed above. The PCA was then applied to each of these smaller datasets separately. Figure 2 shows the eigenvalues of the covariance matrix, normalized so that the largest is one, for each group of variables. Clearly, there exists a large degree of (linear) correlation between the measurements within each group. As can be seen in figure 2, the 36 measurements

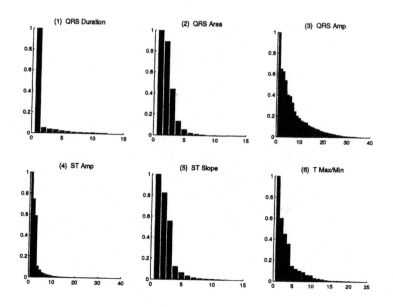

Figure 2: Principal component analysis of the measurements from the dataset of current ECGs. The plots show the (normalized) eigenvalues of the covariance matrix for the variables i each of the groups.

of the ST-amplitudes are well described by only 3 PCA variables. The PCA analysis of the other datasets showed a similar behavior with large linear correlations. Table 1 summarizes the variables used as inputs for the neural network classifiers.

4   Artificial Neural Networks

Feed-forward ANN have turned out to be a very powerful approach for classification problems. A general introduction to the subject can be found elsewhere [3]. We used a standard multilayer perceptron architecture with one hidden layer of 10 nodes. The output layer consisted of one neuron that coded whether the patient suffered from acute myocardial infarction (1) or not (0). A summed-square error function was used together with a Langevin extension [4] of the back-propagation updating rule. Langevin updating consists of adding a random Gaussian component to the gradient, which has the effect of speeding up the minimization procedure. In order to avoid over-training a weight elimination [5] regularization term was used, i.e.

$$E \rightarrow E + \nu \sum_i \frac{w_i^2}{\tilde{w}^2 + w_i^2} \tag{1}$$

where $\tilde{w}$ was set to 1. The sum over weights does not include the threshold weights since they should not be part of the regularization.

The ν parameter was set using a 5-fold cross-validation scheme on the training set. Finally, a committee of 20 networks was trained on the full training set using this ν.

Table 1: Summary of the variables used as inputs to the neural networks.

	Number of PCA components used		
	Current ECGs	Previous ECGs	Previous - Current ECGs
QRS-duration	1	1	1
QRS-area	1	1	1
QRS-amplitudes	2	0	0
ST-amplitudes	5	3	3
ST-slope	2	0	0
T-max/min	5	3	3
Total	16	8	8

The classification performance on the test set was calculated using the average output of the committee.

5   Results

Figure 3 shows the result of the networks on the test set. It it presented as the *receiver-operating characteristic* (ROC) curve. The area under this curve is a measure of the performance of the classifier. The area for the network committee trained on the dataset with only the current ECGs was 0.858. The committees trained with additional information from a second ECG got 0.863 using the previous ECG itself and 0.878 for the committee with the additional dataset of the difference between the current and the previous ECG. A statistical analysis gives a p-value of 0.03 between ROC areas 0.858 and 0.878, a small but statistically significant difference. There are of course many ways of representing the difference between two ECGs. The one employed in this paper increased the classification performance slightly i.e. a previous ECG, recorded on the same patient, helped the classifier for the task of diagnosing acute myocardial infarction.

To validate that the PCA reduction was good several committees was trained using direct QRS- and ST-T measurements as inputs. The number of inputs varied from 40-84 and the obtained (test) ROC areas was between 0.83-0.86.

6   Conclusion

A neural network classifier were constructed to detect acute myocardial infarction, using the 12-lead ECG. Key indicators were extracted with principal component analysis on common measurements of the ECG. When diagnosing an acute myocardial infarction ECG a previously recorded one (if available) is often used as a reference. The presence of such a previous ECG appears to have a small, but positive, effect on the performance of the network classifier.

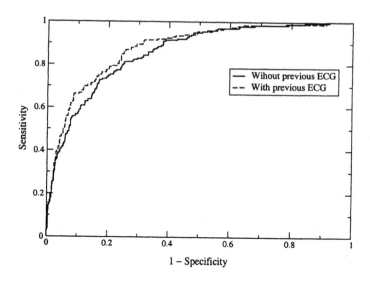

Figure 3: ROC curves for the classification with and without previous ECGs. The previous ECG is represented as the difference of the measurements between the current and the previous ECG.

Acknowledgments

This study was supported by grants from the Swedish Medical Research Council (K99-14X-09893-08B), The Swedish Foundation for Strategic Research and the Swedish National Board for Industrial and Technical Development, Sweden.

References

[1] Hedén B, Ohlsson M, Rittner R et al. Agreement between artificial neural networks and human expert for the electrocardiographic diagnosis of healed myocardial infarction. J Am Coll Cardiol 1996;28:1012-1016

[2] Hedén B, Öhlin H, Rittner R, Edenbrandt L. Acute myocardial infarction detected in the 12-lead ECG by artificial neural networks. Circulation 1997;96:1798-1802

[3] Hertz J, Krogh A and Palmer RG. Introduction to the Theory of Neural Computation. Addison-Wesley, Redwood City, Ca, 1991

[4] Rögnvaldsson T. On Langevin updating in multilayer perceptrons. Neural computation 1994;6:916-926

[5] Hanson SJ and Pratt LY. Comparing biases for minimal network construction with back-propagation. In: D. S. Touretzky (ed) Advances in Neural Information Processing Systems. Morgan Kaufmann, San Meteo CA, 1989, pp 177-185 Morgan Kaufmann (1989)

Bayesian Neural Networks used to Find Adverse Drug Combinations and Drug Related Syndromes

Roland Orre

Mathematical Statistics, University of Stockholm,
Stockholm, Sweden

Andrew Bate and Marie Lindquist

WHO Coll. Centre for Internl. Drug Monitoring, Uppsala Monitoring Centre,
Uppsala, Sweden

Abstract

The data mining task we are interested in is to find associations between variables in a large database. The method we have earlier proposed to find outstanding associations is to compare estimated frequencies of combinations of variables with the frequencies that would be predicted assuming there were no dependencies. The method we now propose use the same strategy as an efficient way of finding complex dependencies, i.e. certain combinations of explanatory variables, mainly medical drugs, which may be highly associated with certain outcome events or combinations of adverse drug reactions (ADRs). Such combinations of ADRs may also be recognized as syndromes.

The method we use for data mining is an artificial neural network architecture denoted Bayesian Confidence Propagation Neural Network (BCPNN). To decide whether the joint probabilities of events are different from what would follow from the independence assumption, the "*information component*" $\log(P_{ij}/(P_i P_j))$, which is a weight in the BCPNN, and its variance plays a crucial role. We also suggest how this method might be used in combination with stochastic EM to analyse conditioned dependencies also between real valued variables, e.g. to consider the amount of each drug taken.

1 Introduction

We are studying an international data base of case reports, each one describing a possible case of adverse drug reactions (ADRs), which is maintained by the Uppsala Monitoring Centre (UMC), for the WHO international program on drug safety monitoring. Each case report, which can be seen as a row in a data matrix, consists of a number of variables, like drugs used, which amounts of each drug and how the drugs were taken, observed ADRs and other patient data like sex, age and resulting outcome events for the patient [1]. The fundamental problem is to find significant dependencies which might be signals of potentially important ADRs, to be investigated by clinical experts. The estimates of significance are obtained with a Bayesian approach via the variance of posterior probability distributions. The posterior is obtained by fusing a prior Dirichlet distribution with a batch of data, which is also the prior used when the next batch of data arrives.

2 Method

The Bayesian Confidence Propagation Neural Network (BCPNN) [2],[3], can be seen as one way of rewriting Bayes theorem into a form which is reminiscent of other feed forward artificial neural network architectures. It works by propagating probabilistic belief values for a set of inputs or explanatory variables into a set of outputs which are the beliefs that the given input represents one of a set of mutually exclusive classes, which are the response variables. In the work presented here the inputs constitute the drugs suspected of causing the adverse reactions and the outputs are the observed set of adverse reactions or the outcome event suspected to be caused by the reported drug or drug combination.

In the following, let a_j denote the j:th compononent of composite output A of a set of mutually exclusive outcome events. That is, output A could represent a certain adverse reaction [*true, false*] or it could be a set of events like [*alive, coma, death*] where a_j would represent one of these outcomes. In a similar way input D may represent the presence of a suspected drug on a report. More generally, let D, denote a multiple variable input event, where d_i is the input variable i of a set of conditionally independent $(P(D|A) = P(d_1|A) \cdot P(d_i|A) \cdots P(d_n|A))$ variables, and d_{ik} is the k:th mutually exclusive component of one of these input variables. Then $\pi_{d_{ik}}$ is the current "belief" on input event k of variable i. If we only have discrete input belief values $(\pi_{d_{ik}} \in \{0, 1\})$ and none of the input states overlap, then the feed forward neural network like expression to produce posterior beliefs for a_j given the input belief values of $\pi_{d_{ik}}$ can be written

$$P(a_j|D) \propto \exp\left[\log P(a_j) + \sum_i \sum_k \log\left[\frac{P(d_{ik}, a_j)}{P(d_{ik})P(a_j)}\right]\pi_{d_{ik}}\right]. \tag{1}$$

The weight expression

$$IC_{ijk} = \log\frac{P(d_{ik}, a_j)}{P(d_{ik})P(a_j)} = \log\frac{P(d_{ik}|a_j)}{P(d_{ik})} \tag{2}$$

in (1) we denote the "information component" between state j and state k of variable i. Often we leave the k index out and just write IC_{ij} because the explanatory variables are usually binary and we are most often only interrested in the positive occurances, *i.e.* the *true* states of the variables. The motivation for the notation "information component" is that mutual information [4] in its discrete form can be regarded as a weighted sum of *information components*:

$$I(Y; X) = \sum_j \sum_k P(x_k, y_j) \log\frac{P(x_k, y_j)}{P(x_k)P(y_j)}, \tag{3}$$

which defines the amount of information passed on from one variable to another. The IC_{ijk} in particular is a measure of the information migrating from state k of variable i to state j of the other variable. Due to the properties of the logarithmic function the expectation and variance for the IC_{ij} kan be expressed as

$$E(IC_{ij}) \quad = \quad E(\log \frac{p_{ij}}{p_i p_j}) = E(\log p_{ij}) - E(\log p_i) - E(\log p_j), \tag{4}$$

$$V(IC_{ij}) \quad = \quad V(\log p_{ij}) + V(\log p_{ij}) + V(\log p_{ij})$$
$$\qquad - 2cov(\log p_{ij}, \log p_i) - 2cov(\log p_{ij}, \log p_j) + 2cov(\log p_i, \log p_j).$$

A reasonable model distribution for $P(d_{ik}, a_j)$ is Dirichlet [3]. However, here will the p_{ij}, p_i and p_j all become a special case of the Dirichlet, the Beta. When p being Beta(a, b) distributed there is an exact form [5] of the expectation and variance

$$E(\log p) \quad = \quad \frac{b}{a(a+b)} - b \cdot \sum_{n=1}^{\infty} \frac{1}{(a+n) \cdot (a+b+n)}, \tag{5}$$

$$V(\log p) \quad = \quad \sum_{n=0}^{\infty} \frac{b^2 + 2ab + 2bn}{(a+n) \cdot (a+b+n)^2}. \tag{6}$$

From the expectation and variance values (4) a probability interval for the IC_{ij} can be calculated, which is used as one signalling criterion when searching for unexpected associations between drugs and adverse reactions.

2.1 Combinations of Variables

As described above, the IC_{ij} is a useful measure to find new unexpected single drug ADR associations. The focus of interest for this paper is, however, extending this to analyse also combinations of variables. We want to find variables which interact conditionally, *i.e.* given a certain outcome a set of drugs may show an unexpected interaction, alternatively when a certain drug or combination of drugs is given as input we may find that a set of adverse reactions interact. The latter form of interaction between adverse reactions may lead to detection of *syndromes*. Earlier [3] we have indicated a way of finding such syndrome interactions by looking at conditioned probabilites for combinations of adverse reactions. By looking at, *e.g.* the quotient

$$\log \frac{P(A_1, A_2, A_3 | D_1)}{P(A_1, A_2, A_3)} \quad = \quad IC(A_1, A_2, A_3; D_1), \tag{7}$$

where A_j stand for an adverse drug reaction and D_i for a medical drug, we may find conditionally interacting triplets of adverse reactions. We demonstrated this earlier [3] by looking for a specific syndrome and sorting the results on the $IC(A_1, A_2, A_3; D_1)$ according to (7) and found that we got very high rankings on the combinations of adverse reactions known to appear within the syndrome picture. There are, however, certain limitations, this method will highlight strong dependencies between three states, but they may be due to strong lower order dependencies. The purpose of our search for interacting combinations is to find those where the interaction may not be explained by lower order interactions. Assume that we are looking for pairs where $P(A_1, A_2 | D) >> P(A_1, A_2)$, further assume that $P(A_1 | D)$ and $P(A_2 | D)$ are independent as well as $P(A_1)$ and $P(A_2)$ being independent.

$$\frac{P(A_1, A_2 | D)}{P(A_1, A_2)} = \frac{P(A_1, A_2, D)}{P(A_1, A_2) P(D)} = \frac{P(A_1, A_2 | D) P(D)}{P(A_1, A_2) P(D)} = \frac{P(A_1 | D) P(A_2 | D)}{P(A_1) P(A_2)} \tag{8}$$

Under this assumption would the joint probability increase when a marginal probability (κ) increases. We were then looking for a measure which would make the combinations stand out despite lower order interactions. An idea for extension of the IC-measure was tu use the IC between a set of ADRs conditioned on a drug

$$IC(A_1;A_2|D) = \log \frac{P(A_1,A_2|D)}{P(A_1|D)P(A_2|D)}, \tag{9}$$

which when compared with an unconditioned IC measuring the general interaction between the same set of adverse reactions

$$IC(A_1;A_2) = \log \frac{P(A_1,A_2)}{P(A_1)P(A_2)} \tag{10}$$

could tell us if the presence of a drug increases or decreases the interaction between the adverse reactions. For a drug related syndrome it could then be expected that when

$$IC(A_1,A_2;D) = \log \frac{P(A_1,A_2|D)}{P(A_1,A_2)} >> 0 \tag{11}$$

would $IC(A_1;A_2) << IC(A_1;A_2|D)$, but the investigation we have done so far, has, however, not given us indications about this. We intend to investigate these measures (9,10) more in the future, but the results we present in this paper are based on the measure in eq. (11) only, which when combined with the variance measure, eq. (5), gives us the ability to sort the obtained results on credibility levels for the IC values.

3 Results

The aim with our experiment was to verify that the algorithm could extract a well known syndrome which is considered drug related. The layer specification for the BCPNN was to consider the two classes "haloperidol" and "other drug" as inputs and let the output layer represent a subset of the power set of all adverse reactions (ADRs) occuring on every report.

The drug "haloperidol" is considered to be the main cause of the Neuroleptic Malignant Syndrome (NMS) and included in the symptom picture of NMS are the following four ADRs *creatine phosphokinase increased, fever, death and hypertonia.*

In the setup of this experiment we generated up to the fourth power of ADR combinations in the output layer. The criterion used to associate the ADR combination with the drug haloperiod was the one given by eq. (11).

The weights inside the BCPNN would then implement the following measures $IC(A_1;D)$, $IC(A_1,A_2;D)$, $IC(A_1,A_2,A_3;D)$ and $IC(A_1,A_2,A_3,A_4;D)$, which are referred to as $IC(A*;D)$ in the table below. In the analysis step we generated lists of these different IC values which were then filtered on two different criteria, either IC being greater than zero or $IC - 2\sigma$ being greater than zero. The latter criterion, $IC - 2\sigma > 0$, gives us an approximate credibility level of 97% for the IC to be positive.

ADR-combinations found where drug haloperidol is suspected					
ADR	#A*	#D-A*	$\#IC(A*;D) > 0$	$\#IC(A*;D) - 2\sigma > 0$	all within
single	2329	623	298	117	91
pair	125791	4952	4458	651	162
triple	496007	6290	6245	256	61
quad	433993	3315	3315	23	0

As could be expected the single term NMS, representing the syndrome, was on the top of all these lists. In the pairs, triplets and quadruples list NMS was also found to be strongly associated with some of the other symptoms which are included in the symptom picture of the selected ADR terms. The reason for this is that the syndrome is not so strictly defined as being composed of these symptoms only, it is enough that the patient has a couple of these symptoms to be diagnosed as NMS. We also found that the selected ADRs were high on all three lists. For the single ADR list all four reactions were among the highest 91 IC values. For the list with ADR pairs combinations of these four ADRs plus the syndrom itself were also found among the highest 162 IC values, and three of these were in the top eight. For the list with triple ADRs, complete combinations were found within the 61 highest. Of the quadruples none of them included a complete symptom picture, on the other hand if we look for the number of terms included in each combination it looks like this:
(3 2 2 2 2 2 3 1 3 2 1 2 3 1 1 3 2 2 1 0 1 1 0), $i.e.$ the part with the highest $IC - 2\sigma$ values also contains most of the terms. For the triplets we obtain a similar picture:
(2 2 3 1 2 1 1 1 2 1 2 1 2 1 2 1 1 2 1 0 2 2 2 1 2 1 2 2 1 0 1 0 1 1 2 0 2 2 1 3 2 0 2 1 0 1
1 1 1 1 0 1 2 2 1 1 0 0 1 1 3 1 0 1 0 1 2 0 1 2 1 0 2 1 2 0 1 2 1 1 1 0 1 0 1 1 1 1 2 0 1 0
1 0 2 1 1 1 1 1 1 1 2 0 0 1 0 1 1 1 2 0 1 1 1 1 0 1 1 2 0 2 1 0 1 1 2 2 1 1 0 0 1 1 1 1 1
1 1 0 1 0 1 0 2 0 1 1 0 0 0 0 0 1 1 1 1 2 1 1 1 0 2 2 0 1 0 2 0 0 2 0 1 1 1 0 0 0 0 1 0 0
1 1 0 2 2 1 0 0 0 0 0 0 0 1 1 1 1 1 1 0 1 1 0 0 0 1 1 1 0 0 1 0 0 1 0 0 0 0 0 0 1 0 1 2 0 1 1
1 0 0 0 1 0 1 0 0 1 0 0 1 1 1 1 1 1 1 0 0 1 1 2 1 1 1)

4 Discussion on Real Valued Dependencies in R^{12000}

This way of serching for discrete value combinations also allows us to further investigate such variables which have real valued attributes. For each drug reported there is also an associated amount, which may be used to give further information about covariances and ranks of the input space. There are about 12000 drugs being reported and there are about 2 million reports where the amount of the drug taken is filled in for about half of the drugs being reported. To search for dependencies in a real space of dimensionality 12000 would be quite a demanding task. However, the dimensionality of the space that need to be considered in this problem is smaller as the there is a limited number of drugs and ADRs which can occur on one single report.

The approach is to adapt a set of Radial Basis Functions (multivaritate gaussians) to the real valued space using a stochastic EM-algorithm [6],[7] by creating a large set of *low dimensional gaussians* representing the density of the subspaces found on the reports as discrete combinations. Further analysis may then be done on these gaussians $N(\mu_i, \Sigma_i)$ which are parameterized by μ_i which is the center and Σ_i is the covariance matrix for the gaussian density function. The *rank* of the inverse covariance matrix gives the dimensionality of the dependency, the *non diagonal elements* in the covariance matrix tell about partial covariates or colinearities. Linear and also non linear dependencies may be found by doing regression analysis on center values μ_i.

5 Summary

The BCPNN *(Bayesian Confidence Propagation Neural Network)* has shown to be a useful tool for data mining large data bases and is used on a regular basis for signalling of adverse drug reactions. The *IC (information component)*, which is the weight in a BCPNN, and its variance is an efficient and intutive measure of the strength and significance of a dependecy relation, which relates to information theory. Our approach to use the *IC* to find interactive discrete varible combinations seems promising but is being investigated further. The approach to use this kind of discrete feature detection in combination with multivariate analysis to find dependencies in sparse high dimensional real valued space will also integrate well with the BCPNN methodology.

References

[1] A. Bate, M. Lindquist, I. R. Edwards, S. Olsson, R. Orre, A. Lansner, and R. M. D. Freitas, "A bayesian neural network method for adverse drug reaction signal generation," *European Journal of Clinical Pharmacology*, vol. 54, pp. 315–321, 1998.

[2] A. Holst and A. Lansner, "A higher order bayesian neural network for classification and diagnosis," in *Computational Learning and Probabilistic Reasoning* (A. Gammerman, ed.), ch. 12, John Wiley & Sons, Ltd, 1996. Proc. of ADT, London, England, April 3-5, 1995.

[3] R. Orre, A. Lansner, A. Bate, and M. Lindquist, "Bayesian neural networks with confidence estimations applied to data mining," *Computational Statistics and Data Analysis*, pp. –, 1999. in press.

[4] J. Pearl, *Probabilistic Reasoning in Intelligent Systems: Networks of Plausible Inference*. Morgan Kaufmann, San Mateo, 1988.

[5] T. Koski and R. Orre, "Statistics of the information component in bayesian neural networks," Tech. Rep. TRITA-NA-9806, Dept. of Numerical Analysis and Computing Science, Royal Institute of Technology, Stockholm, Sweden, 1998.

[6] H. G. Tråvén, "A neural network approach to statistical pattern classification by "semiparametric" estimation of probability density," *IEEE Transactions on Neural Networks*, vol. 2, no. 3, pp. 366–118, 1991.

[7] R. Orre and A. Lansner, "Pulp quality modelling using bayesian mixture density neural networks," *Journal of Systems Engineering*, vol. 6, pp. 128–136, 1996.

Monitoring of Physiological Parameters of Patients and Therapists During Psychotherapy Sessions Using Self-Organizing Maps

Thomas Villmann*   Bernhard Badel   Daniel Kämpf

Michael Geyer

Clinic for Psychosomatic Medicine and Psychotherapy, University Leipzig

04109 Leipzig, K.-Tauchnitz-Str. 25, Germany

Abstract

In the present contribution the authors show the application of SOMs for visualization of physiological parameters of patients and therapists in psycho-therapy sessions. Thereby, using the topology preserving property of SOMs a color representation can be generated allowing an easy assessment of the underlying parameter change which can be interpreted by the therapists. To achieve the topology preserving map a growing extension of the SOM was used together with a magnification control strategy which maximizes the mutual information of the network.

1   Introduction

The therapeutical process in psychotherapy usually is organized in a sequence of single therapy sessions. Thereby, in a stationary therapy, the frequency of sessions is daily. During the therapy and, especially during the single sessions, the emotional feeling can vary in dependence on the actual situation (emotional excitements, as the result of the therapeutical discussion) which can be observed by individual asses by the therapist. Beside the individual observations there exists a large pool of instruments to judge the emotional situation of both the patient and the therapist which consist in a spectrum of several questionnaires [6, 16]. On the other hand, it is generally known that emotions influence physiological parameters as for instance heart rate, respiration rate, muscle tension and electrodermal conductivity (skin conductance) which can end in negative cases in physiological complaints called psycho-somatic symptoms which are under treatment in our clinic.

Vice versa, the therapy process also influences the physiological parameters. The *parallel* observation of them during therapy sessions together with a psychological analysis should lead to a better understanding of the psycho-physiological processes. However, it was never tried before in usual therapy sessions or complete therapies. Thereby, the most difficult thing is that the set of physiological parameters is correlated but have to be considered in a parallel

---

*corresponding author, email: villmann@informatik.uni-leipzig.de

way by a therapist to detect relevant changes [7]. For this purpose a suitable tool is needed for an easy parallel assessment.

2 Data Description

In this initial investigation the therapy of one patient was considered containing $t_{max} = 37$ single sessions. Each of them usually takes approximately 45 min.. During all the sessions for both the patient and, additionally, for the therapist the following parameters were simultaneously measured: *heart rate, muscle tension (by surface electromyogram), skin conductance level (SCL), skin conduction reaction (SCR)*. The values are determined as averages over time intervals of $30s$. In the above outlined approach we are interested in the variation of the parameters. Hence, we investigate only the difference of the parameters for a certain time point in comparison to the previous one. In this way we are independent on the basic level of data in each session which can be influenced by climatic conditions, nourishment, etc. Yet, only the actual development is considered.

We used 3105 data vectors for both the patient and the therapist. Hence we have $n_{max} = 6210$ data vectors for SOM learning. Subsequently, the data were normalized such that for the resulted vectors $\mathbf{v} \in \mathcal{V} \subseteq \Re^4$ we have for all components j: $\sum_{i=1}^{n_{max}} v_j = 1$, i.e., all measured parameters are assumed to be of the same importance [11].

3 Self-Organizing Maps (SOMs)

3.1 SOMs for Topology Preserving Mapping

Self-organizing maps (SOM) [8] as special kind of neural maps project data from some (possibly high-dimensional) input space $\mathcal{V} \subseteq \Re^{D_\mathcal{V}}$ onto a position in some output space, such that a continuous change of a parameter of the input data should lead to a continuous change of the position of a localized excitation in the neural map.[1] This property of *neighborhood preservation* depends on an important feature of the SOM: the a priori defined output space topology usually chosen as a $D_\mathcal{A}$-dimensional rectangular grid $\mathcal{A}$ (hypercube) of N neurons situated on positions $\mathbf{r} = (r_1, r_2, r_3, ..., r_{D_\mathcal{A}})$, $1 < r_j < n_j$ with $N = n_1 \times n_2 \times ... \times n_{D_\mathcal{A}}$.[2] Associated to each neuron $\mathbf{r} \in \mathcal{A}$, is a weight vector $\mathbf{w_r}$ in $\mathcal{V}$. The map $\Psi_{\mathcal{V} \to \mathcal{A}}$ is realized by a winner take all rule

$$\Psi_{\mathcal{V} \to \mathcal{A}} : \mathbf{v} \mapsto \mathbf{s} = \underset{\mathbf{r} \in \mathcal{A}}{\operatorname{argmin}} \|\mathbf{v} - \mathbf{w_r}\| \tag{1}$$

whereas the back mapping is defined as $\Psi_{\mathcal{A} \to \mathcal{V}} : \mathbf{r} \mapsto \mathbf{w_r}$. Both functions determine the map $\mathcal{M} = (\Psi_{\mathcal{V} \to \mathcal{A}}, \Psi_{\mathcal{A} \to \mathcal{V}})$ realized by the network. All data points $\mathbf{v} \in \mathcal{V}$ which are mapped onto the neuron $\mathbf{r}$ perform its (masked) receptive

---

[1] In this way the SOM determines the non-linear principle components of the data.

[2] Yet other arrangements are also admissible which can be described by a connectivity matrix. Here we only consider hypercubes.

field $\Omega_{\mathbf{r}}$. To achieve the map $\mathcal{M}$, SOMs adapt the pointer positions with respect to a presented sequence of data points $\mathbf{v} \in \mathcal{V}$ selected according to the data distribution $\mathcal{P}(\mathcal{V})$:

$$\triangle \mathbf{w_r} = \epsilon h_{\mathbf{rs}} (\mathbf{v} - \mathbf{w_r}) \tag{2}$$

$h_{\mathbf{rs}}$ is the neighborhood function depending on the best matching neuron according to (1), usually chosen to be of Gaussian shape: $h_{\mathbf{rs}} = \exp\left(-\frac{\|\mathbf{r}-\mathbf{s}\|^2}{2\sigma^2}\right)$.

Topology preservation in SOMs is understood as preserving of the continuity of the mapping from the input space onto the output space, exactly spoken: it is equivalent to the *continuity* of $\mathcal{M}$ between the *topological spaces* with properly chosen metric in both $\mathcal{A}$ and $\mathcal{V}$. Because of the lack of space we refer to [14] for a detailed consideration. If the topology of $\mathcal{A}$ does not match that of the shape of the data distribution, neighborhood violations are inevitable [14]. On the other hand, a higher degree of topology preservation, in general, improves the accuracy of the map [3]. Several approaches were developed to judge the degree of topology preservation for a given map [2].

3.2 Variants of the SOM for Suitable Data Representation

To overcome the problem of prior specified lattices one can adapt also the shape of the grid in addition to the usual weight vector dynamic. One approach si the *growing* SOM (GSOM) [4] which realizes an data driven structure adaptation, however, always remaining a hypercube structure of the grid which allows a simple post-processing. During the learning scheme both the lattice dimension and the edge length ratios are adapted (in addition to the weight vectors), i.e. we allow a variable overall dimensionality and variable dimensions along the individual directions in the hypercube. It can be taken as a non-linear principle component analysis.

The GSOM starts from an initial 2-neuron chain, learns according the regular SOM-algorithm, adds neurons to the output space with respect to a certain criterion to be described below, learns again, adds again, etc., until a prespecified maximum number $N_{\max}$ of neurons is distributed. During this procedure, the output space topology remains to be of the form $n_1 \times n_2 \times ...$, with $n_j = 1$ for $j > D_{\mathcal{A}}$, where $D_{\mathcal{A}}$ is the current dimensionality of $\mathcal{A}$.[3] From there it can grow either by adding nodes in one of the directions which are already spanned by the output space or by initilizing a new dimension. For a detailed study of the algorithm we refer to [13].

A further extension of the basic SOM concerns the so-called magnification: the usual SOM distributes the pointers $\mathbf{W} = \{\mathbf{w_r}\}$ according to the input distribution: $\mathcal{P}(\mathcal{V}) \sim \mathcal{P}(\mathbf{W})^{\alpha}$ with the magnification factor $\alpha = \frac{2}{3}$ [10].[4] BAUER

---

[3] Hence, the initial configuration is $2 \times 1 \times 1 \times ...$, $D_{\mathcal{A}} = 1$.

[4] This result is valid for the one-dimensional case and higher dimenional ones which separate.

ET AL. in [1] introduced a local learning parameter ϵ_r with $\langle \epsilon_r \rangle \propto \mathcal{P}(\mathcal{V})^m$ in (2) which now reads as

$$\triangle \mathbf{w_r} = \epsilon_s h_{rs} (\mathbf{v} - \mathbf{w_r}) \tag{3}$$

This local learning rule leads to a relation $\mathcal{P}(\mathcal{V}) \sim \mathcal{P}(\mathbf{W})^{\alpha'}$ with $\alpha' = \alpha(m+1)$ and, hence, allows a magnification control. Especially, one can achieve a resolution $\alpha' = 1$ which maximizes the mutual information (corresponding to a maximization of the entropy) [9, 15].

4 Application and Results

The GSOM was applied to the above described data set extended by the magnification control scheme to obtain a maximal mutual information, i.e. the control parameter m was chosen as $m = \frac{1}{2}$.

From technical point of view, we have judged the degree of topology preservation of the resulted GSOM $7 \times 5 \times 7$–lattice using a special kind of the topographic product [12, 2] and obtain a value $\tilde{P} = 0.007$ referring to a good topology preservation. The magnification control scheme has lead to an improvement of the entropy: without control the entropy was 94.8% of the maximal possible entropy [15] value in comparison to 96.4% using the control scheme.

As mentioned before, the GSOM-result is a $3d$-lattice and, hence, we can interpret the neuron positions as colors. In this way we are able to code each data vector as a color pixel. In Fig.1 we plotted all time points for each session according to the outlined method. Now, the following interpretation is possible: We observe a clear change in the distribution of colors after the 20th session for the therapist (the occurrence of yellow typed colors is increasing) which is related to a change in the treatment concept regarding the therapeutical interactions. A second alteration in the color distribution of the therapist picture can be recognized for the last 5 sessions which are indicated by the detachment of the patient from the therapist [5]. On the other hand, investigating the initial phases of therapy sessions we observe that the frequency of blue-green colors for the patient decreases. We have found that these colors correspond to a decreasing heart rate. This is in agreement to the assessment by therapist where is stated that in the first sessions there was a big emotional pressure and expectation from the patient in the beginning phase which is lost during the following ones.

Moreover, it is possible to assign characteristic pattern of parameter changes to specific colors by a deeper analysis. For instance, as mentioned above we found that green-blue colors are related to an decreasing heart rate whereas red indicates an increase. Yellow-green colors symbolize increasing muscle tensions. The rise of electrodermal conductivity is coded by magenta-like colors. However, for a detailed considerations further investigations are necessary.

In conclusion, we can say that the demonstrated visualization technique of physiological parameters during psychotherapy sessions can help the therapist to understand the psycho-physiological feeling of the patient.

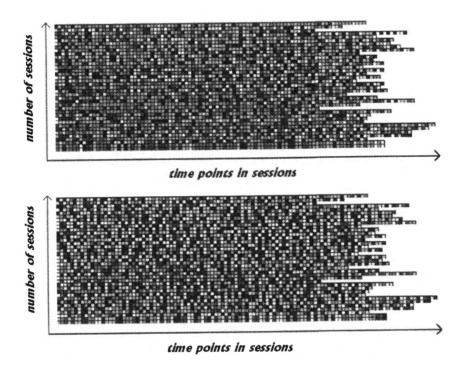

Figure 1: Color description of the variation of the physiological parameters of both the patient (above) and the therapist (below) for the whole therapy for each single session. Each colored pixel codes a characteristic pattern. Global color texture changes within the pictures correspond to variations in the therapy (see text for details). Single colors can be related to characteristic patterns. The color representation is resulted from a GSOM generated $7 \times 5 \times 7$ SOM including a magnification control scheme for maximizing the mutual information. (Regrettably, only a greyscale image can be provided in the printed paper. A color image is available on request from the corresponding author or on http://www.phil.gu.se/ann/villmann)

References

[1] H.-U. Bauer, R. Der, and M. Herrmann. Controlling the magnification factor of self–organizing feature maps. *Neural Computation*, 8(4):757–771, 1996.

[2] H.-U. Bauer, M. Herrmann, and T. Villmann. Neural maps and topographic vector quantization. *Neural Networks*, 12(4–5):659–676, 1999.

[3] H.-U. Bauer and K. R. Pawelzik. Quantifying the neighborhood preservation of Self-Organizing Feature Maps. *IEEE Trans. on Neural Networks*, 3(4):570–579, 1992.

[4] H.-U. Bauer and T. Villmann. Growing a Hypercubical Output Space in a Self–Organizing Feature Map. *IEEE Transactions on Neural Networks*, 8(2):218–226, 1997.

[5] H. Hess. *Untersuchungen Zur Abbildung Des Prozeßgeschehens und der Effektivität in der Intendiert-Dynamischen Gruppenpsychotherapie.* PhD thesis, Humbold University Berlin (Germany), 1986.

[6] G.-J. Hogrefe. *Testkatalog.* Hogrefe-Verlag, Göttingen, 1996/97.

[7] W. Hubert and R. de Jong-Meyer. Psychophysiological response patterns to positive and negative film stimuli. *Biological Psychology,* 31(1):79–93, 1991.

[8] T. Kohonen. *Self-Organizing Maps.* Springer, Berlin, Heidelberg, 1995. (Second Extended Edition 1997).

[9] R. Linsker. How to generate maps by maximizing the mutual information between input and output signals. *Neural Computation,* 1:402–411, 1989.

[10] H. Ritter and K. Schulten. On the stationary state of Kohonen's self-organizing sensory mapping. *Biol. Cyb.,* 54:99–106, 1986.

[11] P. Tyror and M. Lader. Central and peripheral correlates of anxiety: A comparative study. *Journal Nerve Mental Disorder,* 162(2):99–104, 1976.

[12] T. Villmann. Benefits and limits of the self-organizing map and its variants in the area of satellite remote sensing processing. In *Proc. Of European Symposium on Artificial Neural Networks (ESANN'99),* pages 111–116, Brussels, Belgium, 1999. D facto publications.

[13] T. Villmann and H.-U. Bauer. Applications of the growing self-organizing map. *Neurocomputing,* 21(1-3):91–100, 1998.

[14] T. Villmann, R. Der, M. Herrmann, and T. Martinetz. Topology Preservation in Self–Organizing Feature Maps: Exact Definition and Measurement. *IEEE Transactions on Neural Networks,* 8(2):256–266, 1997.

[15] T. Villmann and M. Herrmann. Magnification control in neural maps. In *Proc. Of European Symposium on Artificial Neural Networks (ESANN'98),* pages 191–196, Brussels, Belgium, 1998. D facto publications.

[16] G. Westhoff. *Handbuch Psychosozialer Meßinstrumente.* Hogrefe-Verlag, Göttingen, 1993.

Biomolecular Applications

and

Biological Modelling

Neuronal Network Modelling of the Somatosensory Pathway and its Application to General Anaesthesia

A. Angel

Centre for Research into Anaesthetic Mechanisms
Department of Biomedical Science
The University of Sheffield, Western Bank, Sheffield, S10 2TN, UK

D.A. Linkens and C.H. Ting

Intelligent Systems Laboratory
Department of Automatic Control & Systems Engineering
The University of Sheffield, Mappin Street, Sheffield, S1 3JD, UK

Abstract

The sensory responses evoked from the cortex by stimulating peripheral sensory nerves act as a sensitive index of anaesthetic effect. The whole question of consciousness, awareness and depth of anaesthesia is both timely, little understood, and deeply challenging. Models of the underlying neural pathway mechanisms/dynamics are necessary for understanding the interaction involved and their structure and function. A neuronal network of the somatosensory pathway is proposed in this paper based on experimental information and physiological investigation in anaesthesia. Existing mathematical neuronal models from the literature were modified and employed to describe the dynamics of the proposed pathway network. Effects of anaesthetic agents on the cortex were simulated on the model from which evoked cortical responses were obtained. By comparison with responses from anaesthetised rats, the model's responses are able to describe the dynamics of realistic responses. Thus, the proposed model promises to be valuable for investigating the mechanisms of anaesthesia on the cortex and brain lesions.

1 Introduction

Electrophysiological signals such as electroencephalograms and evoked responses have been shown to be an effective tool in the assessment of the state of the central nervous system[1]. The somatosensory nervous system, from which somatosensory evoked responses (SER) can be elicited, performs the role of interpreting external somatic stimuli. Lesions or malfunctions of the system can abolish an individual's sensation of external somatic stimuli. Anaesthesia is a typical example which attenuates the activity in the somatosensory pathway and hence a patient is unable to feel during surgery.

The alpha-rhythm EEG activity can be well represented by combining the thalamus and cortex as a single thalamocortical relay cell with an inhibitory interneuronal feedback [2]. This is a spatial distributed model since the distribution of neurons in a

230

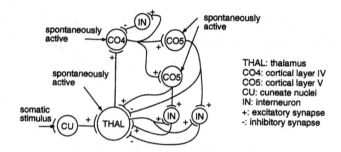

Figure 1: Proposed network model of the somatosensory pathway.

specific area is considered. Lumped modelling which treats physiological properties as several single parameters was adopted in the model. This methodology is acceptable because of the lack of sufficient physiological knowledge and a high degree of redundancy in the cortex and has also been applied to model the visual nervous system [3, 4]. This modelling approach was also employed in this paper.

Anaesthesia simulation results show that the dose versus latency change curve replicates the Hill equation as well as results from animal experiments. The simulation results are interesting not only because SERs respond realistically to changes in the model parameters, but also because the mechanisms of the individual neurons respond characteristically to the parameter variation.

2   Somatosensory Neuronal Pathway

Figure 1 illustrates the proposed network model of the somatosensory pathway in the brain based on physiology and experimental information [1]. In this model each cortical cell at its respective cortical layer is treated as an independent module but networked via interneurons or collateral direct connections. The number of either excitatory or inhibitory synapses per neuron is represented using a connectivity coefficient [5] which can be partially derived from histological literature[4]. The evoked response source is not from a single neurological structure but is the response of a large section of the cortex to a specific afferent point. Hence, for simplicity, each neuronal module represents a population in a specific region, as the figure legend describes.

All cortical cells or interneurons are assumed to have identical physiological properties because of the lack of sufficient detailed physiological knowledge [2]. Interneurons interconnecting different cortical cells are supposed to have different physiological properties to the cortical cells [4]. Many sensory cells show spontaneous activity in the absence of external inputs [1]. Thus it is reasonable to assume that every cortical cell in the network receives spontaneous activity from the surroundings.

Somatic stimuli from a sensory receptor are transmitted along the axonal pathway ascending through the cuneate nuclei to the thalamus. The sensory inputs further ascend to layer IV and then layer V of the cortex (primary somatosensory area). The real contribution of one of the neurons of cortex layer V is not well understood, yet. The presence of inhibitory inputs as well as excitatory inputs affects the network properties considerably. It works as a negative feedback to a cortical cell and introduces stability to the network. Physiologically, the signals presenting at each neuron project to the scalp and can be be picked up on the scalp with suitable electrodes. In simulation, we are able to select membrane potentials of a specific cortical cell.

3 Modelling the Neuron Cells

A neuron can be seen as the construction of synapses, cell body, and axon hillock using the concept of lumped modelling [3]. The synapses convert incoming signals into membrane potentials. Membrane potentials, either temporally or spatially distributed, are integrated by the cell body. The integration is then converted to nerve impulses at the axon hillock if the voltage reaches a threshold potential. Finally, the nerve impulses are transmitted through the axons.

Physiological experiments show that the EPSPs and IPSPs have impulse responses and can be lumped as follows [3]:

$$h_e(t) = \begin{cases} Aate^{-at}, & t \geq 0 \\ 0, & t < 0 \end{cases} \tag{1}$$

$$h_i(t) = \begin{cases} Bbte^{-bt}, & t \geq 0 \\ 0, & t < 0 \end{cases} \tag{2}$$

with $A = 3.25$mV, $B = 22$mV, $a = 100$s^{-1}, and $b = 50$s^{-1}. A and B represent the amplitude gains of the PSP functions and a and b the transmission lag constants.

The interneuron has its own transfer function [4]:

$$h_d(t) = \begin{cases} Aa_dte^{-a_dt}, & t \geq 0 \\ 0, & t < 0 \end{cases} \tag{3}$$

where $a_d \approx \frac{a}{3}$ means that the interneuron has a latency 3 times longer than the cortical cells. The axon hillock model is lumped as [6]:

$$\sigma(v) = \frac{2e_0}{1 + e^{\gamma(v_0 - v)}} \tag{4}$$

with $\gamma = 0.56$mV^{-1}, $e_0 = 2.5$Hz, and $v_0 = 6$mV. Because of the lack of sufficient physiological information, both cortical cells and interneurons are assumed to have the same hillock model.

4 Block Diagram of the Somatosensory Pathway

A block diagram of the physiological pathway of Figure 1 is obtained by applying the linear synaptic transfer functions and the nonlinear hillock activation function to each cell in the pathway and is shown in Figure 2. Constants $C_\#$ represent the connectivity coefficients which account for the average number of synaptic contacts between the correlating cells [5].

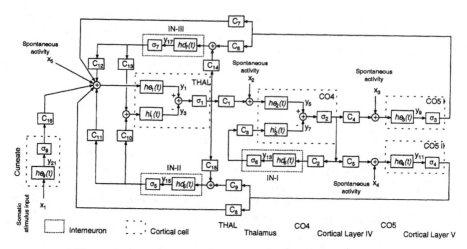

Figure 2: Block diagram of the proposed somatosensory pathway.

The block diagram can be solved using numerical techniques. Responses recorded on the scalp represent the integrated membrane potential fluctuations of the cortical cells. Hence, the measurements are obtained by collecting the results $y_5 - y_7$ (SERs) for the output of the primary somatosensory cortex cells in layer IV.

Referring to the proposal of [4] and our simulation studies, we adopted the following connectivity coefficients: $C_1 = C$, $C_2 = 0.25C$, $C_3 = 0.1C$, $C4 = C$, $C_5 = C$, $C_6 = C$, $C_7 = C$, $C_8 = 0.25C$, $C_9 = C$, $C_{10} = 0.1C$, $C_{11} = 0.8C$, $C_{12} = 0.1C$, $C_{13} = 0.1C$, $C_{14} = C$, $C_{16} = C$, and $C_{16} = C$ with C the base constant to be decided by trial-and-error. $C = 135$, which gives the best simulation results [4], is used here. Since large changes of the connectivity coefficients can make the system unstable, the above selection was maintained constant throughout the whole course of stimulation. This is physiologically reasonable since the number of synaptic contacts should be a constant in normal conditions [7].

5 Application to General Anaesthesia

Anaesthetic agents acting on the somatosensory nervous system either attenuate or block the signal transmission from the peripheral sensory receptors to the somatic sensory cortex [8]. At the cellular level, anaesthetics modify transmission at synapses [9]. Hence, the effect of an anaesthetic agent is simulated by varying the amplitude gains and delay constants of all cortical cells of the model. The integrated membrane potentials of the CO4 are recorded and averaged to form SERs.

A monotonic function was used to mimic the somatic stimulus [4]:

$$u(t) = q \cdot \left(\frac{t}{w}\right)^n \cdot e^{-\frac{t}{w}} \tag{5}$$

with $n = 7$, $w = 0.005$, and $q = 0.5$. Stimuli generated using the above function were applied to the cortical model at the 50th sample time of each sweep. Sponta-

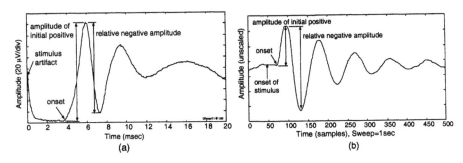

Figure 3: A simulation response is analogue to a realistic SER. (a) from an anaesthetised rat. (b) simulation result.

neous activities applied to the cortical cells from the surroundings were assumed to be a random disturbance with a magnitude range of (120, 320) impulse per second representing the instantaneous frequencies.

In analogy to SERs recorded from living subjects, we are able to define the usual terms relating to SER analysis as shown in Figure 3. Panel (a) is a response recorded from a urethane-anaesthetised rat. In panel (b), the onset, initial positive peak, and initial negative trough are well defined. Hence we can easily estimate the onset latency and the relative amplitudes and use them as an index of anaesthetic effect.

The nervous network is now assumed to be anaesthetised with different intensities of doses. The intensity is represented by a percentage change in the nominal parameter values from 100% to 5% in steps of 5%. This was modelled by attenuating the amplitude gain (A) and the time constant (a) of the EPSP function (1) of all cortical cells by the same percentage. The lower the percentage, the bigger is the intensity of the dose, i.e. 10% of the nominal values is more intense than 20% of the nominal values in light of attenuating synaptic transmission. The simulation results are shown in Figure 4(a). The onset latency increases and the relative amplitude decreases as the dose increases. This result coincides with experimental results on humans and animals with most anaesthetic agents [10, 1]. For model parameters less than 25% of the nominal values, there are almost no oscillations within the sweep duration. This can be interpreted to be that the subject is over anaesthetised and might be dead.

We define the anaesthetic effect as the multiplication of the latency (θ) and the amplitude (ΔP) of the initial positive wave, i.e. $E = \theta(\Delta P_0 - \Delta P)/\Delta P_0$ with ΔP_0 the amplitude at the control period. The normalised contour together with its fitted Hill equation is depicted in Figure4(b).

6 Conclusions

Although the model gives exciting results in anaesthesia simulation, the attempt to use it to quantify the depth of anaesthesia is infeasible at this moment because of insufficient information about the nervous system and also the lack of a clear relationship between the evoked potentials and anaesthesia. However, the model would be valuable as a scenario for hypothesising how brain lesions and anaesthetic drugs affect the

234

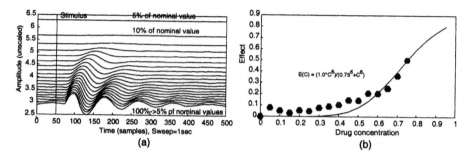

Figure 4: Application to anaesthesia simulation. (a) responses to various levels of anaesthesia. (b) anaesthetic effect vs. drug concentration .

evoked responses and could lead therefore to clinical diagnosis and prediction.

More realistic models, with consideration of characteristics such as adaptation, refractory period, and warm-up period, would be necessary for more profound analysis. However, a major drawback of those models is the lack of known model parameters and the complexity of the constructed network. To put the models in practice, we will need more knowledge and real-world information of neuronal physiology.

References

[1] A. Angel. Central neuronal pathways and the process of anaesthesia. *Brit J Anaesth,* 71:148–163, 1993.

[2] F. H. Lopes da Silva, A. Hoeks, H. Smits, and L. H. Zetterberg. Model of brain rhythmic activity: the alpha-rhythm of the thalamus. *Kybernetik,* 15:27–37, 1974.

[3] A. van Rotterdam, F. H. Lopes da Silva, J. van der Ende, M. A. Viergever, and A. J. Hermans. A model of the spatial-temporal characteristics of the alpha rhythm. *B Math Biol,* 44(2):283–305, 1982.

[4] B. H. Jansen and V. G. Rit. Electroencephalogram and visual evoked potential generation in a mathematical model of coupled cortical columns. *Biol Cybern,* 73:357–366, 1995.

[5] H. R. Wilson and J. D. Cowan. Excitatory and inhibitory interactions in localized populations of model neurons. *Biophys J,* 12:1–24, 1972.

[6] W. J. Freeman. Simulation of chaotic EEG patterns with a dynamic model of the olfactory system. *Biol Cybern,* 56:139–150, 1987.

[7] A M. Thomson and J. Deuchars. Temporal and spatial properties of local circuits in neocortex. *Trends Neurosci,* 17(3):119–126, 1994.

[8] A. Angel. How do anaesthetics work? *Curr Anaesth Crit Care,* 4:37–45, 1993.

[9] D. A. McCormick, H. C. Pape, and A. Williamson. Actions of norepinephrine in the cerebral cortex and thalamus: implications for function of the central noradrenergic system. *Prog Brain Res,* 88:293–305, 1991.

[10] P. S. Sebel, C. P. Heneghan, and D. A. Ingram. Evoked responses – a neurophysiological indicator of depth of anaesthesia? *Brit J Anaesth,* 57(9):840–842, 1985.

A Hybrid Classification Tree and Artificial Neural Network Model for Predicting the *In vitro* Response of the Human Immunodeficiency Virus (HIV1) to Anti-Viral Drug Therapy

W.R. Danter[1], D. Gregson[1], K.A Ferguson[1], M.R. Danter[1] and J. Bend.[2]

Departments of Medicine[1] and Pharmacology and Toxicology[2], University of Western Ontario, London Ontario, Canada.

Abstract

Artificial Neural Networks (ANN) are pattern recognition and prediction tools modeled on the biologic nervous system. Chemical structures can be decomposed to critical elements that can be used to teach an ANN to relate structure to function. This study was carried out to evaluate a hybrid ANN system for predicting the *in vitro* anti-HIV1 activity of potential anti-viral drugs based on their chemical structures.

Structure and activity data were obtained from an AIDS anti-viral screen database for 371 drugs. A proprietary algorithm was used to decompose molecules into critical elements that were used as input variables for a probabilistic artificial neural network (PNN). Initial modeling and input variable selection was carried out using Classification and Regression Tree analysis (CART™, 1999).

Important variables together with the terminal node outputs from CART™ were used to train the hybrid ANN to predict *in vitro* anti-HIV1 activity of each drug. Cross-validated testing results for the ANN model using only the important variables as inputs were compared with hybrid ANN predictions. The final hybrid ANN model correctly classified 96.8% (96.1-99.2%) of the chemical structures re anti-HIV1 activity. The area under the receiver operator curve (ROC) for the hybrid model was 0.97 (0.95-0.99), for the conventional ANN 0.89. The hybrid ANN model performed better for sensitivity, specificity, positive/negative predictive values (p<0.01). We conclude that elements of chemical structure can be used to train a hybrid ANN to accurately predict *in vitro* activity of drugs against wild type HIV1.

1   Introduction

In general the effects of drugs result from their interactions with macromolecules present in biologic organisms. Frequently these macromolecules are receptors residing on the cell surface. The affinity of a molecule for a specific macromolecule and the intrinsic activity of the molecule are determined by its chemical structure. This relationship is usually quite precise and even subtle changes in molecular structure (e.g., stereoisomerism) may result in significant changes in pharmacologic or biologic properties. Such changes may produce desirable properties or decreased toxicity but the converse may occur. Modeling the relationship between structure and activity using critical elements of molecular structure could result in (i) a general methodology for screening known molecules for specific biological activities and (ii) the synthesis of new molecules with improved therapeutic activity.

Artificial neural networks (ANN) are software programs modeled after the biological nervous system that are capable of recognizing complex patterns in data based on experience. They are useful in solving complex problems because they can evaluate large numbers of linear and non-linear variables without the operator making assumptions about the relationships between the variables. Neural networks are trained by presenting a set of patterns with their outcomes that the trainer wishes the network to learn. The trained ANN can then be evaluated by inputting similar, but previously unseen, data. This artificial intelligence (AI) approach for outcome prediction has been used for several medical applications including the diagnosis of myocardial infarction [1], pulmonary embolism [2], and obstructive sleep apnea [3], and in the prediction of survival following ICU admission [4]. ANNs equal or exceed traditional statistical methods in predicting certain outcomes [5,6]. The role of ANNs in medicine continues to develop with a broad range of applications [7] and are most likely to impact clinical medicine by their use as complex pattern recognition tools [8]. To the best of our knowledge ANNs have not been applied to molecular modeling for the purpose of predicting biological activity.

ANNs have been linked to symbolic/production rule systems [9], fuzzy logic [9,10,11], genetic algorithms [9,12,13], logistical regression models [14] and classification trees [11,14,15]. In general, hybrid modeling approaches using several AI tools produce more accurate and robust prediction systems when compared to single methodologies [9].

Anti-HIV1 therapies were a good option for the development of a molecular modeling system because a large database is available detailing the molecular structures of compounds that have been tested for *in vitro* anti-HIV1 activity. This study was carried out to develop and evaluate a hybrid ANN system for predicting the *in vitro* response of HIV1 to potential anti-viral therapies based upon their chemical structure. A secondary study objective was to compare the hybrid ANN system to more conventional ANN modeling.

2   Methods

Complete chemical structure, pharmacologic and 2 or 3 dimensional molecular image data for 178 Active and 193 Inactive molecules were obtained from the National Cancer Institute (NCI)/Developmental Therapeutics Programme (DTP) and National Institutes of Health (NIH) AIDS Anti-Viral Screen databases. All molecules had been tested in an *in vitro* HIV1 tissue culture assay and determined to be either Active or Inactive. The *in vitro* assay has been described in detail [16].

A patent pending algorithm (CHEMSAS™ v3.5, 1999) was used to decompose each molecule into critical element variables (e.g., presence of a halogen). The variables were entered into a database for further evaluation. The decomposition process is limited to molecules with molecular weights ≤1600 daltons.

Initial model development and variable selection was done with Classification and Regression Tree software (CART™, v 3.6 (Salford Systems, 1999). CART™ performed an extensive binary partitioning of the data and produced an optimal model by pruning and v-fold cross-validation. The input variables (i.e., critical structural elements) generated by CART™ were used in the standard ANN and the input variables from CART™ along with the terminal node outputs of the optimal classification tree were used as inputs to develop the hybrid ANN model.

Neuroshell Classifier v 2.0 (Ward Systems Group, 1999) was used for ANN modeling. In order to use all data, the ANN was trained and validated using a genetic optimization algorithm with a leave one out training and testing strategy. A leave one out strategy meant that 370 molecules were used to train and one molecule was held out to test. This was done repeatedly for each molecule such that the PNN used all 371 cases for training and testing. The test case was never in the training set at testing (N test set 371). Training and genetic optimization were stopped when 1000 generations had passed without further improvement. Many models were generated in the training process and the model with the best predictive ability was chosen.

The ANN models were asked to predict the *in vitro* activity of each molecular structure against HIV1 in tissue culture as outlined above. The optimal standard PNN model with critical element variables as inputs was compared with those of the hybrid PNN model which used the important variables plus the terminal node outputs from the optimal classification tree model. Cross-validation testing results are presented as data, proportions or percentages and their 95% Confidence Intervals (i.e. 95% CI) and the two tailed Fisher exact test p values where appropriate.

3   Results

Pre-processing and initial modeling with CART™ reduced the number of input variables from 76 to 28. This process resulted in a final ratio of learning patterns to input variables of 371/28 or >13 patterns per ANN input variable. Using the leave one out training and testing strategy the final hybrid ANN model accurately classified 359 of the 371 chemical structures as Active or Inactive against HIV1 *in vitro*, resulting in an overall testing accuracy of 96.8% (96.1-99.2%) (Table 1). The sensitivity of ANN predictions was 96.1% and the specificity was 97.4%. The positive predictive value (PPV) that the molecule would be active against HIV1 was 97.2% while the negative predictive value (NPV) that molecule would be inactive was 96.4%. The area under the receiver operator curve (ROC) was 0.971.

Table 1: Summary results for 178 active and 193 inactive chemical structures for the optimal hybrid ANN model for prediction of *in vitro* susceptibility

ANN Prediction	Active	Inactive
Active	171	5
Inactive	7	188
Sensitivity	96.1% (93.2-98.9%)	Specificity 97.4% (95.2-99.6%)
PPV	97.2% (94.7-99.6%)	NPV 96.4% (93.8-99.0%)

Predictions and receiver operating characteristics for the optimal hybrid ANN model (identified critical elements and terminal node outputs) were better than those from the standard ANN which used only the critical elements. Both models employed identical training, genetic optimization and testing methods (Table 2).

Table 2: Summary Results for 178 active and 193 inactive chemical structures comparing ANN with Hybrid ANN Models

Outcome Measure	ANN	Hybrid ANN	p value
Accuracy (N=371)	88.7%	96.8%	0.0003
Sensitivity	88.2%	96.1%	0.0094
Specificity	89.1%	97.4%	0.0018
Positive Predictive Value	88.2%	97.2%	0.0017
Negative Predictive Value	89.1%	96.4%	0.0059
Area under ROC	0.892	0.971	

4 Discussion

This study evaluated a hybrid ANN system with genetic optimization for modeling the relationship between molecular structure and specific anti-HIV activity as determined by an *in vitro* test [16]. A proprietary decomposition algorithm was used to identify 76 critical elements of molecular structure. Identification of the important input variables is a crucial step in the ANN modeling process [15]. In this study classification tree analysis (CART™) reduced the size of input space by 63.2% by identifying a subset of 28 variables including the terminal node outputs of the optimal tree. The ANN accurately distinguished between Active and Inactive molecules and the hybrid model was more accurate, with better receiver operating characteristics, than the ANN model which used only the important critical structure elements as input variables. These results support the contention that hybrid AI modeling systems usually outperform single methods.

Poor generalizability due to overfitting data is a potential problem with small data sets and large numbers of input variables. Several steps were taken to minimize overfitting. The initial classification tree analysis developed an overly large model which intentionally overfit the data and then by successive pruning and 10 fold cross-validation found the optimal smaller tree with the lowest cross-validated relative cost. The process identified a subset of input variables which improved the ratio of data patterns to input variables from less than 5 to 1 to more than 13 to 1 in the final model. While there is no best ratio of data patterns to input variables, optimizing the ratio has been recommended since reducing the dimensions of the input space results in models that are more accurate, more robust with less risk of overfitting [15]. A PNN model was chosen in order to minimize spurious correlations between irrelevant inputs and the output. Bayesian methods used assigned large weight decay rates to irrelevant variables that minimized overfitting that may be caused by those variables [17,18]. Finally, the leave one out method of testing and training helped to minimize overfitting since each of the data patterns being evaluated was never in the training set at the time of testing.

While these results are promising, the number of structures in this study is relatively small (N=371). The molecular structure database continues to grow with the inclusion of new compounds representing a range of structures, activity levels, and mechanisms of action. The next model will include more than 500 molecules. Increasing the training data and representing all of the currently recognized classes of drugs with *in vitro* anti-HIV activity should improve the generalizability of future models. Larger databases of >1000 moleucles would allow for validation using a separate out of sample test set in addition to the leave one out validation strategy.

The decomposition algorithm has some limitations. The algorithm is applied manually and is labour intensive. The process is limited to compounds with a maximum molecular weight of 1600 daltons and does not distinguish well between very close stereoisomers. The next generation algorithm deals with the problem of stereoisomers. Future models will use additional validation methods to those used in this study such as a committee of experts approach using bootstrap resampling to improve the classification tree modeling process. This approach generates hundreds of different classification trees and the final classification of the committee of trees is derived from a plurality voting systems. Future models will also explore a more comprehensive AI approach to molecular structure-activity modeling.

That an ANN can be trained to recognize similarities and differences between molecular structures is not surprising. The hybrid ANN system is impressive in it's ability to distinguish active from inactive compounds (sensitivity, specificity, PPV, NPV all >95%). Many inactive molecules are structurally similar to active agents but are still identified as inactive by the ANN. In addition, the drugs used for training and evaluating the network work at several different steps in the viral replicative process. The structures of molecules active at these different sites are radically different and the network is able to compare a given compound for likeness to at least 8 different families of functional structures.

Drug development is a costly process. If proven to be reliable, the present hybrid ANN could screen any compound with a molecular weight ≤1600 daltons for anti-HIV activity. Drugs predicted to be effective against HIV1 could then be entered into the drug development pipeline with formal *in vitro* testing, animal studies, etc. This could reduce the entry of ineffective drugs into the development process.

The availability of anti-HIV therapy is presently limited to affluent societies and individuals. Use of an ANN could potentially reduce drug development costs resulting in new drugs that are less costly. Of the over 30 million people infected with HIV worldwide [20] less than 10% can afford effective treatment. A further use of this ANN would be to screen existing pharmaceuticals for anti-HIV activity. Once identified as having activity a licensed compound could be readily tested in humans allowing a more rapid entry to market. Less expensive alternatives to current multi-drug regimens might be identified with this technology resulting in less expensive therapies available to many more people. The present research represents the early stages of an exciting new interface between artificial intelligence and biomedical research that we have termed Molecular Mining.

References

[1] Baxt WG, Skora J. Prospective validation of artificial neural network trained to identify acute myocardial infarction. Lancet 1996; 347:12-5.

[2] Patil S, Henry JW, Rubenfire M, et al. Neural network in the clinical diagnosis of acute pulmonary embolism. Chest 1993; 104:1685-9.

[3] Kirby SD, Danter W, George CFP, et.al. Neural network prediction of obstructive sleep apnea from clinical criteria. Chest 1999; 116:409-15.

[4] Danter WR, Wood T, Morrison NJ, et al. Artificial neural network prediction of survival or death following admission to ICU. Clin Invest Med 1996; 19:S14.

[5] Pilon S, Tandberg D. Neural network and linear regression models in residency selection. Am J Emerg Med 1997; 15:361-4.

[6] Kennedy RL, Harrison RF, Burton AM, et. al. An artificial neural network system for diagnosis of acute myocardial infarction (AMI) in the accident and emergency department: evaluation and comparison with serum myoglobin measurements. Comput Methods Programs Biomed 1997; 52:93-103.

[7] Baxt WG. Application of artificial neural networks to clinical medicine. Lancet 1995; 346:1135-8.

[8] Guerriere RJ, Detsky AS. Neural networks: what are they? Ann Intern Med 1991; 115:906-7.

[9] Kasabov KK. Foundations of neural networks, fuzzy systems and knowledge engineering. MIT Press, Cambridge MA 1998; 421-74.

[10] Fraleigh S. Fuzzy logic and neural networks. PC AI 1994; 8:16-21.

[11] O'Rourke BM. Analyzing financial neural networks performance: applying fuzzy clustering and tree classification. PC AI 1999; 13:37-40.

[12] Murray D. Tuning neural networks with genetic algorithms. AI Expert 1994; 9:27-31.

[13] Reed RD, Marks RJ. Neural smithing: supervised learning in feedforward artificial neural networks. MIT Press, Cambridge MA 1999; 185-95.

[14] Steinberg D, Cardell NS. Achieving results with next generation data mining techniques. Salford Systems 1998.

[15] Dwinnell W. Modeling methodology 2: model input selection. PC AI 1998; 12:23-6

[16] Weislow OS, Kiser R, Fine DL, et.al. New soluble formazan assay for HIV-1 cytopathic effects: application to high flux screening of synthetic and natural products for AIDS antiviral activity. J Nat Cancer Inst. 1989; 81:577-86.

[17] MacKay DJC. Bayesian non-linear modeling for the energy prediction competition. ASHRAE Transactions 1995; v100 pt.2:1053-62.

[18] Sarle WS. Stopping training and other remedies for overfitting. Proc. 27th Symposium, Interface of Computing Science and Statistics. 1995; 352-60.

Neural Unit Sensitive to Modulation*

Anatoli Gorchetchnikov[†]
Al Cripps
Department of Computer Science, Middle Tennessee State University
Murfreesboro, TN 37132, USA

Abstract

Neuromodulation is an important mechanism of information processing control in the brain. Implementation of this mechanism in Artificial Neural Nets (ANNs) will enable intelligent structures to control their own cognitive function. What follows is a description of a unit created for parallel synchronous neural networks. It was designed to include the neuromodulation in the network functionality. Experiments showed that the model's behavior corresponds to the behavior of neurons in the brain.

1 Introduction

In the application of neural networks there are parameters that must be set according to some criterion. These criteria may be based upon sophisticated mathematical calculations or simple trial and error, but in any case the process requires the experimenter's interference. Often the same neural network require different parameter settings for different tasks. In this situation it would be convenient (and more biologically realistic) if the network will adjust these parameters by some process of self regulation and control. In the brain one of the means of such control is neuromodulation. It is based on the long-term action of several simple neurotransmitters: acetylcholine (ACh), norepinephrine (NE), γ-aminobutyric acid (GABA), dopamine (DA), and 5-hydroxytryptamine (5HT). Major effects of these neuromodulators on the firing activity of the cell were listed in [1]. Three of them that were simulated in this study are discussed in the following subsections.

1.1 Depolarization of the Pyramidal Cells (ACh) and Some Interneurons (ACh, NE, DA, 5HT)

Depolarization of the cell increases its excitability. In the case of pyramidal cells, high concentration of neuromodulator allows the relatively weak afferent input to produce some spiking activity in the cell. This is important in learning new information since it increases the cell's 'attention' to afferent signals. The low concentration of neuromodulator can enhance recall since only the signal through the strong, well-trained connection will produce the output activity. Indeed, in the brain high levels of ACh

---
*This research is partially supported by MTSU grant#2-21026.
[†]Please send correspondence to ag2a@mtsu.edu

were recorded during exploratory behavior and low levels during quiet waking and slow-wave sleep [2, 3].

In the case of interneurons, increase of excitability leads to stronger lateral inhibition, and, therefore, increases the selectivity of the network and reduces the overlap between the internal representations of input patterns [4]. This overlap reduction have been proven to reduce catastrophic interference in neural networks [5]. Reducing the interneurons' excitability (and therefore lateral inhibition) by the means of neuromodulation allows more related patterns to be active during initial phase of recall, and therefore broadens the memory search.

1.2 Suppression of Neuronal Adaptation (ACh, NE)

Barkai and Hasselmo [6] discussed the modulatory suppression of neuronal adaptation and provided some additional evidence of slower suppression rates in the presence of cholinergic agonist carbachol. The influence of this effect on learning and memory can be very important. Neuronal adaptation can change the behavior of the neural net during recall from convergence to a single attractor to movement between several centers of attraction [7]. This process allows the recall of chunks of related information, and, with some additional guidance, the recall of temporal and/or spatial sequences. Thus, high rate of adaptation provided by the low concentration of modulators can enhance recall.

A high concentration of modulator reduces the rate of adaptation and lets the signal persist for a longer time. This effect increases the active learning time, and, therefore, improves learning. At first glance there is a contradiction with the reported in [8] increase of selectivity in the neural net due to the presence of adaptation during learning. This contradiction can be resolved by phasic changes in neuromodulation [9]. For the increase of selectivity, it is not necessary to have a high level of adaptation throughout learning. At the end of the learning cycle, when the signal is still present, and modulation is low, the adaptation will not only increase the selectivity (as shown in [8]), but also trigger the reset of the neurons preparing the net for the next pattern.

1.3 Increase of LTP by the Means of EPSP (ACh, NE) or IPSP (GABA) Height

Hebbian learning in the brain requires the postsynaptic cell to fire within 50 ms after the presynaptic cell in order to enhance the synaptic connection between the two [10]. How does the cell know which synapses received signals 50ms ago? It is only possible, if there is still some trace of activity in the synapse, and postsynaptic potential (PSP) is this trace. As long as it persists in the synapse, the cell knows there was a signal coming to this synapse some time ago. Therefore, the higher postsynaptic potential the longer time it will persist, the higher chance of learning in this synapse. High concentration of neuromodulators have been shown to increase the PSP amplitude about threefold [11]. Again, the presence of neuromodulators enhances learning.

All three discussed effects consistently drove us to the idea that neglecting the neuromodulatory effects in artificial neural networks most likely impairs their learning ability. To include these effects following model was developed.

2 Method

2.1 Model Description

The model presented here consists of two parts: dendritic queue, and cell body. Temporal discretization is set to 0.01 ms of simulation time. The dendritic queue is an implementation of linear compartmental dendritic tree and has a length of 500. Input can be provided to any location (synapse) along the queue. Therefore, the distinction between distal and proximal inputs can be made. On each simulation step, the synaptic potential is multiplied by the decay constant and shifted one position closer to cell body. Next, the post-synaptic potential (PSP) in the location is calculated according Cohen-Grossberg [12] activation dynamics used in Hopfield nets. We removed the summation from the original formula, because there is only one input to each location, and received:

$$\frac{\partial x}{\partial t} = -\frac{1}{C}(Ax - I - mS(y)) \tag{1}$$

where x is activation ($x(0) = 0$); m and $S(y)$ are the weight of the synaptic connection and signal through this connection respectively. C and A represent to some extent membrane capacitance and permeability respectively; and I is an external input. Permeability linearly depends on the synapse location, decreasing with the distance from the cell body to implement longer time constant for distal synapses. Both A and C dynamically depend on the level of modulator (henceforth we will use ACh as an example) to implement the influence of ACh on PSP height. C increases with increase of ACh, while A decreases. Both dependencies were made linear for the sake of simplicity. Calculated current PSPs are added to potentials received during shift.

The cell body receives the signal popping out of the queue and generates the action potentials according to the canonical model of type 1 excitability [13]:

$$\frac{\partial S}{\partial t} = 1 - cos(S) + (1 + cos(S))\left(r + \sum_j c_j \delta(S_j - \pi)\right), \tag{2}$$

from which we removed the summation of incoming signals using total activation x instead. To implement the effect of neuronal adaptation we added adaptive activational threshold θ, so the final formula is:

$$\frac{\partial S}{\partial t} = 1 - cos(S) + (1 + cos(S))(r + x - \theta). \tag{3}$$

Initial condition is $S(0) = 0$. Original bifurcation parameter r in our model is the concentration of ACh and ranges from -0.75 with no ACh to +0.25 with 100% ACh. We call 100% concentration the level of ACh when it reaches the saturation and any further increase does not noticeably affect the process. At 75% concentration of ACh our model experiences bifurcation ($r = 0$) ant transition from the excitable state to periodic spiking.

The adaptive threshold θ also depends on the ACh level. After every spike, the threshold increases by the constant value 0.3, which was determined empirically to achieve the adaptation time of about 400 ms. After these increases, the threshold decays back to zero according to equation suggested in [7]:

$$\frac{\partial \theta}{\partial t} = -\alpha\theta \tag{4}$$

where in our model α depends on the ACh level. Again, for simplicity we used linear dependency with coefficient 0.0002.

2.2 Experimental Settings

We ran two experiments using the single unit. In the first study we provided 3 ms pulses that arrived every 225 ms at 150th synaptic location (counting from the cell body). The level of ACh was increasing in each consecutive simulation by 1% from 0% to 100%. For the second study the input signal was changed to constant potential with the amplitude 10% of the original pulse. All other settings remained the same. Additionally, we measured the PSP in the 250th synapse receiving the actual spike through the connection with weight 1.0 for ACh levels of 0% and 100%.

3 Results

The results of the two experiments are summarized in Figure 1. It shows the signal 'recorded' from the output of equation 3. Figure 2 shows the excitatory post-synaptic potential (EPSP) in the synapse as a reaction on the spike in presynaptic cell for two extreme levels of modulation.

4 Discussion

Excitability of the cell obviously increased as ACh level increased. For constant stimulation, the modulation below 17%, does not provide enough depolarization for action potentials to be generated. This situation is normal for biological neurons, where many input signals are usually required to activate the cell. For pulse stimulation we used overkill amplitude, so it would invoke the spike even with no ACh present. That make easy to observe the suppression of neuronal adaptation. With ACh of 10% and lower adaptation is so strong, that the cell is not able to recover after first spike to generate the other two. If ACh is in-between 10% and 15% only one of the consecutive spikes was skipped. With ACh approaching 100% the neuronal adaptation became hardly noticeable, but did not completely disappear, which is also biologically realistic. EPSP height in reaction to the spike increased about threefold with the increase of modulation from 0% to 100%. This corresponds to the data observed in natural neurons [11].

Thus, all three studied modulatory influences on the spiking behavior of the model showed high biological realism and justified further experiments with the networks of described units. Same adjustments can also be applied to other neural models increasing their functionality and plausibility. The complexity of the single unit that we introduced should not affect the simulation time for the network, because the network

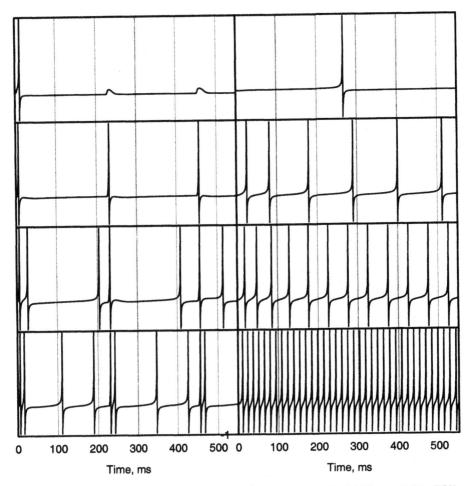

Figure 1: Left panel shows the response on pulse. The level of ACh was 15%, 75%, 85%, and 100% from top to bottom. Right panel shows the response on constant signal. The level of ACh was 17%, 35%, 50%, and 100% from top to bottom.

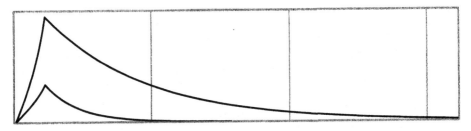

Figure 2: EPSP resulting from spike arriving at 250th synaptic location for levels of ACh 0% and 100%. Gridlines are drawn every 10 ms.

model in our research is designed for massive parallel architecture. This allows the addition of up to 2048 units (number of processors of our MP1) without affecting the simulation time. Our future research will use the neural unit described here in the neural net design for both scientific and commercial applications.

5 Acknowledgments

We thank M. Hasselmo and anonymous reviewers for useful comments.

References

[1] Hasselmo M. Neuromodulation and cortical function. Tutorial at IJCNN'99, in press.

[2] Kamentani H, Kawamura H. Alterations in acetylcholine release in the rat hippocampus during sleep-wakefulness detected by intracerebral dialysis. Life Science, 1990; 47(5): 421–426.

[3] Marrosu F, Portas C, Mascia M et al. Microdialysis measurement of cortical and hippocampal acetylcholine release during sleep/wake cycle in freely moving cats. Brain Research, 1995; 671: 329–332.

[4] Linster C, Hasselmo M. Modulation of inhibition in a model of olfactory bulb reduces overlap in the neural representation of olfactory stimuli. Behavioral Brain Research, 1997; 84: 117–127.

[5] French R. Dynamically constraining connectionist networks to produce distributed, orthogonal representations to reduce catastrophic interference. In: Proceedings of the 16th Annual Cognitive Science Society Conference. NJ:LEA, 1994, pp 335–340.

[6] Barkai E, Hasselmo M. Modulation of the input/output function of rat piriform cortex pyramidal cells. Journal of Neurophysiology, 1994; 72: 644–658.

[7] Horn D, Usher M. Neural networks with dynamical thresholds. Phys. Review, 1989; A40: 1036–1044.

[8] Gorchetchnikov A. Introduction of threshold self-adjustment improves the convergence in feature-detective neural nets. Neurocomputing, in press.

[9] Hasselmo M. Neuromodulation and the hippocampus: memory function and dysfunction in a network simulation. In: Reggia J, Ruppin E and Glanzman D (eds) Progress in Brain Research, v121. Elsevier Science BV, 1999, pp 153–161.

[10] Larson D, Lynch G. Theta pattern stimulation and the induction of LTP: the sequence in which synapses are stimulated determines the degree to which they potentiate. Brain Research, 1989; 489: 49–58.

[11] Hasselmo M, Barkai E. Cholinergic modulation of activity-dependent synaptic plasticity in the piriform cortex and associative memory function in a network biophysical simulation. Journal of Neuroscience, 1995; 15(10): 6592–6604.

[12] Cohen M, Grossberg S. Absolute stability of global formation and parallel memory storage by competitive neural networks. IEEE Transactions on Systems, Man, and Cybernetics, 1983; v. SMC-13: 815–826.

[13] Hoppensteadt F, Izhikevich, E. Canonical neural models. In: Arbib, M (ed) Brain Theory and Neural Networks. The MIT Press, Cambridge, MA, 2001, in press.

On Methods for Combination of Results from Gene-Finding Programs for Improved Prediction Accuracy

Cecilia Hammar

Department of Computer Science, Bioinformatics Group, University of Skövde
Skövde, Sweden
cissi@ida.his.se

Dan Lundh

Department of Computer Science, Bioinformatics Group, University of Skövde
Skövde, Sweden
dan@ida.his.se

Abstract

Gene-finding programs available today give far from 100% correct predictions. The idea of combining a number of different gene-finding programs to achieve an improved overall prediction accuracy was recently brought forward. Murakami and Takagi [1] published the first attempt to combine results from different gene-finding programs. Their discrete combination methods based on logical expressions showed improved prediction accuracy (by 3-5 %) compared to the best performing individual program (GENSCAN).

This paper shows an initial attempt to use an ANN to combine the results of the three well-known gene-finding programs GRAILII, FEXH, and GENSCAN. The results show a considerable improvement by 11% in prediction accuracy (on the nucleotide level) compared to GENSCAN and better prediction accuracy than the logic based combination methods (AND, OR and Majority).

1 Introduction

Finding genes in DNA is a difficult and time consuming task and there is a great need for computer support. The gene-finding programs are used to pinpoint regions in the DNA that are likely to contain exons[1]. It is shown by a number of evaluations that most gene-finding programs are far from being powerful enough to elucidate the genomic structure completely [2] [3]. Research have showed that different programs consider different gene-features, and often predictions for the same sequences are different [2]. The idea behind this project, which is inspired by [1], is that a combination of gene-finding programs could improve the prediction accuracy if the strenghts of the programs could be captured and the erroneous effects avoided.

---

[1] Exons represent coding regions.

Murakami and Takagi [1] tested four well-known gene-finding programs (FEXH, GRAILII, GENSCAN, and GeneParser) and evaluated five methods for combining the results (AND, OR, HIGHEST, RULE and BOUNDARY). The AND and OR methods were used as logical operators to combine the output of the programs. The HIGHEST method selected the highest probability to be the prediction. The RULE method used a priority order for the programs and the BOUNDARY method used more biological information to make the overall exon prediction. Murakami and Takagi [1] demonstrated that the approximate correlation (AC) was improved by 3-5% by two of the methods (HIGHEST and BOUNDARY) compared to the best performing program (GENSCAN). They also showed that three of the methods (HIGHEST, OR, and BOUNDARY), improved AC as the number of programs combined increased.

The methods used here can be classified into logical and function approximation. The AND, OR, and MAJORITY (MAJ)[2] methods are logical methods with discrete input, while the ANN has the possibility of learning fuzzy situations. In other words the ANN will approximate a function for the combination of the results, if there is one. The type of ANN used is a basic feed-forward neural network, trained with back-propagation. In this paper we demonstrate that the ANN method is the most promising approach.

2 DNA Data Set and Gene-Finding Programs

In this project we use the data set Murakami and Takagi [1] collected from GenBank[3] release 100 (April 1997). The data set consists of human DNA sequences with at least one coding sequence (CDS), which means that sequences can contain parts of genes as well as complete genes. The transcriptions of the standard splice site conservative dinucleotides (i.e. GT-AC) have to be experimentally confirmed. Only sequences that have been registered after June 1996 are included. Immunglobin genes are discarded due to their complex structure. Sequences that contain exons which code for proteins that are homologous to already known proteins are also discarded. The data set consists of 219 loci divided in a training set of 146 loci and a test set of 73 loci. These data sets are available upon request from Murakami and Takagi [1].

Three well-known gene-finding programs are evaluated to keep the similarity to related studies [1]. In this study we use GRAILII [4], GENSCAN[5], and FEXH[6]. These programs use different techniques that should capture different features of the exon characteristics. GRAILII uses a neural network [3] [4]. GENSCAN is based on a hidden Markov model [5] [6]. FEXH uses a linear discriminant function [7]. The output of the programs is specified regions of DNA containing exons, with associated scores (probabilities or weights). The scores are normalized between 0.2

---

[2] MAJORITY can be decomposed into simple logical AND and OR operator.
[3] GenBank http://www.ncbi.nlm.nih.gov
[4] GRAILII e-mail server. GRAIL@ornl.gov (option "-2" to indicate GRAILII)
[5] GENSCAN http://CCR-081.mit.edu/GENSCAN.html
[6] FEXH. http://genomic.sanger.uk/gf/gf.shtml

and 0.8. For information about real exons GenBank data files are used. The exon regions (CDS) are assigned the value 0.8, while other nucleotides are assigned the value 0.2. In the tests done by Murakami and Takagi [1] the scores, i.e. predictions, were transformed into probabilities.

3 Performance Measures

To evaluate the gene-predicting programs on the nucleotide level we use *Sensitivity, Specificity, and Approximate Correlation* [8]. *Sensitivity* is defined as the proportion of coding nucleotides that are correctly predicted as coding. *Specificity* is the proportion of non-coding nucleotides that are correctly predicted as non-coding. *Approximate correlation* (AC) is a widely used accuracy measure that reflects the association between the prediction and the reality. To measure sensitivity and specificity at the exon level we use the measures developed by Burset and Guigo [2], i.e. *Missing exons* and *Wrong exons*. *Missing exons* (ME) is the proportion of actual exons with no overlap of predicted exons. *Wrong exons* (WE) is the proportion of predicted exons with no overlap to any actual exons.

The results from the different gene-finding programs are analysed separately for comparison. The AND, OR, and ANN method are used on the three possible combinations of two programs and the combination of all three programs. The MAJ method is used on the combination of all three programs.

4 Experiments and Results

The AND and OR methods require that all respectively one of the participating programs predict a nucleotide as coding for the overall prediction to be coding. The MAJ method requires at least two programs to predict a nucleotide as coding for the overall prediction to be coding, see Figure 1 for an illustration.

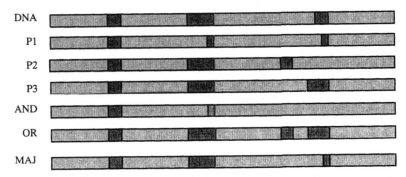

Figure 1. The top sequence represents the actual DNA sequence, where the dark regions are the actual exons. P1, P2, and P3 are the predictions by the three different programs. The resulting overall prediction of the combination methods AND, OR, and MAJ are shown in the lower sequences.

The topology of the ANN is shown in Figure 2. The output of the programs (coded as described above) is fed to the input layer of the network nucleotide by nucleotide. The output from the ANN is a prediction if a nucleotide is coding or not. A threshold to divide the output into coding and non-coding predictions is set to 0.5.

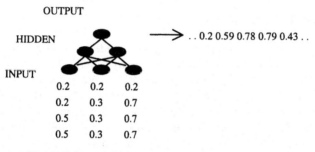

GRAILII, FEXH, GENSCAN

Figure 2. The topology of the ANN. In the figure the ANN combines the predictions from three programs.

The AND, OR, and ANN combination methods are used on all combinations of two programs (i.e. all permutations of programs) and three programs (i.e. GRAILII, GENSCAN, and FEXH). The MAJ method is only used on all the programs.

Table 1. Prediction accuracy results for GENSCAN *(GS)*, GRAILII *(GR)*, and FEXH *(FE)* and the results of the different combinations of programs, see text. The results are derived from using the test set. *Undef* is the number of sequences for which the AC is undefined.

Programs	Method	Undef.	AC	Sn	Sp	WE	ME
GENSCAN	-	18	0.77	0.77	0.91	0.09	0.09
GRAILII	-	16	0.64	0.66	0.80	0.17	0.18
FEXH	-	8	0.62	0.63	0.80	0.22	0.13
GR+FE	AND	23	0.66	0.59	0.90	0.11	0.24
GR+GS	AND	23	0.73	0.67	0.94	0.04	0.18
GS+FE	AND	22	0.73	0.65	0.97	0.03	0.15
GR+GS+FE	AND	27	0.71	0.61	0.97	0.02	0.23
GR+FE	OR	5	0.66	0.72	0.78	0.22	0.06
GR+GS	OR	12	0.70	0.75	0.81	0.17	0.09
GS+FE	OR	6	0.68	0.74	0.78	0.23	0.08
GR+GS+FE	OR	5	0.68	0.76	0.76	0.24	0.05
GR+GS+FE	MAJ	16	0.73	0.71	0.91	0.09	0.10
GR+FE	ANN	17	0.68	0.64	0.88	-	-
GR+GS	ANN	19	0.77	0.75	0.92	-	-
GS+FE	ANN	19	0.77	0.87	0.74	-	-
GR+GS+FE	ANN	22	0.85	0.99	0.80	0.01	0.11

The results on the nucleotide level are shown in Table 1 (above). The logical methods were also tested on the training set, however the results do not differ considerably. As seen in Table 1, using an ANN to combine predictions from all programs results in a higher AC than the best performing program (GENSCAN), i.e. an emerging effect. This effect is not shown by the other combination methods. However, we can also see in Table 1 that the sensitivity in general increases for combinations while specificity decreases. In Figure 3 the approximate correlation is shown with respect to the number of programs that are combined using the different methods. The GENSCAN result is shown for comparison. Figure 3 demonstrates clearly that the ANN method is the only method that considerably improves the AC as the number of programs increase. The ANN captures the specific gene-finding features of the programs which results in a more accurate overall prediction compared to the separate programs.

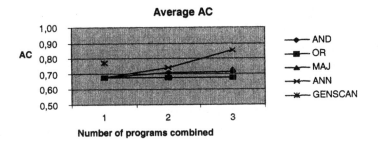

Figure 3. The average approximate correlation (AC) for the different methods in relation to the number of programs combined with the methods. The results are averaged over the sequences in the test set and the tests done with the different number of programs. The 'combination of one program' is the average AC of the three programs. The ANN method shows a clear improvement in AC as the number of programs increase, and the best AC measured (0.86). The difference between GENSCAN and the ANN combination of three programs is 11%.

The idea of combining different gene-finding programs is that different programs might be better at finding different types of exons or better at finding exons in different sequence structures. If the aim is to capture these valuable differences between the programs, a combination using the AND method would fail because of the large number of missing exons. The OR method is not sufficient for the problem either, as long as any program predicts wrong exons. The OR method show very few missing exons, but a large number of incorrectly predicted ones.

The logical functions for combining prediction results do generally improve the prediction performance compared to the average performance of the separate programs. This conclusion agrees with Murakami and Takagi [1]. However, the results show that functions adopted by the ANN are better than the logical functions for this problem. The ANN used has approximated a function that gives a more correct overall prediction than AND, OR, and MAJ. The AC of the best performing program (i.e. GENSCAN) is 0.77. This can be compared to the ANN result (when combining all three programs), which was as high as 0.85. That is, the ANN used in this project improve the prediction performance considerably compared to the results of the different programs.

It is possible that some features of the different programs are lost when combining the predictions and evaluating the overall prediction on the nucleotide and exon levels. It would be interesting to evaluate the programs and combination methods on the complete gene level like Burge and Karlin [6] did with GENSCAN, i.e. the proportion of actual genes which are predicted exactly.

In future work the input to the ANN can also include information about the predictions, i.e. exon, intron, splice site, promoter etc.

In this paper we show one type of ANN (i.e. architecture). It is our belief that an exploration of architectures, tuned training, and incorporation of more gene-finding programs would improve the prediction accuracy further.

References

[1] Murakami K, Takagi T. Gene recognition by combination of several gene-finding programs. Bioinformatics 1998; 8:665-675

[2] Burset M, Guigo R. Evaluation of Gene Structure Prediction Programs. Genomics 1996; 34:353-367

[3] Henderson J, Salzberg S, Fasman KH. Finding genes in DNA with a hidden Markov model. Journal of Computational Biology 1997; 2:127-141

[4] Uberbacher EC, Mural RJ. Locating protein-coding regions in human DNA sequences by multiple sensor-neural network approach. Proceedings of the National Academy on Intelligent Systems for Molecular Biology 1991; 88:11261-11265

[5] Xu Y, Uberbacher EC. Gene prediction by pattern recognition and homology search. Proceedings of the Fourth International Conference on Intelligent Systems for Molecular Biology. World Scientific, 1996, pp 241-251

[6] Burge C, Karlin S. Prediction of complete gene structures in human genomic DNA. Journal of Molecular Biology 1997; 268:78-94

[7] Solovyev VV, Salamov AA, Lawrence CB. The prediction of human exons by oligonucleotide composition and discriminant analysis of spliceable open reading frames. Altman R, Brutlag D, Karp P, Lathrop R, Searls D eds. Proceedings of the Second International Conference on Intelligent Systems for Molecular Biology. AAAI press, 1994, pp 354-362

[8] Sternberg MJE. Protein structure prediction- A practical approach. Oxford University Press, Oxford, 1996

A Simulation Model for Activated Sludge Process Using Fuzzy Neural Network

Taizo Hanai*, Shuta Tomida, Hiroyuki Honda,

Takeshi Kobayashi

*Corresponding author

Postal address: Furo-cho, Chikusa-ku, Nagoya 464-8603, Japan

E-mail address: taizo@nubio.nagoya-u.ac.jp

Department of Biotechnology, Graduate School of Engineering, Nagoya University

Nagoya, Japan

Abstract

In order to construct a simulation model for estimating the effluent chemical oxygen demand (COD) value of an activated sludge process in a "U" plant, fuzzy neural network (FNN) was applied. The constructed FNN model could simulate periodic changes in COD with high accuracy. Comparing the simulation result obtained using the FNN model with that obtained using the multiple regression analysis (MRA) model, it was found that the FNN model had 3.7 times lower error than the MRA model. The FNN models corresponding to each of the four seasons were also constructed. Analyzing the fuzzy rules acquired from the FNN models after learning, the operational characteristic of this plant could be elucidated.

1 Introduction

The activated sludge process has been widely used for water treatment of both municipal and industrial wastewaters. Activated sludge can convert various organic compounds in wastewater to carbon dioxide by the oxidative activities of aerobic microorganisms. Since many microorganisms are associated with the removal of organic compounds, the kinetics of this process is normally more complex than that of industrial microbial processes, in which only one microorganism is used. To date, many models have been proposed to describe the dynamic characteristics of the process, such as the activated sludge models (ASMs) no.1 [1] and no.2 [2] presented by the International Association on Water Quality (IAWQ). These models are constructed based on Monod's equation for growth of microorganisms. However, the ASMs no.1 and no.2 have 19 and over 50 parameters, respectively, and it is difficult to decide these parameters exactly. In recent years, the load for treatment of municipal wastewater has increased due to increasing municipal population, and the criterion for the effluent from the wastewater treatment has become more stringent from the viewpoint of environmental preservation. Therefore, a sophisticated control system of the process needs to be established. In the previous papers [3,4], we applied a fuzzy neural network (FNN) to control fermentation processes, such as the Japanese sake mashing and beer brewing, and a highly accurate FNN model was constructed. In the present study, we apply FNN for constructing a model to estimate chemical oxygen demand (COD) of the effluent from municipal wastewater treatment.

2 Materials and Methods

2.1 Measurement items

The data used in this study were collected at the "U" wastewater treatment plant in "N" city applying the activated sludge method. One-week data collected in May, August and October 1994 and in February, 1995, representing spring, summer, autumn and winter seasons, were used for the analysis. Figure 1 shows the flowsheet of the "U" plant. The following items were measured at the points from (b) to (h) every hour, *i.e.*, (b) primary effluent; SS (suspended sludge), COD, and flow rate of the effluent, (c) the middle portion of the aeration tank; mixed liquor suspended solid (MLSS), dissolved oxygen concentration (DO), and added amount of sodium hypochlorite, (d) the exit of the aeration tank; MLSS, and DO, (e) effluent; COD, pH, and flow rate of the effluent, (f) line of excess sludge; flow rate of the excess sludge, (g) line from blower; air flow rate, (h) line of return sludge; flow rate of the return sludge, and MLSS. These items were used as the input variables for modeling.

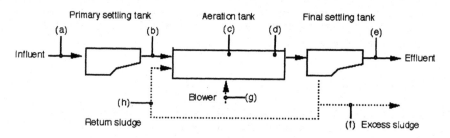

Fig.1 Flowsheet of wastewater treatment plant "U".

2.2 Input and output variables for modeling

The average hydraulic retention times for the aeration tank and the final settling tank were approximately 6 h and 4 h, respectively. Assuming a plug flow, the wastewater passing through point (b) will reach 3 h later at point (c), 6 h later at point (d), 10 h later at point (e). Therefore, the measured values at point (b) at time T h, at point (c) at T+3 h, at point (d) at T+6 h were used for the estimation of the COD value at point (e) at T+10 h. Since the air flow rate in aeration tanks varied, the average air flow rate during 6 h was also added to the input variable. Temperatures of the atmosphere and point (e) were measured twice a day and their average values were used for analysis, but the temperature in the aeration tank was not measured. Therefore, the temperature at point (e) at time T+6 h was used as the temperature in the aeration tank.

The total number of measurable variables was 34. If a variable has a correlation coefficient above 0.8 with another variable, only one of the two variables should be used for the modeling. Therefore, the 16 variables were used to construct the model and these are listed in Table 1. Using these 16 variables, the FNN model estimating the COD value in the effluent, in which the COD value is the output variable, was constructed.

Table 1. Input variables for the model estimating COD.

Number	Variables	Measured point in Fig.1
1	Operation time	
2	Hydraulic retention time	Aeration tank
3	COD	(b)
4, 5	MLSS	(c) and (d)
6	COD / MLSS	Aeration tank
7, 8	DO	(c) and (d)
9	pH	(e)
10	Temperature	(e)
11, 12	Flow rate of water	(b) and (e)
13, 14	Flow rate of sludge (F_s)	(f) and (h)
15	Mass flow rate of biomass $(F_s \cdot MLSS)$	(h)
16	Air flow rate	(g)

Among all sets of the input and output data in each season, which totaled to 168 data sets, half the data were used for learning of the FNN and the other half were used for evaluation of the FNN after learning.

2.3 Fuzzy neural network

In this paper, "Type I" of FNNs proposed by Horikawa et al. [5] was used. The FNN realizes a simplified fuzzy inference of which the consequences are described with singletons. Figure 2 shows an example of the configuration of FNN, where the FNN has two inputs x_1 and x_2 and one output y^* and three membership functions in each premise. The symbols of the circle and square in Fig.2 mean units of the neural network and the denotation wc, wg, wf, 1 and -1 are the connection weights. The connection weights wc and wg determine the positions and gradients of the sigmoid functions "f" in the units in (C)-layer, respectively.

If an unnecessary variable is used as the input for the FNN model, the fuzzy rules of the model can hardly be understood and the accuracy of this model would be lower than that using only necessary variables. Therefore, the parameter increasing method (PIM) was used for the optimization of the input variables [5]. The PIM identifies a fuzzy model by increasing such parameters as the number of membership functions and/or the number of input variables step by step. The fuzzy modeling by PIM was carried out in every combination of input variables at every step. The most

256

suitable combination of input variables which has the smallest error is selected at every step. Considering the relationship between the number of learning data and the scale of the model, the maximum steps of PIM were determined to be six for the model of "U" plant.

Multiple regression analysis (MRA) was also used for comparison with the FNN model. The input variables were optimized by the PIM method and the maximum number of input variables was set to the same maximum step number of the FNN modeling.

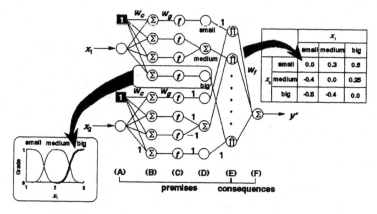

Fig.2   Structure of fuzzy neural network Type I.

3   Result and Discussion

3.1   Selected input variables and the estimated result

Modeling was carried out for the data sets collected in each of the four seasons at "U" plant. The results using the data in the summer season are discussed as an example. Selected input variables of FNN and MRA models are listed in Table 2. The reason why the selected variables were different between FNN and MRA models is that MRA can handle only linear relationships, while FNN can deal with more complex relationships.

The simulated results for the evaluation data by the constructed FNN and MRA models using these input variables are shown in Fig. 3. The FNN model estimated the time courses with high accuracy. The average error and the average relative error of this model were 0.094 (mg/l) and 0.69%, respectively. On the other hand, the MRA model could roughly estimate the COD trend. Some points were highly inaccurate. The average error and the average relative error of this model were 0.355 (mg/l) and 2.62%, respectively. The FNN model showed 3.7 times lower error than the MRA model. Modelings in spring, autumn and winter were also carried out independently and the three models were constructed. Average relative error for learning in the MRA model was about 2.9 times higher than that of the FNN model.

Table 2. Selected input variables for the FNN and MRA models using summer data by PIM.

FNN model	MRA model
-COD at point (b)	-COD at point (b)
-Operation time [a]	-DO at point (c)
-Hydraulic retention time	-DO at point (d)
-Temperature at point (e) [a]	-Flow rate of effluent at point (e)
	-COD / MLSS
	-Hydraulic retention time

a the variable was selected twice by PIM.

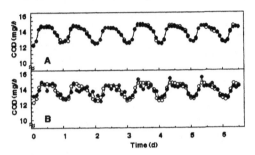

Fig. 3 Simulation result for summer data using the FNN(A) and MRA(B) models. Symbols: open circle, measured value; filled circle, estimated value.

3.2   Analysis of the FNN rules

In FNN, the acquired knowledge can be described in the form of fuzzy rules by analysis of the connection weight in the FNN after learning. Table 3 shows the acquired fuzzy rules of FNN models using the summer data. In this table, "S", "M" and "B" indicate small, medium and big, respectively. If the values in this table become big, the COD value of the effluent becomes large. From this table, the following rules were elucidated after the analysis; (i) If "operation time" becomes medium (the time is about noon), then the COD value becomes small (effluent COD becomes low). If "operation time" becomes small or big (the time is about early in the morning or late at night), then the COD value becomes big (effluent COD becomes high). This rule reflects the effect of hydraulic retention time of this plant (about 12 h) and of day-cycle of domestic wastewater. (ii) In the case of big (long) "hydraulic retention time", if the "temperature in the effluent" becomes small, medium and big (water temperature becomes low, medium and high), then the effluent COD becomes small, medium and big (effluent COD becomes low, slightly high and high), respectively. This rule can be interpreted as follows. In this plant, a nitrifying reaction often occurs in summer at high water temperature. The expert worker in this plant controls the air flow rate at a low level so as to limit the nitrifying reaction, which leads to a decrease in the activities of aerobic microorganisms, and the effluent COD becomes high. (iii) In the case of small (short) "hydraulic retention time", if the "temperature in effluent" becomes small, medium and big (water temperature becomes low, medium and high), then the effluent COD becomes big, medium and small (effluent COD becomes high, slightly low and low), respectively. This means that effluent COD decreases with increase in temperature since the hydraulic retention time is short and the nitrifying reaction does not easily occur.

Table 3. Identified fuzzy rule using summer data.

				Operation time					
				S		M		B	
				COD at point (b)					
				S	B	S	B	S	B
Temperature at point (e)	S	Hydraulic retention time	S	0.93	0.72	0.26	0.20	0.74	0.72
			B	0.01	0.36	0.95	0.08	0.58	0.76
	M		S	1.11	0.57	0.25	0.18	0.62	0.61
			B	0.41	0.81	0.29	0.44	0.74	0.80
	B		S	0.44	1.08	0.10	0.08	0.19	0.38
			B	0.79	0.66	0.26	0.56	0.74	0.74

In these ways, characteristic rules of each season can be elucidated from the FNN model of each season. The simulation method presented in this paper can be easily applied to other wastewater treatment plants, if the data are measured every hour. In those cases, the membership functions will be tuned automatically by learning and the fuzzy rules can be automatically elucidated. Comparing these rules with the ones derived from the present plant, the characteristics of these plants are understood clearly.

References

[1] Henze, M., Grady, C. P. L., Jr., Gujer, W., Marais, G.v.R., Matsuo, T. : Activated sludge model no.1. IAWPRC Scientific and Technical Reports 1987, 1

[2] Henze, M., Gujer, W., Mino, T., Matsuo, T., Wentzel, M.C., Marais, G.v.R. : Activated sludge model no.2. IAWPRC Scientific and Technical Reports 1995, 3

[3] Hanai, T., Katayama, A., Honda, H., Kobayashi, T. : Automatic fuzzy modeling for ginjo sake brewing process using fuzzy neural networks. J. Chem. Eng. Japan 1997, 30, 94-100

[4] Noguchi, H., Hanai, T., Takahashi, W., Ichii, T., Tanikawa, M., Masuoka S., Honda, H., Kobayashi, T. : Model construction for quality of beer and brewing process using FNN. Kagaku Kougaku Ronbunshu 1999, 25, 695-701 (in Japanese)

[5] Horikawa, S., Furuhashi,T., Uchikawa, Y. : A study on fuzzy modeling using fuzzy neural networks. Proc. of International Fuzzy Engineering Symposium'91 1991, 562-573

A General Method for Combining Predictors Tested on Protein Secondary Structure Prediction

Jakob V. Hansen

Department of Computer Science, University of Aarhus
Ny Munkegade, Bldg. 540, DK-8000 Aarhus C, Denmark

Anders Krogh

Center for Biological Sequence Analysis, Technical University of Denmark
Building 208, DK-2800 Lyngby, Denmark

Abstract

Ensemble methods, which combine several classifiers, have been successfully applied to decrease generalization error of machine learning methods. For most ensemble methods the ensemble members are combined by weighted summation of the output, called the linear average predictor. The logarithmic opinion pool ensemble method uses a multiplicative combination of the ensemble members, which treats the outputs of the ensemble members as independent probabilities. The advantage of the logarithmic opinion pool is the connection to the Kullback-Leibler error function, which can be decomposed into two terms: An average of the error of the ensemble members, and the ambiguity. The ambiguity is independent of the target function, and can be estimated using unlabeled data. The advantage of the decomposition is that an unbiased estimate of the generalization error of the ensemble can be obtained, while training still is on the full training set. These properties can be used to improve classification. The logarithmic opinion pool ensemble method is tested on the prediction of protein secondary structure. The focus is on how much improvement the general ensemble method can give rather than on outperforming existing methods, because that typically involves several more steps of refinement.

1   Introduction

Empirically it has proven very effective to average over an *ensemble* of neural networks in order to improve classification performance, rather than to use a single neural network. See [1, 2] for examples in protein secondary structure, [3] for an over-view of applications in molecular biology. Here we present a general ensemble method, and show how the error of an ensemble can be written as the average error of the ensemble members minus a term measuring the disagreement between the members (called the ensemble ambiguity). This proves that the ensemble is always better than the average performance – substantiating the empirical observations.

The ambiguity is independent of the training targets and can thus be estimated from unlabeled data. Therefore an *unbiased* estimate of the ensemble error can be found when combining cross-validation and ensemble training. This error estimate can be used for determining when to stop the training of the ensemble. We test this general approach to ensemble training on the secondary structure prediction problem. This problem is well suited for the method, because thousands of proteins without known structure are available for estimation of ambiguity. It is shown that the estimated error works well for stopping training, that the optimal training time is longer than for a single network, and that the method outperforms other ensemble methods tested on the same data.

2 The ensemble decomposition

An ensemble consists of M predictors f_i, which are combined into the combined predictor F. We consider the case where each predictor outputs a probability vector $\{f_i^1, \ldots, f_i^N\}$, where f_i^j is the estimated probability that input $\vec{x}$ belongs to class c_j. The ensemble has associated M coefficient $\{\alpha_1, \ldots, \alpha_M\}$ obeying $\sum_i \alpha_i = 1, \alpha_i \geq 0$.

It is very common to define the ensemble predictor as $F^{\text{LAP}} = \sum_{i=1}^N \alpha_i f_i$, which is called the linear average predictor (LAP). The combined predictor for the logarithmic opinion pool (LOP) of the ensemble members is

$$F^j = \frac{1}{Z} \exp(\sum_{i=1}^M \alpha_i \log f_i^j), \tag{1}$$

where Z is a normalization factor given by $\sum_{j=1}^N \exp(\sum_{i=1}^M \alpha_i \log f_i^j)$. This combination rule is non-linear and asymmetric as opposed to the LAP. Unless otherwise stated, this is the combined predictor we are considering in this work.

An example set T consist of input-output pairs, where the output is also a probability vector $\{t^1, \ldots, t^N\}$. The examples are assumed generated by a target function t. The difference between the target t and the combined predictor F is measured by the Kullback-Leibler (KL) error function

$$E(t, F) = \sum_{j=1}^N t^j \log \left(\frac{t^j}{F^j} \right). \tag{2}$$

The error is zero if F^j is equal to t^j for all j. For all appearances of this error function the mean over the training set is implicitly taken. If the target probabilities are restricted to one and zero the error function (2) reduces to $E(t, F) = -\log(F^k)$, where t^k is one. This would be the case if the error function is used on a training set consisting of class examples.

The error in (2) can be decomposed into two terms with the LOP in (1)

$$E(t, F) = \sum_{i=1}^M \alpha_i E(t, f_i) - \sum_{i=1}^M \alpha_i E(F, f_i) = \langle E(t, f_i) \rangle - A(f), \tag{3}$$

where $A(f)$ is the ambiguity and $\langle \cdot \rangle$ is the weighted ensemble mean. By using Jensen's inequality it can be shown that ambiguity is always greater than or equal to zero. This implies that the error of the combined predictor always is less than or equal to the mean of the error of the ensemble members. We see that diversity among the ensemble members without increase in the error of each ensemble member will lower the error of the combined predictor. The decomposition in (3) also applies for the LAP ensemble with the quadratic error $E(t,F) = (t-F)^2$ replacing (2) [4]. The LOP decomposition is due to Tom Heskes [5, 6], although it is derived in a slightly different setting.

We see that the ambiguity term in (3) is independent of the target probability, which means that the ambiguity can be estimated using unlabeled data, or if the input distribution is known the ambiguity can be estimated without any data. Assuming we have estimates of the generalization error of the ensemble members and an estimate of the ambiguity, the estimated generalization error comes directly from (3). This can be achieved in the *cross-validation ensemble*, where the training set is divided into M equal sized portions. Ensemble member f_i is trained on all the portions except for portion i. The error on portion i is independent of training and can be used to estimate ensemble member f_i's generalization error. With the estimated ambiguity, this gives us a method for obtaining an unbiased estimate of the ensemble error and still use all the training data.

3 Tests on the protein secondary structure problem

The method is tested on the standard secondary structure problem, in which the task is to predict the three-class secondary structure labels α-helix, β-sheet or coil, which is anything else. There is much work on secondary structure prediction, for overviews see [3]. The currently most used method is probably the one presented in [1].

The LOP ensemble is suitable for the protein problem for several reasons. Firstly the protein problem is a classification problem, and the LOP ensemble method is tailor-made for classification, and secondly there is huge amount of unlabeled data, i.e. proteins of unknown structure, which can be used to estimate the ambiguity.

The ensemble members are chosen to be neural networks. The output of a neural network is post-processed by the SOFTMAX function defined as $f^j = e^{g_j}/\sum_k e^{g_k}$, where g^j is the linear output (the weighted sum of the hidden units) of output unit j. This ensures that the ensemble members obey $\sum_j f_i^j = 1, f_i^j \geq 0$. The ensemble coefficients are uniform. Learning is done with back-propagation and momentum. A window of 13 amino acids is used, together with a sparse coding of amino acid into a vector of size 20, so each network has 260 inputs and three outputs. Each network has a hidden layer with 50 nodes. All examples in the training set are used once and only once during each epoch. Weights are updated after a certain number of randomly chosen examples have been presented (batch-update). During training the batch size is increased every tenth epoch by a number of examples equal to the square root of the training set size. The learning rate is decreased inverse proportional to the batch size. Training is stopped after 500 epochs, or if the validation error is increased by more than 10 % over the best value.

Our data set consists of 650 non-homologous protein sequences with a total of about 130000 amino acid [7]. Four-fold cross-validation is used for each test, which means that the training set is divided into four equally sized sets. Training is done on three sets with 97000 amino acids in total, while testing is done on the remaining set of 33000 amino acids. The test set is rotated for each cross-validation run. The average of the four test runs is calculated.

In the cross-validation ensemble the validation error is the estimate of the generalization error calculated from (3) by using a set of 70000 unlabeled amino acids to estimate the ambiguity. Seven ensemble members are used, which means that each one was trained on about 83000 amino acids and validated on 14000. Apart from the cross-validation ensemble, we also test what we will term a simple ensemble, in which all the ensemble members are trained on the same training set of 83000 amino acids, and the validation error is calculated from an independent set of labeled data containing 14000 amino acids. Note that there is still the four-fold cross-validation as an 'outer loop' for both ensemble methods.

For every tenth epoch the ensemble validation error, the ensemble test error, and the average test error of the individual ensemble members were calculated. The graphs in Figure 1 shows for a single test run these values for the cross-validation ensemble.

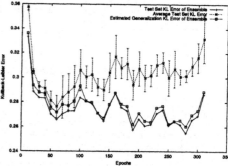

It is clearly seen that an ensemble is better than the average of the individual members, as it should be.

The cross-validation ensemble reaches a lower generalization error (represented by the test error) than the simple ensemble. This can be explained by the fact that the difference in training sets makes the ensemble members differ more than if they are trained on the same data, and this increases the ambiguity, which in turn lowers the generalization error.

Figure 1: Cross-validation LOP ensemble

In a practical application, one would select the ensemble at the training epoch with the lowest validation error. The various errors are shown for the ensemble selected in this manner. Using the validation error to select the ensemble training makes training dependent on the validation error, and therefore the estimate of the generalization error becomes biased. We will call the time of lowest validation error the stopping time. In

Table 1: The Kullback-Leibler errors and misclassification rates at stopping time.

Combination rule	LOP	LOP	LAP
Validation error	Cross-validation	Simple	Cross-validation
Train error function	KL	KL	MSE
Test error	0.2546	0.2585	0.6293
Average of indiv.	0.287 ± 0.010	0.2802 ± 0.0065	1.46 ± 0.12
Misclassification rate	0.3331	0.3385	0.3542
Ambiguity	0.0327	0.0217	0.8320

Table 1 the Kullback-Leibler error and misclassification rate at stopping time for the test runs is given. In these runs the validation error fluctuates by about ±3 % around the test error. However, the oscillations of the validation error follows the oscillations of the test set error, as can be seen in Figure 1, so the validation error can still be used to find the lowest test error. The validation error is not always lowest when the test set error is lowest, so a measure of the usability of validation error for stopping is the average difference between the lowest test set error and the test set error at stopping time. For both types of ensembles this difference is as low as 0.0001 or close to 0.05 %. So the estimated generalization error can very accurately be used to find the right stopping time.

The cross-validation ensemble reaches a test error that is 1.5 % lower than for the simple ensemble, and a misclassification rate that is 1.6 % lower. The explanation is in a larger ambiguity for the cross-validation ensemble, since the average test error of the ensemble members are comparable. The ambiguity of cross-validation ensemble is 1.5 times the ambiguity for the simple ensemble.

As noted in section 2 the error of the combined predictor is always better than the average of the error of the ensemble members. Still, one of the ensemble members can be better than the combined predictor. For the cross-validation ensemble method the test error is 0.2546, while the average of the error of the ensemble members is 0.2873. The difference (the ambiguity) is 0.0327. A gain of 12.8 %, which is substantial. The standard deviation on the test error of the ensemble members is 0.010, so the ambiguity is more than three times larger. It is very unlikely that any ensemble member has a lower generalization error than the ensemble error. For the simple ensemble method the ambiguity is smaller: 0.0217 or a gain of 8.4 %. The standard deviation among the ensemble members is 0.0065, so the ambiguity is more than three times the deviation.

The lowest average error for the ensemble members do not have to happen when the ensemble error is lowest. Typically the lowest average generalization error of the ensemble members will be reached before the lowest generalization error for the ensemble, so the ensemble can actually gain from overfitting in the individual ensemble members. This effect can be seen in Figure 1. Also the optimal architecture for a simple predictor is often smaller than for ensemble members. A number of single predictors with different size hidden layer has been trained. The number of nodes in the hidden layer are varied from 3 to 400. The training must be done with a separate validation set, since there is no ambiguity for a single predictor. The best result is achieved with 10 hidden nodes giving an average generalization error of 0.2598, and misclassification rate of 0.3413, which is respectively 2.0 %, and 2.5 % more than for the cross-validation LOP ensemble.

A *standard* cross-validation ensemble using the MSE error function and LAP combination rule is trained on the same data as the cross-validation LOP ensemble. The validation error is calculated using (3) even though the outputs do not necessary sum to one. The validation error have lost it's meaning as an error, e.g. it can be negative, but it is still valid as an early stopping indicator. This is supported by the fact that the lowest misclassification rate on the test set is only 0.6 % lower than the misclassification rate on the test set at stopping time. The generalization error for the standard ensemble is much higher measured with the KL error function, namely 0.6293 or about 2.5 times more than the cross-validation LOP ensemble. This is not a fair comparison, since the LOP ensemble is trained to minimize the KL error. Another measure is the misclassification rate. For the standard ensemble the misclassification rate is 0.3542, which is 6.3 % more than the misclassification rate of the cross-validation LOP ensemble.

Surprisingly the benefit is not in the combination rule. A test run, where the LOP is replaced with the LAP, while training still uses the KL error, yields a generalization of 0.2543, which is essentially the same as for the LOP combination rule. The misclassification rate for the LAP is 0.3363, which 1.0 % more than the LOP.

4 Conclusion

It was shown how the generalization error of an ensemble of predictors using a logarithmic opinion pool (LOP) can be estimated using cross-validation on the training set and an estimate of the ambiguity from an independent unlabeled set of data. When testing on prediction of protein secondary structure it was shown that this estimate follows the oscillations of the error measured on an independent test set.

The estimated error can be used to stop training when it is at a minimum, and it was shown that the cross-validation LOP ensemble method is superior to single predictors and standard ensemble methods using mean square error function on the protein problem. The benefit is not as much in the combination rule, as in the use of the Kullback-Leibler error function and the target independent ambiguity term.

5 Acknowledgments

We would like to thank Claus Andersen and Ole Lund at CBS for sharing their protein data set. This work was supported by the Danish National Research Foundation.

References

[1] B. Rost and C. Sander. Prediction of protein secondary structure at better than 70 % accuracy. *Journal of Molecular Biology*, 232(2):584–599, Jul 20 1993.

[2] S. K. Riis and A. Krogh. Improving prediction of protein secondary structure using structured neural networks and multiple sequence alignments. *Journal of Computational Biology*, 3:163–183, 1996.

[3] P. Baldi and S. Brunak. *Bioinformatics - The Machine Learning Approach*. MIT Press, Cambridge MA, 1998.

[4] Anders Krogh and Jesper Vedelsby. Neural network ensembles, cross validation, and active learning. In G. Tesauro, D. Touretzky, and T. Leen, editors, *Advances in Neural Information Processing Systems*, volume 7, pages 231–238. The MIT Press, 1995.

[5] Tom Heskes. Bias/variance decompositions for likelihood-based estimators. *Neural Computation*, 10(6):1425–1433, 1998.

[6] Tom Heskes. Selecting weighting factors in logarithmic opinion pools. In Michael I. Jordan, Michael J. Kearns, and Sara A. Solla, editors, *Advances in Neural Information Processing Systems*, volume 10. The MIT Press, 1998.

[7] O. Lund, K. Frimand, J. Gorodkin, H. Bohr, J. Bohr, J. Hansen, and S. Brunak. Protein distance constraints predicted by neural networks and probability density functions. *Protein Engineering*, 10(11):1241–1248, 1997.

A Three-Neuron Model of Information Processing During Bayesian Foraging

Noél M. A. Holmgren

Dept of Natural Sciences
University of Skövde
P.O. Box 408
SE-541 28 Skövde
Sweden
email: noel.holmgren@inv.his.se

Ola Olsson*

Dept of Animal Ecology
Ecology Building
SE-223 62 Lund
Sweden
email: ola.olsson@zooekol.lu.se

*current adress:
Dept. Biological Sciences
Univ. of Illinois at Chicago (M/C 066)
845 W. Taylor Street
Chicago, IL 60607
USA

Abstract

A foraging animal is often confronted with uncertainty of resource abundance. A Bayesian model provides the optimal forgaing policy when food occurrence is patchy. The solution of the Bayesian foraging policy requires elaborate calculations and it is unclear to what extent the policy could be implemented in a neural system. Here we suggest a network architecture of three neurones that approximately can perform an optimal Bayesian foraging policy. It remains to be shown how the network could be self-learned e.g. through Hebbian learning, and how close to to the optimal policy it can perform.

1 Introduction

Foraging is an essential activity for most animals, since it provides them with energy and nutrients that are used for survival and reproduction. The foraging process is therefore expected to be under strong selection, and one can hypothesize that it is optimized [6]. This means that foraging policies have evolved, under given constraints, to serve the end of maximal propagation of the forager's genes in the next generation. During a non-reproductive period, the forager is expected to maximize its survival until the next breeding event.

Food items are often patchily distributed, so a forager needs a policy when to leave a food-patch and go on to the next one. Many food items need to be searched for because they are hidden or are cryptically coloured. In other words, the patch content can not be estimated before the searching for food has begun. However, the forager may, through experience, have a general view of what to expect from a patch (i.e. it knows the general distribution of the food content in patches). The optimal policy for foraging under these conditions has been solved numerically, but involves time-consuming calculations with recursive functions [1, 4]. It may be questioned if a forager actually could follow the optimal policy given by Olsson & Holmgren since it requires such elaborate calculations to be solved. The idea of this paper is to present a simple neuron-based model capable of performing close to optimal in a Bayesian foraging situation as just described.

2 The Model

Data on the number of food items found and the time spent in a patch has been shown being sufficient statistics for the estimate of number of prey in a patch [2]. Patch contents often belongs to a clumped statistical distribution [5], such as the negative binomial distribution. When the prey distribution across patches follow the negative binomial distribution, the estimate of the number of remaining number of prey can be calculated using Bayes theorem [2]:

$$r(n,t) = \frac{\lambda + n}{e^{At}\frac{\alpha+1}{\alpha} - 1}, \tag{1.1}$$

where n is the number of food items found, t is the time spent in the patch, A is a search efficiency parameter, α and λ are parameters of the negative binomial distribution [2]. The estimate of number of prey declines over time but increases with the capture of a food item (Figure 1). In a patch foraging situation, however, the forager has to anticipate the reward of staying in the current patch and weigh it against the reward expected in another patch (i.e. the average patch content). The future in the current patch does not only involve benefits in terms of food-items as estimated, but it also involves information for better patch estimates. The estimate of number of remaining food items will increase in the future if another food-item is found (Figure 1). This implies that the optimal policy is not to leave a patch at a fixed estimate (r; [1, 2]), but to be more reluctant to leave during the early part of the patch visit [1, 3, 4].

The optimal policy is a set of leaving points $\{n, t_n\}$, which could be calculated by an iterative algorithm [1] or by stochastic dynamic programming [4]. These procedures are too extensive to be presented here. The interested reader is referred to the original publications.

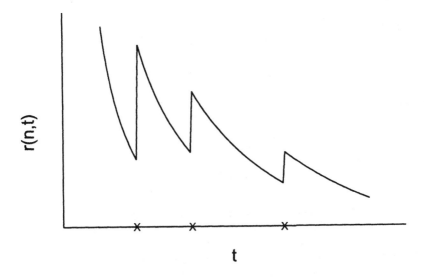

Figure 1. The estimate of number of remaining prey in a patch ($r(n,t)$) as a function of time (t) and number of food-items found (n). The crosses on the abscissa indicate time of food-item capture.

A three-neuron network could at least approximately perform the optimal policy described above. One neuron could produce an "r-signal" (the estimate of number of remaining food items), and another neuron could produce a "reference-signal" responsible for the reluctancy of leaving in the beginning of the stay. The two output signals produced could then be compared in a projection-neuron signalling when it is appropriate for the forager to leave the current patch for a new one (Figure 2).

A foraging cycle starts with the forager entering a food patch. The forager determines its arrival in a patch by its sensory system and neural layers behind (this part is not modelled here), which produce a pulse in the lower left corner of Figure 2. This pulse is an input signal to both inter-neurones in Figure 2, causing them to increase their firing rate. Due to the feed-back connections of the inter-neurones, their firing rate will decline gradually after the patch-arrival pulse has disappeared. Unaffected, as from I_2, this output is a memory of patch time (dotted line in Figure 3).

After some searching, the forager discovers a food item. A more peripheral system signals this event with a pulse entering our model system in the upper left corner of Figure 2. This signal triggers an increase in the firing rate of I_1. The result is an output signal from I_1 that increase when food items are found, and decline with time (hatched line in Figure 3). This is an approximation to the behaviour of the estimate function of remaining number of food items (Figure 1).

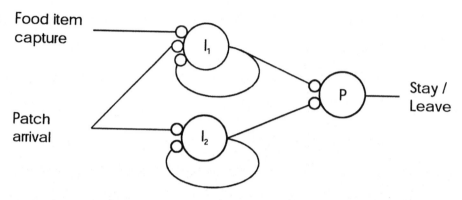

Food item capture

Patch arrival

I₁

I₂

P

Stay / Leave

Figure 2. A three-neuron network able to perform Bayesian foraging routines. A short pulse for the capture of a food item, and a short pulse when the forager arrives to a patch are input channels to the network. The input signals trigger an high output from the inter-neurones, which is slowly decaying due to the feed-back connections. The output of the inter-neuron (I_1) increases sharply with a food-item capture. This output signal could represent an estimate of the number of items remaining in the current patch. The output of the other inter-neuron (I_2) is a reference signal to the comparison in the projection-neuron (P). When the "items remaining" signal becomes too low in relation to the reference signal, the output signal from P changes from signalling "stay" to signalling "leave". The neurones are using the threshold function: $y=1/(1+\exp(-\alpha\ v))$, where y is the output, v is the cell activity (i.e. the sum of weighted inputs) and α is a threshold parameter.

In the final step, the two output signals are inputs to the projection-neuron (P). Here they are summed and passed through a threshold function. As long as the sum is high, which it is in the beginning, and as long as the forager is successful finding food items, the output of P is high. The high output is a signal to search for food in the current patch. After some time with unsuccessful search, the firing of P will be inhibited, and the forager will interrupt the food search in the current patch and leave (solid line Figure 3).

3 Discussion

The current analysis shows that it is theoretically possible to approximately perform a Bayesian foraging policy with very few neurones. Although some analysis of optimal behaviour require elaborate calculations, approximate computations can be performed even by very small animals with comparably low computational capacity.

The model above can be evaluated in comparison with the optimal policy. This would require further analysis to estimate the parameters of the model. This includes the synaptic weights, and slope parameters for the threshold function of each neuron. Even if this can be solved, it still remains to be shown how the parameters could be adjusted unsupervised, for instance through Hebbian learning. It is also possible that some changes in the architecture could enhance the performance of the network.

The simple neural network model is not only useful for understanding foraging behaviour. It may be a useful model for any learning system dealing with optimal resource use.

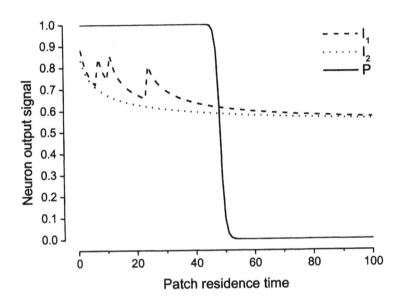

Figure 3. The output signals of the three neurones shown in Figure 2. The parameter settings of the neurones $W=\{\alpha, w_b, w_p, w_i, w_r\}$, where w_b, w_p, w_i, and w_r are the synaptic weights for the bias signal (-1), the patch arrival signal (1 at patch arrival), the food item found signal (1 when food item found), and the recurrence signal, respectively, are: $W(I_1)=\{4.0, 0.5, 1.0, 0.2, 1.0\}$, $W(I_2)=\{4.0, 0.5, 0.9, 0.0, 1.0\}$, and $W(P)=\{1000.0, 0.6^*, 0.5^*, 0.5, 0.0\}$. The weigths marked with * are for the output signal of I_1 and I_2, respectively.

References

[1] Green, R. F. 1988. Optimal foraging for patchily distributed prey: random search. Department of Mathematics & Statistics, University of Minnesota, Duluth.

[2] Iwasa, Y., M. Higashi, and N. Yamamura. 1981. Prey distribution as a factor determining the choice of optimal foraging strategy. The American Naturalist 117: 710-723.

[3] McNamara, J. 1982. Optimal patch use in a stochastic environment. Theoretical Population Biology 21:269-288.

[4] Olsson, O., and N. M. A. Holmgren. 1998. The survival rate maximizing policy of Bayesian foragers - wait for good news! Behavioral Ecology 9:345-353.

[5] Pielou, E. C. 1969. An introduction to mathematical ecology. Wiley-Interscience, New York.

[6] Stephens, D. W., and J. R. Krebs. 1986. Foraging Theory. Princeton University Press, Princeton, New Jersey.

Sensorimotor Sequential Learning by a Neural Network Based on Redefined Hebbian Learning

Karl Theodor Kalveram[1]

Institute for General Psychology, Section Cybernetical Psychology and
Psychobiology, Heinrich-Heine-University
40225 Duesseldorf/Germany, Universitaetsstr.1
e-mail: kalveram@uni-duesseldorf.de
http://www.psycho.uni-duesseldorf.de/kalveram.htm

Abstract

A two-jointed arm is used to discuss the conditions under which a neural controller can acquire a precise internal model of a plant to be controlled without the help of an external superviser. The problem can be solved by a 'modified Hebbian rule' ensuring convergence of the synaptic strengths, a feedforward network called 'power network', and a learning algorithm called 'auto-imitation'. The modified Hebbian rule describes a neuron, that – in addition to a number of inputs with plastic weights – has also a teaching input established with a fixed synaptic weight. The power network can adopt accurate models even of non-linear plants like the two-jointed arm when established with modified Hebbian synapses. The auto-imitation algorithm provides the power network with the values to be achieved by the network after learning. The training must be able to generate arbitrary movements, first of low velocity, then of higher velocity.

1 Introduction

'Hebbian learning', an old neuro-psychological concept describing the process of accomplishing associations at the synaptic level [4], is being increasingly confirmed by neuro-biological explanations especially referring to long term potentiation (LTP) [1], [17], [14], [3]. The question tackled in the present paper is whether an organism using such a mechanism is principally capable of acquiring, for instance, feedforward controllers with the complexity and the high precision necessary for goal-directed low-level motor behaviour. Recently it has been shown that at least for a simulated one-jointed arm such an internal model can be established [12]. However, even this seemingly simple task turns out to require a complex learning environment of a particular stucture. This demands a redefinition of the synaptic update mechanism in Hebbian learning, as well as a special type of a three-layered artificial neural network called "power network" [9], and also the embedding of both network and limb system into a special learning algorithm called "auto-imitation" [7], [10]. The purpose of the present paper is to demonstrate that in this learning environment an internal model also of a two-jointed arm can be acquired. Because of the dynamical inter-limb interactions due to reactive dynamic consequences and gravitational effects, this is a complex problem that is much more difficult to solve than for the simple case of an one-jointed arm.

---

[1] Supported by grant 417/18-2 from Deutsche Forschungsgemeinschaft (DFG).

2 Theory

2.1 Redefined Hebbian learning means adaptive linear filtering

Given a neuron completely behaving linearly, Hebbian learning can be expressed by the 'updating rule' $w'_i = w_i + r \cdot x_i \cdot z$, where x_i represents the (momentary) presynaptic activation at a synapse i, z the postsynaptic potential generated through all n synaptic connections of the neuron with other neurons, w_i the momentary synaptic strength, and w'_i the resulting synaptic strength in the next time step. r controls the learning rate. In terms of continuous time, this can be written as

$$w_i = r \int_0^T x_i \, ?z \, ?dt \; , \text{ with } \; z = \sum_{j=1}^n w_j \, ?x_j \; , \tag{1}$$

where T is the duration of the learning phase. (1) suggests that the strength of Hebbian synapse i is given by the correlation between the two variables x_i and z, and, as a consequence, that z represents the required values to be attained after learning. However, this interpretation leads to a serious problem in that the synaptic strength thus defined cannot converge in those cases where the correlation between the variables x_i and z differs from zero. This means that w_i depends on the learning duration T, and if T approaches infinity, w_i goes either into positive or negative saturation [13]. However, to process continuous data the synaptic weights of a neuron must also attain scaled values between such limits. Different normalization methods can be applied to get convergence [15], but if they fail, equation (1) is not applicable.

Another feature of Hebbian learning is that it is not possible to directly apply desired values to the neuron during learning, except for the non-realistic supposition of an enforcing input that overrides the activation given by all other inputs. However, both problems can be solved if a new synapse with the fixed weight of -1 is added to the neuron. This has the function of a "teaching input" that provides a target variable y to the neuron. This can be expressed by

$$w_i = -r \int_0^T x_i \, ?(z - y) dt \; \text{ with } \; z = \sum_{j=1}^n w_j \, ?x_j \; \text{ and } \; T \blacklozenge \times \; , \tag{2}$$

where $-y$ represents that part of the postsynaptic potential that originates from the teaching input. If weights can be selected such that the weighted sum z of x_1 to x_n finally attains y, the problems of convergence and of teaching are both solved in a biological plausible manner. Equation (2) also formally embodies the delta rule, formerly considered as not deducible from original Hebbian learning. Obviously, (2) imposes that learning and the application of that learning must be assigned to mutually exclusive phases, and in the application phase the teaching input must be such that $y=0$. In original Hebbian learning this teaching input is not provided [1], and thus (2) can be regarded as representing a modified type of Hebbian learning.

Assuming that the transfer characteristic between the postsynaptic potential of the neuron and its output is linear, Fig.1 provides an analogue visualization of (2) in terms of the neuron model of McCulloch and Pitts.

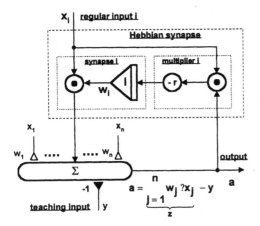

Figure 1: Hebbian synaptic learning redefined. Following (2), during learning the integrator I accumulates the products put out by the multiplier i. After learning, the connection to the synapse i is cut, and y is set to zero. Now I holds the output value w_i representing the synaptic strengths. All other n synaptic units operate in the same manner. If learning ends with a→0, for all combinations of values of the input variables $x_1, ..., x_n$ the output a now automatically attains the desired values of y.

Notice that the teaching input neither enforces the neuron to attain a post-synaptic potential equal to the input value of y, nor serves as a re-enforcing input signalling back an error provided by an external supervisor. The intracellular postsynaptic potential itself mirrors the error z-y, and has to be minimized (relaxation principle).

Referring to adaptive signal processing, the neuron in Fig.1 with teaching input is an "adaptive linear combiner" capable of "adaptive linear filtering". As has been pointed out [16], the adaptation procedure, if conducted appropriately and using finite time intervals, approaches the minimum error step by step in a least mean squares steepest-descent manner, with r regulating the decrease of error per step.

2.2 The power network realizes multi-dimensional power series

Much evidence exists that the central nervous system is both additive and multiplicative, at least at the spinal level. Processes like sensory gating or descending modulation of reflexes [2] suggest that in these cases premotor efferent activity is a product of two or more variables. In order to map this property of the nervous system, the "power network" [9] can be used. This is a three layer SIGMA-PI network with feedforward architecture, fixed synaptic weights in the hidden layer, and plastic weights in the output layer. Thereby, the hidden neurons multiply potentiated variables coming from the input layer, and the output neurons then compute a weighted sum of these products. If M=number of output neurons and K=number of input neurons, the power network represents M abbreviated K-dimensional power series, or even Taylor series, known to approximate any function to every required degree of precision. In the present investigation, such a power network was used, established with 6 input neurons, 10 hidden nodes, 9 modified Hebbian synapses for the first output neuron, and 6 such synapses for the second output neuron.

2.3 The auto-imitation algorithm needs no external teacher

When modelling a plant, an autonomous controller has to solve the problem to get an internally produced teaching signal. An appropriate algorithm has been proposed by Kalveram [7], later on also called "auto-imitation" or "self-imitation" [8].

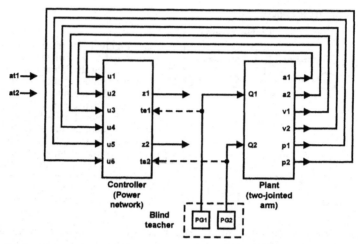

Figure 2: Neural controller learning an inverse model of the plant (two-jointed arm of equation (3)) in the framework of auto-imitation. The plant to be controlled is given by its tool transformation; i.e., the rule governing the behaviour of the plant in the forward direction (from torques to kinematics).

In Fig.2, the two-jointed arm movable in a plane as given in (3) demonstrates how the principle of auto-imitation is applied. In the learning phase, the plant gets arbitrary torques from the "blind teacher". The outputs z1, z2 of the power network are not connected to the inputs Q1, Q2 of the plant. These torques produce kinematic values of angular acceleration (a1, a2), velocity (v1, v2) and position (p1, p2) at joint 1 (shoulder) and joint 2 (elbow) according to (3). The kinematic outputs are proprioceptively measured and fed back into the network's regular inputs u1 to u6. The torque values, also measured proprioceptively, are simultaneously offered also to the teaching inputs te1 and te2. They function as targets to be attained after learning. This setup should enable the modified Hebbian synapses of the output layer to attain the right strengths. After training, the blind teacher's ouputs are removed, the outputs (z1, z2) of the network are linked with the plant's inputs, and the connections between u1, u2 and a1, a2 are cut. The position and velocity variables remain connected with the network, because they are required as state feedback. Now, desired angular accelerations (at1, at2) can be fed into the inputs u1, u2 of the network which computes torque outputs forcing the plant to attain the desired accelerations. It should be noticed, that this auto-imitation is not only a "direct method of inverse modelling" [6], but can, as previously pointed out [7], [11], also immediately be related to the reafference principle [5].

3 Simulation of learning to control a two-jointed arm

The simulation experiment is based on the the two-jointed arm, governed by the coupled differential equations [8]

$$
\begin{aligned}
A\ddot{\varphi}_1 + C\ddot{\varphi}_2 - D\dot{\varphi}_2^2 - 2D\dot{\varphi}_1\dot{\varphi}_2 - E + B_1\dot{\varphi}_1 &= Q_1 = K_1\left(\varphi_{01} - \varphi_1\right) \\
B\ddot{\varphi}_2 + C\ddot{\varphi}_1 + D\dot{\varphi}_1^2 \qquad\quad - F + B_2\dot{\varphi}_2 &= Q_2 = K_2\left(\varphi_{02} - \varphi_2\right)
\end{aligned}
\tag{3}
$$

where A, B, C, D, E, F denote coefficients depending on the masses, lengths, moments of inertia, and angular positions j_i of both the limbs. E and F refer to gravitationally induced torques. B_i resp. K_i (i=1,2) represent the coefficients of viscous damping resp. stiffness caused by muscular activity, which is considered to drive the limbs by displacing the mechanical equilibrium positions φ_{0i} of the joints.

For simulation, the length and mass of both arm segments were set to 0.5 m resp. 1 Kg. Mass was assumed to be symmetrically distributed over the segment lengths yielding moments of inertia of 0.1 Kg·m². The selected values of damping and stiffness were B_i=0.5 N·m·s and K_i= 5 N·m. Initially, all Hebbian synapses were set to zero strength. The training runs endured 300 s. The blind teacher generated two approximately sinusoidal torque patterns, which were sent to the torque inputs of the arm and the teaching inputs of the power net. In condition 'hom', during the whole training the torque frequencies were f_1=0.01 and f_2=0.05. In condition 'spl', the training phase was split: In the first half, frequencies of f_1=0.01 and f_2=0.05 Hz were applied as before, in the second half frequencies of f_1=0.1 and f_2=0.5. The amplitudes in both modes were set to 4 and 2 N·m. All computations where performed using MATLAB and SIMULINK.

4   Results and discussion

Table 1 shows, that very exact identifications of the inverse model could be achieved with split training procedure 'spl'. This uses low velocity torque trajectories to train the network in the first half of a training session, followed by high velocity trajectories in the second half. In the low velocity part, predominantly synaptic connections responsible for compensation of gravitational forces were established, while in the high velocity part synapses necessary for balancing out dynamical interactions between the limb segments were programmed. In the homogeneous condition 'hom', however, model identification was bad. Therefore, unlike to the simultaneous LSQ-rule descripted in [8], [10], the sequential algorithm applied in the present paper didn't produce reasonable synaptic weights in every case.

Table 1: "True" synaptic weights of the synapses S1-S15, and weights finally attained in simulation runs under the training conditions 'hom' and 'spl'.

Syn.	S1	S2	S3	S4	S5	S6	S7	S8
true	.45	.125	.1	-.125	-.25	.125	.5	7.5
hom	.37	.001	-.12	-.16	.23	.362	.38	5.7
spl	.45	.126	.1	-.124	-.25	.126	.50	7.5

Syn.	S9	S10	S11	S12	S13	S14	S15
true	2.5	.1	.125	.1	.125	.5	2.5
hom	4.1	-.6	.475	-.01	.009	.68	1.6
spl	2.5	.1	.125	.10	.125	.5	2.5

The question, whether high precision low level sensorimotor learning is principally possible using a setup including auto-imitation and a power network with synapses from redefined Hebbian type, can be answered by "yes", so far as training is appropriate. Otherwise "life-long" motor deficiencies can occur.

References

[1] Brown, T.H., Chattarji, S. (1995) Hebbian synaptic plasticity. In M. Arbib (Ed) The handbook of brain theory and neural networks. MIT Press (pp. 454-459)

[2] Gossard, J.P. and S. Rossignol, 1990. Phase-dependent modulation of dorsal root potentials evoked by peripheral nerve stimulation during fictive locomotion in the cat. Brain Research 537, 1-13

[3] Hagemann, G., Redecker, C., Neumann, D., Haefelin, T. Freund, H. J. and O. D. Witte, 1998. Increased long-term potentiation in the surround of experimentally induced focal cortical infarction. Ann. Neurol. 44, 255-258

[4] Hebb, D. O., 1949. The organization of behaviour. New York: Plenum Press

[5] Holst, E. von and H. Mittelstaedt, 1950. Das Reafferenzprinzip. Naturwissenschaften 37, 464-476

[6] Jordan, M. I., 1988. Supervised learning and systems with excess degrees of freedom. COINS Technical Report 88-27, 1-41

[7] Kalveram, K.Th., 1981. Erwerb sensumotorischer Koordinationen unter störenden Umwelteinflüssen: Ein Beitrag zum Problem des Erlernens von Werkzeuggebrauch. (Engl. titel: Acquisition of sensorimotor co-ordinations under environmental disturbations. A contribution to the problem of learning to use a tool) In: L. Tent (Ed.): Erkennen, Wollen, Handeln. Festschrift für Heinrich Düker (S. 336-348). Göttingen: Hogrefe

[8] Kalveram, K.Th., 1992. A neural network model rapidly learning gains and gating of reflexes necessary to adapt to an arm's dynamics. Biological Cybernetics 68, 183-191

[9] Kalveram K.Th., 1993a. Power series and neural-net computing. Neurocomp. 5, 165-174

[10] Kalveram K.Th., 1993b. A neural network enabling sensorimotor learning: Application to the control of arm movements and some implications for speech-motor control and stuttering. Psychological Research 55, 299-314

[11] Kalveram K. Th., 1998. Wie das Individuum mit seiner Umwelt interagiert. Psychologische, biologische und kybernetische Betrachtungen über die Funktion von Verhalten (engl. titel: How the individual interacts with its environment. Psychological, biological and cybernetical considerations about the function of behaviour). Lengerich: Pabst

[12] Kalveram, K. Th., 1999. A modified model of the Hebbian synapse and its role in motor learning. Hum. Mov. Science 18, 185-199

[13] MacKay, D.J. and K.D. Miller, 1990. Analysis of Linsker's simulations of Hebbian rules to linear networks. Network 1, 257-297

[14] Pennartz, C.M.A. 1997. Reinforcement learning by Hebbian synapses with adaptive thresholds. Neuroscience 81, 303-319

[15] Shouval, H.Z. and M.P. Perrone, 1995. Post-Hebbian learning rules. In: M. A. Arbib (ed.) The handbook of brain theory and neural networks. Cambridge: The MIT Press (pp. 745-748)

[16] Widrow, B. and S.D. Stearns, 1985. Adaptive signal processing. Englewood Cliffs: Prentice-Hall

[17] Wilson, R.I., J. Yanovsky, A. Gödecke, D.R. Stevens, J. Schrader, and H.L. Haas, 1997. Endothelial nitric oxide synthase and LTP. Nature 386, 338

On Synaptic Plasticity: Modelling Molecular Kinases involved in Transmitter Release

Dan Lundh

Dept. of Computer Science, Bioinformatics Group, University of Skövde
Skövde, Sweden, *email: dan@ida.his.se*

Ajit Narayanan

Dept. of Computer Science, University of Exeter
Exeter, England, *email: ajit@dcs.exeter.ac.uk*

Abstract

The neurotransmitter release depends, in principal, on two neuronal events. One is the influx of Calcium in the active zone due to polarization, and the other is the amount of fusion available neurotransmitter vesicles. We state a model of the phosphorylation of the Synapsins (I and II), which creates a pool of fusion available vesicles. The exocytosis of vesicles is stated by Calcium on and off rates, modelling the complex of the fusion pore. The aim here is to model paired pulse facilitation due to phosphorylation of the Synapsins, and post-tetanic potentiation due to increased kinetic activity of the cAMP complex. In addition we show that the regulation of fusion competent vesicles in post-tetanic potentiation is highly correlated to Calcium residues released by organelles, i.e Calcium buffering.

1   Introduction

The molecular mechanisms behind transmitter release machinery can be divided into four different stages [10]. As an initial stage, it is believed that vesicles in the recruitment pool are attached to Actin by the protein Synapsin, and released by phosphorylation [4, 11]. The Synapsin protein is fairly well studied both by micro-injection [4, 9] and gene knockout experiments [15, 16]. The protein indicated as a transport mechanism from the recruitment pool to the plasma membrane is Myosin II [10, 17], an ATP-dependent motor. The molecular events involved with membrane fusion are not clear, and the suggested roles below should be interpreted as tentative and considered as a working model. The second stage, during or following docking, consist of proteins[1] forming complexes with the preceding dissociation of complexes[2] [10]. In the third and reversible stage SNARE complexes in the docked state are shown to contain NSF and SNAP[3] proteins, which act during ATP-dependent priming to disassemble complexes and promote conformational changes in SNARE proteins [10, 13]. This stage prepares the vesicle for exocytosis. In the last stage, Calcium dependent interaction between Synaptotagamin and Syntaxin and Calcium stimulated interaction between Synaptotagamin and SNAP-25 suggests an initiation of membrane merge and

---

[1] SNARE-proteins; soluble N-ethylmaleimide sensitive factor (NSF) attachment protein receptor.
[2] VAMP/synaptobrevin-synaptophysin.
[3] SNAPS; soluble (NSF) attachment proteins; a cytosolic ATP binding protein.

ultimately fusion-pore formation [10, 13]. In this paper we address two of these stages, the initial stage; recruitment from the vesicle pool and the last stage, the Calcium dependent fusion-pore formation.

2 Model and Methods

The model consist of a set of substances (ratios), and their changes expressed with differential equations (see below). To update substances we use a forward Euler, updating 1000 times per millisecond and plotting each hundred update. A change in a substance ratio reflects changes in other substance updates (differential equations), i.e. changes are propagated to other substance ratios. This results in a more or less continous model with substance reactions linked by a system of differential equations.

2.1 The Synapsins

In dephosphorylated form the binding of Synapsin I and Synapsin II to synaptic vesicles (one equation for each type of molecule) is stated as:

$$\frac{d[synapsin]_{bound}}{dt} = \frac{(B_{max} - [synapsin]_{bound})[synapsin]_{free}}{K_d}$$

where $B_{max} = 800 \ pM/mg$ protein [1], the amount of Synapsin I and Synapsin II is 621 fM and 415 fM (calculated from [1]), respectively, and $K_d = 10 \ nM$ [1]. The amount of vesicle protein, 0.842 μg is based on calculations[4] from [12] and is consistent with observations of dephospho-Synapsin I as 6-7% (Synapsin II 3%) of total vesicle proteins [1]. The amount of Synapsin is consistent with observations of a total of 9% Synapsin [16]. The phosphorylation, which causes the release of the vesicle, is stated as:

$$\frac{[synapsin]_{site \ x}}{dt} = \frac{V_{m_{agent}}[agent]}{K_{m_{agent}} + [agent]} - \frac{[synapsin]_{site \ x}}{K_{dis}}$$

where agent denotes the kinetic agent (i.e cAMP, and Ca/CaM kinase). The values used for site 1 phosphorylation are $K_{m_{cAMP}} = 2.0 \ \mu M$ [1], V_m is approximately 12 pM/min [14], and $K_{dis} = 16 \ nM$ [1]. We here assume that the constants are identical for Calcium/Calmodulin kinase type I [1], with exception of $V_{m_{Ca/CaM \ I}} = 2 \ pM/min$ (from [2], calculated on 10% of total amount of Calmodulin). The phosphorylation of site 2 and 3 of the Synapsin I molecule could be stated in similar way with the difference that we here have a re-association to vesicles of $K_{dis} = 50 \ nM$, a $K_{m_{CaM \ II}} = 0.4 \ \mu M$ and $V_{m_{Ca/CaM \ II}} = 64 \ pM/min$ [1, 7]. The paired pulse facilitation depends on the phosphorylation of site 2 and 3, i.e the CaM II kinase [1]. The rate of exocytosis is an estimation of the formation of the fusion pore. By assuming a Calcium binding of four ions per vesicle we can state the rate as (from R. Heidelberger et. al [3]):

$$\frac{excytosis}{dt} = k_{Ca_a}[Ca] \ free\_pool - k_{Ca_d} \ exocytosis$$

---

[4]The amount of vesicle proteins is calculated on a relation between the isoforms (Ia:Ib and IIa:IIb) as 1:2 with molecular weights of 86000 and 80000 (Synapsin Ia and Ib) and 74000 and 55000 (Synapsin IIa and IIb), respectively [1].

where the maximum secretion rate $k_{Ca_a} = 14 \mu M \ s^{-1}$ and $k_{Ca_d} = 2 \ s^{-1}$ [3].

Due to the duration time of phosphorylation in site 1, it is most likely that a phosphorylation of site 2 and 3 happens on a vesicle whose molecules already are phosphorylated at site 1, i.e. there cannot be any fusion competent vesicles in the pool unless the attached Synapsin I molecules are phosphorylated at site 2 and 3 [1]. The number of secreted vesicles peaks at 43 vesicles (see figure 2.1), resulting in approximately 230 vesicles released per pulse [5, 6]. In the simulations we modelled action potential

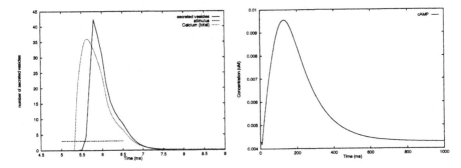

Figure 1: Left, the number of released vesicles from the synaptic terminal, based on 14 Synapsin II and 20 Synapsin I molecules attached to each synaptic vesicle. Right, the cyclic Adenosine monophosphate (cAMP), formed by the balance of its synthetic enzyme Adenylyl cyclase and the degradative enzyme Phosphodiesterase.

by Hodgkin & Huxley equations. The Calcium influx in the active zone is modelled by calculation of the Calcium current [18] which is converted to concentration. The Calcium influx creates three different pools[5] with different dynamics and durations[6].

2.2 The cAMP formation

The cyclic Adenosine monophosphate (cAMP) is formed by the balance of its synthetic enzyme Adenylyl cyclase and its degradative enzyme Phosphodiesterase. We express the Adenylyl cyclase formation according to R. Kötter et. al [8]:

$$+\frac{d[cAMP]}{dt} = \frac{pf\ hf\ V_{m_{AC}}[ATP]\left(1+\frac{b[CaMCa_4]}{EC_{50_{CaMCa_4}}} + \frac{a[Ca]}{IC_{50_{Ca}}} + \frac{ab[CaMCa_4][Ca]}{EC_{50_{CaMCa_4}}IC_{50_{Ca}}}\right)}{K_{m_{AC}} \quad denominator}$$

$$denominator = 1 + \frac{[ATP]}{K_{m_{AC}}} + \frac{[CaMCa_4]}{EC_{50_{CaMCa_4}}} + \frac{[Ca]}{IC_{50_{Ca}}} + \frac{[CaMCa_4][Ca]}{EC_{50_{CaMCa_4}}IC_{50_{Ca}}} +$$

$$\frac{[ATP]}{K_{m_{AC}}} \times \left(\frac{[CaMCa_4]}{EC_{50_{CaMCa_4}}} + \frac{[Ca]}{IC_{50_{Ca}}} + \frac{[CaMCa_4][Ca]}{EC_{50_{CaMCa_4}}IC_{50_{Ca}}}\right)$$

where $pf = 20000 \ mg/l$ (assuming 20 g per liter), $hf = 1.85$ (correction factor for low substrate saturation), $V_{m_{AC}} = 1.3 \times 10^{-6} \ \mu mole \ s^{-1} \ mg^{-1}$, $[ATP] = 900 \ \mu M$,

---

[5]Calcium influx through the active zone, Calcium diffusion through the terminal, and Calcium buffering (and release) by organelles.

[6]Created by having different on and off rates to each pool.

$K_{m_{AC}} = 320\,\mu M$, $EC_{50_{CaMCa_4}} = 1.32 \times 10^{-9}\mu M$, $IC_{50_{Ca}} = 0.9\mu M$, $a = 0.046$, $b = 3.55$, $c = 2.23$, $ab = 0.0034$, $ac = 0.13$, and $bc = 5.32$ [8]. The activity of the degradative enzyme (Phosphodiesterase) is stated in a similar way [8]:

$$-\frac{d[cAMP]}{dt} = \frac{\frac{pf\,V_{m_{PDE}}[cAMP]}{K_{m_{PDE}}} + \frac{pf\,bV_{m_{PDE}}[cAMP][CaMCa_4]}{aK_{m_{PDE}}bEC_{50_{CaMCa_4}}}}{denominator}$$

$$denominator = 1 + \frac{[cAMP]}{K_{m_{PDE}}} + \frac{[CaMCa_4]}{bEC_{50_{CaMCa_4}}} + \frac{[cAMP][CaMCa_4]}{aK_{m_{PDE}}bEC_{50_{CaMCa_4}}}$$

where $V_{m_{PDE}} = 0.028\,\mu mole\,s^{-1}mg^{-1}$, and $K_{m_{PDE}} = 68\mu M$ [8]. The other values are $bV_{m_{PDE}} = 0.168\,\mu mole\,s^{-1}\,mg^{-1}$, $aK_{m_{PDE}} = 12\mu M$, and $bEC_{50_{CaMCa_4}} = 0.011\mu M$ [8].

The careful balance of these enzymes gives the cAMP a lifespan of several hundreds of milliseconds (by a single stimulus pulse (depolarization)), see Figure 2.1.

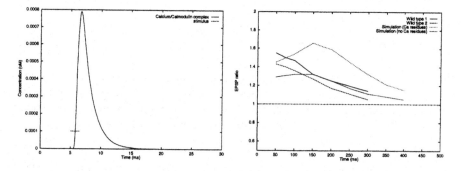

Figure 2: Left, The Calcium/Calmodulin complex, binding step four. Right, simulated paired pulse facilitation in comparison to wild type.

2.3   The Calcium/Calmodulin complex

The four-step binding of Calcium to Calmodulin is stated by [8]:

$$\frac{d[CaM]}{dt} = k_{-1}[CaMCa] - k_{1a}[CaM][Ca]$$

$$\frac{d[CaMCa]}{dt} = k_{1a}[Ca][CaM] + k_{-1}[CaMCa_2] - $$
$$k_{-1}[CaMCa] - k_{1b}[Ca][CaMCa]$$

$$\frac{d[CaMCa_2]}{dt} = k_{1b}[Ca][CaMCa] + k_{-2}[CaMCa_3] - $$
$$k_{-1}[CaMCa_2] - k_{-2}[Ca][CaMCa_2]$$

$$\frac{d[CaMCa_3]}{dt} = k_2[Ca][CaMCa_2] + k_{-2}[CaMCa_4] - $$
$$k_{-2}[CaMCa_3] - k_2[Ca][CaMCa_3]$$

$$\frac{d[CaMCa_4]}{dt} = k_2[Ca][CaMCa_3] - k_{-2}[CaMCa_4]$$

where the values of the constants are $k_{1a} = 2.67 \ /s/\mu M$, $k_{1b} = 7.41 \ /s/\mu M$, $k_{-1} = 20 \ /s$, $k_2 = 20 \ /s/\mu M$, and $k_{-2} = 600 \ /s$. Initial CaM was set to $10 \ \mu M$. The end product, $[CaMCa_4]$, takes part in the formation as well as degradation of cAMP.

3 Results

Our aim was to model molecular kinases underlying paired pulse facilitation, and post-tetanic potentiation. Paired pulse facilitation (PPF) occurs when a stimulus is followed by a second stimulus with a short intermediate time. The short intermediate time results in an increased response at the second stimulus if the intermediate time between stimuli pulses is small (some hundred milliseconds), see Figure 2.2. With some discrepancy to short intermediate time between pulses, our model coinsides with PPF data. The reason for this is believed to be the accuracy in modelling Calcium residues and transmitter diffusion. A tetanic stimulation results in intracellular build-

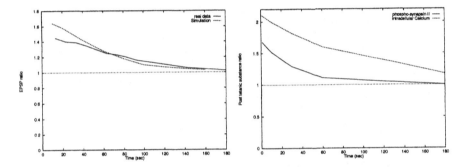

Figure 3: Left, Post-tetanic potentiation as a result of 1 second 100 Hz stimulus, compared to wild type (real data). Right, substance ratio for phospho-Synapsin II, and Calcium residues, baseline amounts are 0.0022 aM, and 0.024 μM, respectively.

up and buffering of Calcium by organelles. This buffering and release is believed to underly post-tetanic potentiation[7], see Figure 3. Here we see a match between the model and real data. The post-transactional effect, e.g. after $100 \ Hz$ 1 *second* stimuli, is highly correlated with the Calcium residues, see Figure 3. This correlation gives us reason to believe that the underlying dynamics of Synapsin site I phosphorylation, e.g. Synapsin II phosphorylation, play a crucial role in post-tetanic potentiation. Small amounts of intracellular Calcium do not give any lasting change in the lifespan of the Calcium/Calmodulin complex. The behavior (post-tetanic potentiation) is almost due to increased activity of cAMP, i.e increased phosphorylation of site 1.

---

[7]The residues must be present for some time, i.e. it is not sufficient with Calcium residues caused by one or a few depolarizations.

282

References

[1] P.D. Camilli, F. Benfenati, F. Valtorta, and P. Greengard. The Synapsins. *Annual Review of Cell and Developmental Biology*, 6:433–460, 1990.

[2] D. Chin, K.E. Winkler, and A.R. Means. Characterization of substrate phosphorylation and use of calmodulin mutants to address implications from the enzyme crystal structure of calmodulin-dependent protein kinase I. *Journal of Biological Chemistry*, 272(50):31235–31240, 1997.

[3] R. Heidelberger, C. Heinemann, E. Neher, and G. Matthews. Calcium dependence of the rate of exocytosis in a synaptic terminal. *Nature*, 371:513–515, 1994.

[4] H.C. Hemmings, A.C. Nairn, T.L. McGuinnes, R.L. Huganir, and P. Greengard. Role of protein phosphorylation in neuronal signaling transduction. *FASEB*, 3:1583–1592, 1989.

[5] B. Katz. Neural Transmitter Release: from quantal secretion to exocytosis and beyond. *Journal of Neurocytology*, 25:677–686, 1996.

[6] F. Kawasaki and H. Kita. Two Excitatory Motorneurons Differ in Quantal Content of Their Junctional Potentials in Abdominal Muscle Fibres of the Cricket, Gryllus bimaculatus. *Journal of Insect Physiology*, 43(2):167–177, 1997.

[7] M.B. Kennedy, T. McGuinness, and P. Greengard. A Calcium/Calmodulin-dependent protein kinase from mammalian brain that phosphorylates synapsin I: partial purification and characterization. *Journal of Neuroscience*, 3(4):818–831, 1983.

[8] R. Kötter, D. Schirok, and K. Zilles. Concerted regulation of cyclic AMP by calcium, calmodulin, and dopamine: A kinetic modelling approach. In R. Paton, editor, *Information Processing in Cells and Tissues (IPCAT)*, pages 199–210. Liverpool: Univ. of Liverpool, 1995.

[9] R. Llinas, T.L. McGuinnes, C.S. Leonard, M. Sugimori, and P. Greengard. Intraterminal injection of synapsin I or calcium/calmodulin-dependent protein kinase II alters neurotransmitter release at the squid giant synapse. *Proceedings of the National Academy of Sciences of USA*, 82:3035–3039, 1985.

[10] T.F.J. Martin. Stages of regulated exocytosis. *Trends in Cell Biology*, 7:271–276, 1997.

[11] A.C. Nairn, H.C. Hemmings, and P. Greengard. Protein Kinasis in the Brain. *Annual Review of Biochemistry*, 54:931–976, 1985.

[12] E.J. Nestler and P. Greengard. *Protein Phophorylation in the Nervous System*. New York: Wiley, 1984.

[13] B.J. Nichols and H.R.B. Pelham. SNAREs and Membrane Fusion in the Golgi Apparatus. *Biochimica et Biophysica Acta*, 1404:9–31, 1998.

[14] R.B. Pearson, S. Forrest, M. Davis, T.J. Martin, and B.E. Kemp. Comparison of substrate specificity of myosin kinase and cyclic AMP-dependent protein kinase. *Biochim Biophys Acta*, 786(3):261–266, 1984.

[15] T.W. Rosahl, M. Geppert, D. Spillane, J. Herz, R.E. Hammer, R.C. Malenka, and T. Südhof. Short-Term Synaptic Plasticity Is Altered in Mice Lacking Synapsin I. *Cell*, 75:661–670, 1993.

[16] T.W. Rosahl, D. Spillane, M. Missler, J. Herz D.K. Selig, J.R. Wolff, R.E. Hammer, R.C. Malenka, and T. Südhof. Essential functions of synapsins I and II in synaptic vesicle regulation. *Nature*, 375:488–493, 1995.

[17] J.L. Stow and Kirsten Heinmann. Vesicle budding on Golgi membranes: regulation by G proteins and myosin motors. *Biochimica et Biophysica Acta*, 1404:161–171, 1998.

[18] W.M. Yamada, C. Koch, and P.R. Adams. Multiple Channels and Calcium Dynamics. In C. Koch and I. Segev, editors, *Methods in Neuronal Modeling*, pages 97–133. Cambridge, Massachusetts: MIT Press, 1989.

Self-Organizing Networks for Mapping and Clustering Biological Macromolecules Images

Alberto Pascual[1], Montserrat Barcéna[1], J.J. Merelo[2], José-María Carazo[1]

[1]Centro Nacional de Biotecnología. Universidad Autónoma, 28049. Madrid. Spain.

E-mail: pascual@cnb.uam.es

[2]GeNeura Team, Dpto Arquitectura y Tecnología de Computadores, Facultad de Ciencias, Campus Fuentenueva, s/n, 18071 Granada, Spain;

E-mail: geneura@kal-el.ugr.es

Abstract

In this work we study the effectiveness of the Fuzzy Kohonen Clustering Network (FKCN) in the unsupervised classification of electron microscopic images of biological macromolecules. The algorithm combines Kohonen's Self-Organizing Feature Maps (SOM) and Fuzzy c-means clustering technique (FCM) in order to obtain a clustering technique that inherits their best properties. Two different data sets obtained from the *G40P* helicase from B. Subtilis bacteriophage SPP1 have been used for testing the proposed method, one composed of 2458 rotational power spectra of individual images and the other composed by 338 images from the same macromolecule. Results of FKCN are compared with Self-Organizing Maps (SOM) and manual classification. Experimental results have proved that this new technique is suitable for working with large, high dimensional and noisy data sets.

1 Introduction

Image classification is a very important step in the three-dimensional study of biological macromolecules using Electron Microscopy (EM) because three-dimensional reconstruction methods require that the set of images that is going to be considered correspond to different views of the same biological specimen. Obtaining homogeneous set of images corresponding to views from the same direction of the same biological specimen is a hard task due to several factors, mainly the low signal/noise ratio of the images obtained in the electron microscope and the potential structural heterogeneity of the biological specimen under study.

In the context of Pattern Recognition and Classification in Electron Microscopy, different approaches have been previously used: classical statistical methods and clustering techniques (see [1] for a review). Kohonen's self-organizing feature maps have also been successfully applied in classification of Electron Microscopic images [2], creating a new and powerful methodology for classification in this field. One of the advantages of this approach is that it does not need any prior knowledge of the number of clusters present in the data set. In the present work we describe the application in Electron Microscopy image classification of a hybrid approach using the integration of Kohonen´s Self-organizing Feature Maps (SOM) [3] and Fuzzy c-means clustering algorithm (FCM) [4]: the Fuzzy Kohonen Clustering Network (FKCN). This method takes advantage of the self-organizing structure of SOM and the Fuzzy-clustering model of FCM. It was first proposed in [5] and has also been successfully applied in the field of image processing [6].

2 Materials and methods

2.1 Experimental data sets

We have used a set of 2458 images of negatively stained hexamers of the SPP1 G40P helicase obtained in the electron microscopy [7]. The images have a very low signal/noise ratio, making impossible visual classification. The images were translationally and rotationally aligned and their rotational power spectra were calculated [8]. For experimental purposes, we created two data sets: one composed of 2458 rotational power spectra (up to 15 harmonics) and the other composed of 338 50x50 pixels images that were extracted from a visually homogeneous 6-fold symmetric population.

2.2 Fuzzy kohonen clustering network (FKCN)

Fuzzy Kohonen clustering network [11] is a type of Neural Network that combines two well-known and widely used algorithms: Kohonen's Self-Organizing Maps (SOM) [3] and Fuzzy c-means (FCM) [4]. The structure of this self-organization feature-mapping model consists of two layers: input and output. The input layer is composed of p nodes, where p is the number of features and the output layer is formed by c nodes, where c is the number of clusters to be found. Every single input node is fully connected to all output nodes with an adjustable weight v_i (cluster center) assigned to each connection. Given an input vector, the neurons in the output layer update their weights based on a pre-defined learning rule α. This approach integrates the fuzzy membership u_{ik} from the FCM in the following update rule:

$$v_{i,t} = v_{i,t-1} + \alpha_{ik,t}(x_k - v_{i,t-1})$$

Where the learning rate α is defined as:

$$\alpha_{ik,t} = (u_{ik,t})^m t; \qquad m_t = (m_0 - t\Delta m) \qquad \text{and} \qquad \Delta m = (m_0 - 1)/t_{max}$$

And $u_{ik,t}$ is the Fuzzy membership matrix calculated by Fuzzy c-means, m_0 is any positive constant greater than one, t is the current iteration and t_{max} is the iteration limit.

This method possesses several interesting properties:

- The learning rate is a function of the iteration t and its effect is to distribute the contribution of each input vector x_k to the next update of the neuron weights inversely proportional to their distance from x_k. The winner node updates its weight favored by the learning rate as the iteration increases and in this way the Kohonen concept of neighborhood size and neighborhood updating are embedded in this new learning rate.

- FKCN is a truly Kohonen-type algorithm, because it possesses a well defined method for adjusting both the learning rate distribution and update neighborhood as function of time. Hence, FKCN inherits the "self-organizing" structure of SOM-types algorithms and at the same time, is stepwise optimal with respect to a widely used fuzzy clustering model.

- FKCN is not sequential; different sequence of the data do not alter the result of the algorithm.

- For a fixed $m_t >1$ FKCN is a truly Fuzzy c-means algorithm and in the limit ($m_t=1$), the update rule reverts to Hard c-means (winners take all).

3   Results

3.1   Experiment using rotational power spectra

In this example, 2458 images were used for analysis. The rotational power spectrum [8] of each particle was calculated yielding a 2458 15-dimensional data set. A 7x7 SOM was applied to the data set and the resulting map was manually clustered in four classes as described in [7]. The results are shown in Figure 1. Group A shows a predominant 6-fold symmetry with a small but noticeable component on harmonic 3. Group B represents 3-fold symmetry images. Group C is closely related to 2-fold symmetry particles and Group D showed only a prominent harmonic in 1, which can be interpreted as a lack of symmetry in this group of particles.

FKCN was applied to the whole set of particle's spectra with the following configuration: 15 input nodes (each node representing a component in the spectrum) and 4 nodes in the output layer (representing four clusters). For comparison purposes with results already obtained using this data set [7], 4 clusters were used. Fuzzy constant m was set to 1.5 and 500 iterations was used The resulting code vectors (cluster centers) are shown in Figure 2.

As can be seen in the cluster centers, the four groups visualized by the SOM were also successfully extracted by FKCN. Quantitative results of coincidence (with respect to the SOM groups) are shown in Table 1, however, a major difference in the sets extracted by both algorithms (SOM and FKCN) should be noticed. When SOM output was manually clustered, a set of code vectors that were bounding the groups were no considered in order to avoid a erroneous classification in the borders. FKCN considered the whole data, so an unavoidable difference will be reflected in the results. Cluster 2 obtained by FKCN seems to be composed by these group of spectra associated to the code vectors that were eliminated from the SOM for being part of the borders between the four hypothetical clusters. Furthermore a large set of noisy spectra as well as the non-symmetry ones are also included in this cluster.

286

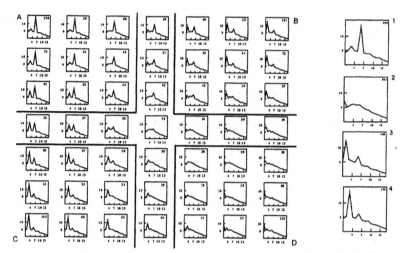

Figure 1: 7x7 SOM output manually clustered in four regions. Group A: particles with a prominent 6-fold component and a small but noticeable 3-fold component. Group B: 3-fold symmetry. Group C: 2-fold symmetry and Group D: lacks of a predominant symmetry.

Figure 2: Four cluster centers obtained from FKCN.

3.2 Experiment using 3-fold and 6-fold symmetry images

In this experiment 338 images were used for testing the algorithms in the presence of high dimensional and very noisy data, a situation very common in Electron Microscopy Image Analysis. These images share similar rotational symmetry (6-fold with a minor 3-fold component) See [7] for details.

For comparison purposes, a 7x7 SOM was applied to the data set and the resulting map was manually clustered in two different classes that apparently exhibited an opposite handedness [7]. Figure 3 shows the map clustered in two classes that seem to reflect essentially the same type of macromolecular structure. In the first experiment we clustered the data into 2 groups, however, results (not shown here) showed that the method could not find two clusters related only to variations in handedness of the particle as obtained by SOM. Only a small but noticeable difference in the rotational spectra was detected. We then clustered the data into 3 groups using $m=1.5$ and 500 iterations. Cluster centers obtained are shown in figure 4. Classification accuracy is also shown in Table 2. Analyzing the results of the clustering algorithm it is clear that FKCN correctly clustered Group A in SOM (Cluster 3), however, Group B need further analysis. The images from the three classes were independently realigned to obtain their average. Figure 5 shows the average image and rotational spectra of subsets belonging to each cluster. In the case of cluster 3, a counterclockwise handedness is clearly observed as expected from Group A of SOM. In the case of clusters 1 and 2, the average images show the same handedness (clockwise) as expected from Group B of SOM, however, it is also clear the differences in rotational symmetry between both clusters. Both of them have a predominant 6-fold symmetry, but Cluster 2, as oppose to Cluster 1, is influenced by a noticeable 3-fold component. This very subtle difference was not detected in SOM. This symmetry variation was maybe the cause of misclassification when using two clusters: symmetry was influencing more than handedness.

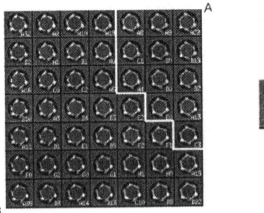

Figure 3: 7x7 SOM output manually clustered in two regions with different handedness.

Figure 4: Three cluster centers obtained from FKCN.

Table 1: Comparison between FKCN and SOM using spectra. ($m = 1.5$, $t = 500$)		
SOM	FKCN	Coincidence
Group A	Cluster 1	92.87%
Group D	Cluster 2	90.97%
Group B	Cluster 3	85.86%
Group C	Cluster 4	96.24%

Table 2: Comparison between FKCN and SOM using images. ($m = 1.5$, $t = 500$)		
FKCN	SOM	Coincidence
Cluster 1	Group B	93.23%
Cluster 2	Group B	96.15%
Cluster 3	Group A	89.3%

4 Discussion and conclusion

In this paper, a new fuzzy classification technique have has been applied to the study of biological specimens by Electron Microscopy. This technique uses a special type of Neural Network named Fuzzy Kohonen Clustering Network (FKCN) that successfully combines the well-known Self-Organizing feature maps (SOM) and the Fuzzy c-means clustering technique (FCM). The need of classification tools suitable for working with long sets of noisy images is evident in electron microscopy field. Here we have proposed a new approach which is in the middle somewhere between two methods already applied in this context: SOM and FCM.

The proposed method combines the ideas of fuzzy membership values for learning rates, the parallelism of FCM and the structure of the update rules of SOM, producing a robust clustering technique with a self-organizing structure. FKCN have been fully tested in this work using two kinds of data sets that are very common in any Electro Microscopy study: the rotational power spectra and the images of individual particles from a protein (In this case the G40P helicase was used). In both

288

cases FKCN was able to discriminate not only evident but subtle variations in the data set. The results demonstrate the suitability of this method for working with these kind of high dimensional and noisy data sets. Comparing this clustering approach with others previously proposed in the field for structure-based classification, we should emphasize that this method directly performs classification (assignment of data to cluster), while at the same time offers a direct visualization by inspecting cluster centers directly. A number of future research topics remain open, especially the automatic determination of number of clusters. We think that this type of neural computation approaches (self-organizing networks) can be successfully employed for exploration of data in the Electron Microscopy field.

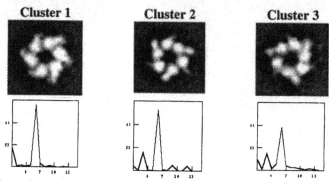

Figure 5: Average Images and rotational spectra of clusters.

References

[1] Bonnet N.: Multivariate statistical methods for the analysis of microscope images series: applications in materials science. J. Microsc. 190 (1998) 2-18.

[2] Marabini, R., Carazo, J.M.: Pattern Recognition and Classification of Images of Biological Macromolecules using Artificial Neural Networks. Biophysical Journal 66 (1994) 1804-1814.

[3] Kohonen, T.: Self-Organizing Maps, 2nd Edition, Springer-Verlag (1997).

[4] Bezdek, J. C.: Pattern Recognition with Fuzzy Objective Function Algorithms. Plenum Press, New York. (1984).

[5] Chen.Kuo Tsao, E., Bezdek, J. C., Pal, N. R.: Fuzzy Kohonen Clustering Networks. Pattern Recognition 27 (1994) 757-764.

[6] Jin-Shin Chou, Chin-Tu Chen, Wei-Chung Lin: Segmentation of Dual-echo MR Images using Neural Networks. Image Processing 1998 (1993) 220-227.

[7] Barcena, M., San Martín, C., Weise, F., Ayora, S. Alonso, J.C., Carazo, J.M.: Polymorphic quaternary organization of the Bacillus subtilis bacteriophage SPP1 replicative helicase (G40P). Journal of Molecular Biology (1988) (in press).

[8] Crowther, R.A., Amos, L.A.: Harmonic analysis of electron microscope images with rotational symmetry. J. Mol. Biol. 60 (1971) 123-130.

Neural Network Model for Muscle Force Control Based on the Size Principle and Recurrent Inhibition of Renshaw Cells

Takanori Uchiyama

Fac. Science and Technology, Keio University

Yokohama, Japan

uchiyama@appi.keio.ac.jp

Kenzo Akazawa

Fac. Engineering, Kobe University

Kobe, Japan

akazawa@in.kobe-u.ac.jp

Abstract

A neural network model for muscle force control was constructed. The model contained a single motor-cortex output cell, the actual number of α motoneurons found in human muscles, Renshaw cells and muscle units. The size of the motor units (motoneurons and muscle units) was distributed as the human brachialis muscle, the extensor digitorum muscle and the first dorsal interosseous muscle. The relationship between the model's muscle force and the firing rate of α motoneurons was investigated. The relationship depended on the absolute refractory time of α motoneurons, RIPSP by Renshaw cells and the firing pattern of Renshaw cells. When these parameters were selected appropriately, the model showed a relationship similar to that observed in isometric contraction of human skeletal muscles. The size distribution of the motor units had a dominant effects on the relationship.

1 Introduction

We proposed a neural network model for the force control of skeletal muscles. The model consisted of a single motor-cortex output cell, α motoneurons (MNs), Renshaw cells and muscle units. The model was based on the size principle and had recurrent inhibition by Renshaw cells. The purpose of this study was to construct a neural network model in which the relationship between the firing rate of α MNs and force was similar to that observed in the isometric contraction of human muscles.

2 Method

2.1 Model

Figure 1 shows the schematic illustration of the model. The model has a structure similar to that reported previously [1], but motor units are not classified into segments

(Fig. 1). α MNs are simply numbered according to their size (from zero to the maximum size). Renshaw cells (RC) are also numbered. The input is discharge impulse train X of motor-cortex output cell, and the output is muscle force F. We made the following assumptions and simplifications: A single cortico motor neuron (CMN) innervates all α MNs, and the Renshaw cell RC_M receives exitatory impulses from R α MNs, and the Renshaw cell RC_M sends inhibitory impulses back to the same R α MNs. Afferent signals are ignored.

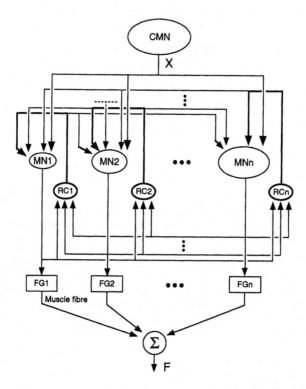

Figure 1: Neural network model for control of muscle force. X is discharge impulses of CMN; F is force of muscle. MN_j, RC_j and FG_j represent the j th α MN, Renshaw cells and muscle fibres, respectively.

The motoneuron MN_j generates a spike when the membrane potential $P_j(t)$ reaches the threshold $H_j(t)$. The potential $P_j(t)$ and the threshold $H_j(t)$ are based on Segundo's model [2]. The potential $P_j(t)$ is expressed by

$$
\begin{aligned}
P_j(t) = {} & P_\infty + (P_0 - P_\infty)e^{-(t-t_0)/\tau_{pj}} \\
& + EP_j \sum_{i=1} e^{-(t-t_i)/\tau_{ej}} - K_s \sum_{i=1} e^{-(t-t_{rji})/\tau_{ej}} \\
& - K_j \sum_{M=1, M \neq j}^{n} \sum_{i=1} e^{-(t-t_{rMi})/\tau_{ej}},
\end{aligned} \tag{1}
$$

where $P_0 = -72\text{mV}$, $P_\infty = -65\text{mV}$. EP_j is peak EPSP of MN_j generated by a single CMN impulse. K_s is peak RIPSP of MN_j generated by a single RC_j impulse ($K_s = 0.5\text{mV}$). K_j is peak RIPSP of MN_j generated by a single RC_M impulse ($M \neq j, K_j \ll K_s$). t_i, t_{rji} and t_{rMi} are i th clock time for receiving single discharge impulses of CMN, RC_j and RC_M, respectively; τ_{pj} and τ_{ej} are time constants. These parameters are similar to those of the previous report [1] which were determined empirically by referring to the electrophysiological data of motoneurons [3].

The threshold potential $H_j(t)$ after a spike firing is expressed as Eq. (2).

$$H_j(t) = H_\infty + (H_0 - H_\infty)e^{-(t-t_0)/\tau_{hj}}, \tag{2}$$

where $H_0 = -50\text{mV}$, $H_\infty = -55\text{mV}$ and τ_{hj} is the time constant. The Renshaw cell RC_M generates a burst of impulses when it receives a single MN_j impulse.

When α MNs fire, impulses are sent to muscle fibres. Force is calculated as Eq. (3).

$$F(t) = \sum_{j=1}^{N} S_j \sum_{i=0}^{P_j} g(t - t_{ji}), \tag{3}$$

where S_j is the size of α MN and $g(t)$ is the twitch tension of each muscle.

2.2 Parameters

We constructed the models of three muscles: the brachialis muscle (BRA), the extensor digitorum muscle (ED) and the first dorsal interosseous muscle (FDI). The size of α MNs are relative values and are distributed as shown in Table 1.

Table 1: Size and number of MN_j.

size S_j	0.5–1.5	1.5–2.5	2.5–3.5	3.5–4.5	4.5–5.5
BRA	21	22	22	16	9
ED	11	11	11	11	11
FDI	27	32	15	6	6
size S_j	5.5–6.5	6.5–7.5	7.5–8.5	8.5–9.5	9.5–10.5
BRA	5	3	1	1	1
ED	11	10	9	8	6
FDI	3	3	3	3	3

The parameters EP_j [mV], τ_{pj} [ms], τ_{ej} [ms] and τ_{hj} [ms] were calculated by interpolating the parameters given in the previous report [1] with following equations:

$$EP_j = \frac{6.4 - 10.0}{10.0 - 1.0}(S_j - 1.0) + 10.0 \tag{4}$$

$$\tau_{pj} = \frac{12.8 - 20.0}{10.0 - 1.0}(S_j - 1.0) + 20.0 \tag{5}$$

$$\tau_{ej} = \frac{4.3 - 7.1}{10.0 - 1.0}(S_j - 1.0) + 7.1 \tag{6}$$

$$\tau_{hj} = \frac{22.8 - 30.0}{10.0 - 1.0}(S_j - 1.0) + 30.0 \tag{7}$$

K_j [mV] was expressed in Eq. (8).

$$K_j = \frac{K_{j1} - K_{j2}}{10.0 - 1.0}(S_j - 1.0) + K_{j2}, \tag{8}$$

where K_{j1} (5μV) and K_{j2} (28μV) were the minimum and the maximum values of RIPSP, respectively. These values were determined empirically because researchers have reported quit different values of RIPSP [4–7].

The absolute refractory time of α MN is calculated by Eq. (9).

$$t_0 = \frac{S_j - 1.0}{10.0 - 1.0}(1.5 - 5.5) + 5.5 \tag{9}$$

The Renshaw cell sends three inhibitory impulses back to α MNs at 2, 3 and 5 ms after receiving an exitatory impulse from α MN.

All of these parameters were employed for three muscles.

3 Results

Figure 2 shows the force-firing rate relationships of three muscles. Left panels (a), (c) and (e) are the relationships obtained by our models. Right panels (b), (d) and (f) are the relationships shown in previous reports [8–10].

The left top panel (a) shows the force-firing rate relationship of the brachialis muscle model. The closed circles, open circles and squares denote the firing rate of the smallest motoneuron, intermediate motoneuron and largest motoneuron, respectively. The firing rate of the small- and medium-sized α MNs increased rapidly after the α MNs began to fire. Then they increased gradually. Subsequently, the firing rate of the medium-sized α MNs increased rapidly once again.

The left middle panel (c) shows the force-firing rate relationship of the model of the extensor digitorum muscle. The firing rate of α MNs increased, first rapidly and then gradually, as muscle force increased. These findings agree with those for human muscles.

The left bottom panel (e) is the force-firing rate relationship of the model of the first dorsal interosseous muscle. The model provided an almost linear relationship between the firing rate of α MNs and force.

For all muscles, only small α MNs fired when the force was small. Large α MNs began to fire as the force increased. This was consistent with the size principle. The force-firing rate relationships of our models were close to those observed in human skeletal muscles.

The differences of the parameters among our models of these three muscles were the distribution of α motoneurons' size. Other structures and parameters have no difference. As a result, the size distribution of motor units has a dominant effect on the force-firing rate relationships.

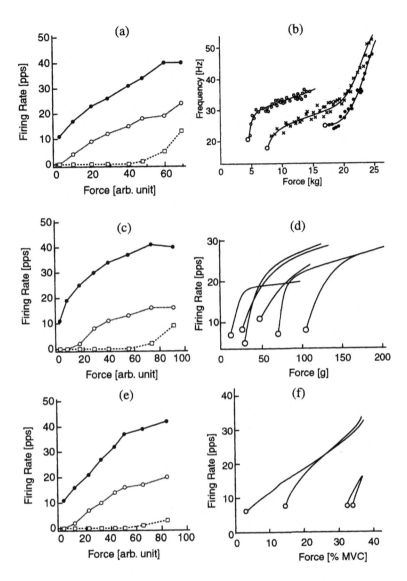

Figure 2: Force-firing rate relationship between the models and human muscles: (a), (c), (e) are the relationships of the models. Closed circles, open circles and squares denote the firing rate of the smallest, the intermediate and the largest motoneuron. (b), (d), (f) are relationships given in previous reports. (a) brachialis muscle, (b) brachialis muscle [8], (c) extensor digitorum muscle, (d) extensor digitorum muscle [9], (e) first dorsal interosseous muscle, (f) first dorsal interosseous muscle [10].

4 Conclusion

In conclusion, our study indicates that the size distribution of motor units has a dominant effect on the relationship between firing rate of α MN and force. The details of the relationships were inconsistent with those observed in human muscles. In constructing the model, we need to take into account the afferent input from muscle spindle and tendon as well as the nonlinear inhibition by Renshaw cells. Further study is needed.

References

[1] Akazawa K and Kato K. Neural network model for control of muscle force based on the size principle. Proc. IEEE 1990; 78:1531–1535

[2] Segundo JP, Perkel DG, Wyman H, Hegstad H and Moore GP. Input-output relations in computer-simulated nerve cells, Infuluence of the statistical properties, strength, number and inter-dependence of exitatory pre-synaptic terminals. Kybernetik 1968; 4:157–171

[3] Burke RE: motor units: anatomy, physiology, and functional organization. In: Handbook of Physiology, vol 2, Motor Control, American Physiology Society, Maryland,1981, pp 345–422

[4] Friedman WA, Sypert GW, Munson JB and Fleshman JW. Recurrent inhibition in type-identified motoneurons, J Neurophysiol 1981; 46:1349–1359

[5] McCurdy ML and Hamm TM. Topography of recurrent inhibitory postsynaptic potentials between individual motoneurons in the cat, J Neurophysiol 1994; 72:214–226

[6] van Keulen L. Autogenetic recurrent inhibition of individual spinal motoneurones of th cat, Neurosci Lett 1981; 21:297–300

[7] Hultborn H, Katz R and Mackel R. Distribution of recurrent inhibition within a motor nucleus. II. Amount of recurrent inhibition in motoneurones to fast and slow units, Acta Physiol Scand 1988; 134:363–374

[8] Kanosue K, Yoshida M, Akazawa K and Fujii K. The number of active motor units and their firing rates in voluntary contraction of human brachialis muscle. Jpn J Physiol 1979; 29:427–443

[9] Monster AW and Chan H. Corticospinal neurons with a special role in precision grip. J Neurophysiol 1977; 40:1432–1443

[10] DeLuca CJ, LeFever MP, McCue MP and Xenakis AP. Control scheme governing concurrently active human motor units during voluntary contractions. J Physiol 1982; 329:113–128

Prediction of Photosensitizers Activity in Photodynamic Therapy Using Artificial Neural Networks: A 3D–QSAR Study

Rozália Vanyúr

Institute of Chemistry, Chemical Research Center, Hungarian Academy of Sciences
1525 Budapest, P.O. Box 17, Hungary

Károly Héberger

Institute of Chemistry, Chemical Research Center, Hungarian Academy of Sciences
1525 Budapest, P.O. Box 17, Hungary

István Kövesdi

EGIS Pharmaceutical Company Ltd.
1475 Budapest P.O. Box 100, Hungary

Judit Jakus

Institute of Chemistry, Chemical Research Center, Hungarian Academy of Sciences
1525 Budapest, P.O. Box 17, Hungary

Abstract

Biological activity in photodynamic therapy was predicted from the molecular structure of pyropheophorbide derivatives using artificial neural networks (ANN). First, the structure of molecules was optimized and various descriptors were calculated. ANN architecture was optimized while suitable descriptors were selected applying a novel variable selection method.

The reliability of models was tested by cross–validation and randomization of biological activity data. Models are able to predict biological activity from the molecular structure of the phorbide derivatives with a leave-one-out crossvalidation Q^2 of 0.956. The size of the substituents is decisive in the third direction (perpendicular to the main plain of the molecules).

1 Introduction

Photodynamic therapy (PDT) is one of the most promising techniques in cancer therapy. It uses a combination of photosensitizing agent and visible light for therapy of many kinds of solid tumors. The light activated sensitizer produces active oxygen species (e.g. singlet oxygen) which results in tumor destruction.

A widely used photosensitizer in clinical trials is Photofrin, which is a complex mixture of several porphyrins not characterized fully. Photofrin is not an ideal material because it is unstable during storage and/or after administration. Moreover, it has a relatively low optical extinction coefficient in the 600-900 nm range, where light exhibits optimal penetration through tissue, and its excretion from tissues is quite slow. Therefore, there is a need for development of new photosensitizers. Testing a sensitizer in biological systems requires a lot of time and live animals, therefore the selection of promising molecules should be done preferably on the basis of their structures.

2 Theory

The structure of pyropheophorbide derivatives is given below.

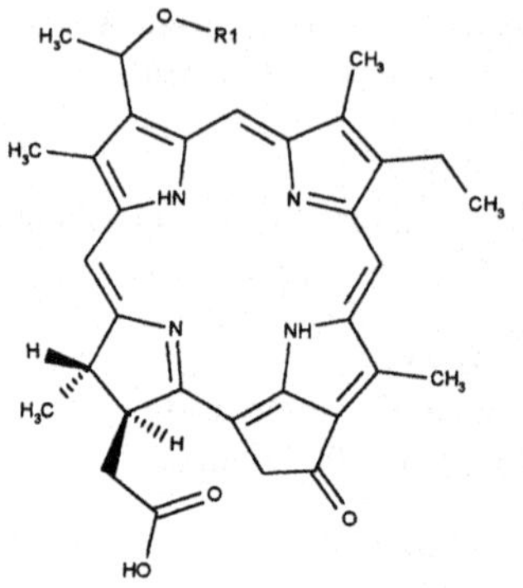

M1	R1=methyl
M2	R1=1-propyl
M3	R1=1-pentyl
M4	R1=1-hexyl
M5	R1=cis-3-hexenyl
M6	R1=tr-3-hexenyl
M7	R1=cyclohexyl
M8	R1=2-hexyl
M9	R1=1-heptyl
M10	R1=1-octyl
M11	R1=1-nonyl
M12	R1=1-decyl
M13	R1=1-dodecyl

Figure 1

The geometry optimization was done by Hyperchem program [1] using MM+ molecular mechanics and conformational search for the global minimum conformer. During the conformational search all dihedral angles around single bonds of the substituents were varied. A minimum of 256 iterations was completed for each compound with 0.1 kcal/angstrom convergence criteria for the gradient. We used global minimum conformers to represent the 3D structures of the molecules in this paper. The 2D, 3D molecular structures and the biological activity data were stored in MDL IsisBase format [2].

2.1 Methods applied

Because of the strong non-linearity between descriptors and activity data to be predicted artificial neural network was chosen to build suitable models. ANN is superior to other methods as principal component regression, partial least squares, and locally weighted regression [3]. For the calculation of 3D QSAR descriptors and for the neural network computations we used 3DNET program [4] that is designed for fast and effective 3D quantitative structure-activity relationship calculations.

The program calculates many 2D, 3D QSAR descriptors from MDL SDF format molecular data sets. We systematically checked 96 descriptors in this work.

In Table 1 we listed the best descriptors of our models.

Table 1

Name of descriptors	Reference
Todeschini descriptors : 3rd mass moment, 3rd pos. moment, 3rd EN moment	[5]
Wiener index	[6]
Calculated lipophilicity	[7]

The program contains a fully connected, three-layer, feed-forward computional neural network with back-propagation training.

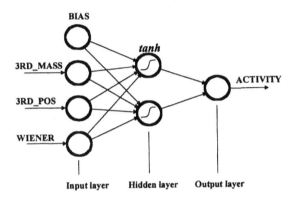

Figure 2

The back-propagation algorithm uses a gradient method with continuously decreasing learning rate during the learning epochs. In this neural network the input and output layers are linear and the hidden layer has neurons with tangent hyperbolicus (tanh) transfer function. The leave-one-out cross-validation procedure is automatic.

The 3DNET program applies a quick and effective algorithm that calculates the relative significance of the input descriptors [8]. It consists of adding a surplus input layer and pushing the input values towards zero in a stepwise manner. During this the back-propagation algorithm tries to decrease the growing error of the calculated outcomes by restoring those inputs via increasing their weights, that are relevant for the calculation of the outcomes.

The extra network weights determined for each input were sorted, and the largest one is taken as having 100% of relative importance on a linear scale. Descriptors with less then 5% relative importance were left out. Then the calculation was done again without these descriptors. The remaining descriptors were selected according to their importance one by one until six of them remained. The last 6 descriptors were tested in every combination.

The promising models were selected by leave-one-out cross-validation procedure. The predictive ability of the best model was checked by leave-n-out cross-validation.

2.2 Predicted activity and computer code used

In this study the 3DNET computer program [4] was applied to build a 3D-QSAR model for a congeneric series of pyropheophorbide photosensitizers.

Experimental data on tumor growth delay have been reported by Barbara W. Henderson *et al.* [9]. The biological activity that they studied was the logarithm of the median time required for the tumor re-growth to a 400 mm^3 after PDT treatment on radiation-induced fibrosarcoma tumor. They found that experimental drug lipophilicity (log$P$) was highly predictive for PDT activity.

3 Results and discussion

In Table 2 we listed the best three models resulting from a systematic descriptor selection for networks containing 1 to 3 hidden neurons. For comparison, the table contains the two "one-descriptor" models using calculated and experimental log$P$, respectively. Here, unlike in the article [9] where two outliers were not included in the fit, we included all data into our calculations.

Table 2

No. of hidden neurons	Descriptors in the model	Leave-one-out Q^2
2	Calculated log$P$ Wiener index	0.932
2	3rd mass moment 3rd pos. moment Wiener index	**0.956**
2	3rd EN moment 3rd mass moment Wiener index	0.928
5	Calculated log$P$	0.316
5	Experimental log$P$	0.256

The best one-descriptor model uses calculated lipophilicity as a descriptor variable. However, the prediction quality of this model, Q^2 = 0.316 is far from satisfactory. In agreement with the results in [9] the experimental lipophilicity descriptor itself yields a practically equivalent model to the previous one: Q^2 = 0.256. However, the best network architecture performs much better using three descriptors ($Q^2 = 0.956$). The ratio of the number of molecules and the network parameters is 1.3 for this model.

The best predictive model by multiple linear regression (MLR) resulted Q^2 of 0.843 using 10 descriptors. Using the same method (MLR) for three descriptors of the best ANN model the Q^2 has been calculated to be equal to 0.631.

The best ANN model was also tested by leave-n-out cross-validation. During this cross-validation 10 random selected molecules were used for the model building. Then, the activities of the three remaining molecules were predicted applying this model. The random molecule selection was done four times and resulted in an average Q^2 of 0.943 while the worst Q^2 was 0.827.

Shuffling of the activity values three times yields the average leave-one-out cross validated Q^2 as bad as –4.5. This shows a definite role of the selected descriptors in describing the studied biological activity.

Among the descriptors the Wiener index provides essentially the same structural information for the ANN as the experimental log$P$. The 3rd moments of mass and atomic position characterizes the smallest diameter of the molecules.

Since pyropheophorbide accumulation in cells increases monotonically with log$P$ while PDT activity shows a parabolic dependence from the same [9], we suggest that there must be a physical impediment to PDT activity at high lipophilicity values. This barrier, however, cannot be the transport through the cell membrane, because pyropheophorbide bioavailability in cells is linear with log$P$. It is worth noting that the pyropheophorbide molecules in this study differ only in the alkyl ether side chain R_1. So our results do not shed light onto the main molecular features responsible for their biological activity. In fact, we only investigated how the effect of varying the side chain can be explained and modeled in QSAR terms.

4 Conclusions

From a robust pool of QSAR descriptors the neural network itself can select the information-rich ones even for a limited, although structurally closely related, number of molecules. The model calculated solely from the 3D molecular structure is able to predict the antitumor activity of some phorbide derivatives. The important descriptors describe how the substituents modify the molecular size in the direction perpendicular to the main plain of the molecules.

References

[1] HyperChem version 4.0, HyperCube Inc., Canada, 1997.

[2] IsisBase1.2 for desktop, Molecular Design Limited, 1996.

[3] Héberger K., Borosy A. Comparison of Chemometric Methods for Prediction of Rate Constants and Activation Energies of Radical addition reactions. J. Chemometrics 1999; 13:473-489.

[4] 3DNET version 1.0, CompElit Ltd., Hungary, 1998.

[5] Todeschini R., Gramatica P. New 3D Molecular Descriptors: The WHIM theory and QSAR Applications Perspectives in Drug Discovery and Design, 1998; 9/10/11:355-380.

[6] Kier LB. Sharpe Indices of Other of One and Three from Molecule Graf. Quant. Struct.-Act. Relat, 1986; 5:1-7.

[7] Broto P., Moreau G., Vandycke C., Molecular-structures – Perception, auto-correlation descriptor and SAR studies – System of atomic contributions for the calculation of the normal-octanol water partition-coefficients. Eur. J. Med. Chem. 1984; 19:71-78.

[8] Masters T., Practical neural network recipes in C++. Academic Press, Inc., 1996.

[9] Henderson BW., Bellnier DA., Greco WR., Sharma A., Pandey RK., Vanghan LA., Weishaupt KR. and Dougherty TJ. An in vivo quantitative structure-activity relationship for a congeneric series of pyropheophorbide derivatives as photosensitizers for photodynamic therapy. Cancer Res. 1997; 57:4000-4007.

Learning Methods

and

Hybrid Algorithms

Case-Based Explanation for Artificial Neural Nets

Rich Caruana

Department of Radiological Sciences, University of California, Los Angeles
Center for Automated Learning and Discovery, Carnegie Mellon University
caruana@cs.cmu.edu

Abstract

Sometimes the most accurate models are the least intelligible. We show how to generate case-based explanations for non–case-based machine learning methods such as artificial neural nets. The method uses the neural net as a distance metric to determine which cases in the training set are most similar to a test case that needs to be explained.

1 Introduction

Artificial neural nets (ANNs) often yield better accuracy than other learning methods. Unfortunately, their complexity makes them difficult to understand. This poses a serious obstacle to using ANNs for medical applications. An advantage of case-based methods is that they can explain their reasoning by presenting to users the cases in the training set that are most similar to the test case being explained [1]. We present a method that allows case-based explanations to be provided for ANNs. The method uses the ANN as a distance metric for kNN. The k cases in the training set that the ANN considers most similar to the test case are returned as the explanation.

Section 2 reviews k-nearest neighbor. Section 3 reviews explanations for kNN. Section 4 shows how to use a neural net as a distance metric for kNN. Section 5 discusses how to evaluate case-based explanations. Section 6 presents an empirical evaluation of case-based explanation of a neural net trained to predict pneumonia risk. Section 7 is a discussion of issues that arise in case-based explanation systems.

2 Case-Based Learning

Consider a training set consisting of N training cases, $X^1...X^N$. Each case X^i is a vector consisting of M input features X^i_1-X^i_M, and a label Y^i. The input features are measurements for each patient such as age, gender, blood pressure, and white blood cell count. The label Y^i is the outcome we want to predict from those inputs. kNN uses the training set to predict an outcome Y^{test} for a test case from the M input features for that test case, X^{test}_1-X^{test}_M.

Nearest neighbor (1NN) finds the case in the training set whose input features are most similar to the input features of the test case. It returns as a prediction for the test case the outcome of that most similar training case. Similarity in the input features is measured using a distance such as unweighted Euclidean distance:

$$\text{Euclidean Distance (A,B)} = \text{Sqrt} \left(\sum_1^M (X_i^A - X_i^B)^2 \right)$$

More complex distance metrics such as weighted Euclidean distance, which weights each input dimension with a weight W_i, often yield better performance, though learning what weights to use for each input dimension can be difficult [2]. A extension of 1NN that usually performs better is k-nearest neighbor (kNN). kNN finds the k training cases closest to the test case. The prediction kNN returns is the majority label of the k nearest neighbors (or an average if the labels are continuous).

3 Case-Based Explanation

Case-based methods use the training set and a distance metric to make predictions. Non-case-based methods such as artificial neural nets train a separate model (the ANN) on the training set, and use only that model to make predictions for test cases. ANNs ignore the training set once the model has been trained.

To make a prediction, kNN must determine which cases in the training set are most similar to the test case. A natural way to explain kNN's predictions is to present to users the k cases in the training set that are most similar to the test case. This explanation accomplishes two things. It explains the distance metric by showing the user what cases the method considers most similar to the test case. And it explains the data by presenting the user with cases similar to the test case.

Because medical training and practice emphasizes case evaluation, practitioners are adept at understanding explanations provided as cases. Case-based explanations allow users to apply their own expertise to judge how similar the explanation cases are to the test case, and how the empirical evidence in the training set affects the decision for this test case. Case-based explanations also help users learn from the data.

4 Using a Neural Net as a Distance Metric

A typical ANN contains thousands of real-valued parameters that interact in a complex, non-linear way. Understanding ANNs is difficult, and is the subject of much research [3]. Consider the ANN in Figure 1. It is trained to predict pneumonia mortality from 65 demographic, history and physical, and hospital measurements [4]. The ANN has 65 inputs, 64 hidden units, and one output (a total of 4,289 weights), and is trained with backprop. The complexity of the ANN makes it difficult to understand both the ANN and the predictions the ANN makes.

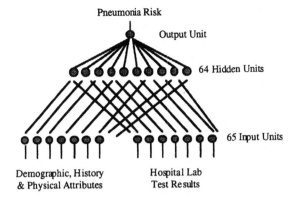

Figure 1. Artificial Neural Net to Predict Pneumonia Mortality

If ANNs are unintelligible, how can we safely employ them in critical applications such as health care? One approach is to focus not on understanding the ANNs themselves, but on understanding the predictions the ANNs make.

The hidden units in an ANN are internal features learned by the net. All of the information from the inputs that the ANN uses to make a prediction at its output passes through the hidden units (assuming no jump connections). The pattern of activation on the hidden layer indicates how the ANN models each case. We can compare the patterns of activation the ANN makes for each of the training cases with the pattern of activation for the test case that needs to be explained. The cases in the training set with patterns of hidden unit activation similar to the test case are the cases in the training set that the ANN models similarly to the test case.

Similarity in hidden unit activation can be measured with Euclidean distance on the hidden unit activations of the two cases:

$$\text{NeuralNetDistance } (A,B) = \text{Sqrt } (\sum_{1}^{H} (HU_i^A - HU_i^B)^2)$$

where A and B are two cases, H is the number of hidden units, and HU_i^A is the activation of hidden unit i for case A. Small distance between the hidden unit activations of two cases suggests that the two cases are modeled similarly by the ANN. By keeping the training set after the ANN is trained, and presenting to users the k training cases with smallest NeuralNetDistance to the test case, we can explain the ANN's reasoning the same way we provide explanations for case-based methods.

One might consider using kNN on input features to find explanations, instead of kNN on hidden unit activations. That is, give users explanations from kNN over inputs, even though predictions are made by an ANN. But this does not help the user understand the *ANN's* predictions. The ANN might provide little or no weight to some inputs, give large weight to other inputs, or combine some inputs in a complex, non-linear way. KNN on *input features* fails to capture how the *ANN* uses the features. To explain the ANN, we must use the ANN in the distance metric.

When generating case-based explanations for ANNs, weights can be used to insure that the hidden units that are most important to the net's prediction receive more weight in the distance metric. These can be estimated from the output layer of the ANN. A large positive or negative connection from a hidden unit to the output unit suggests that the hidden unit is important. A connection near zero indicates that the hidden unit is not important. Taking into account the average activation of the hidden units also helps. For simplicity, we use unweighted Euclidean distance on hidden unit activation for the experiments reported here.

5 Evaluating Case-Based Exaplanation Systems

How do we know if the cases the method above selects as an explanation are appropriate? This section proposes one necessary (but insufficient) measure for judging the effectiveness of any case-based explanation system.

Consider the predictions made by an ANN for a test case, and the predictions it makes for each of the cases in the explanation. If the explanation cases are similar to the test case *from the ANN's point-of-view*, the ANN should make similar predictions for the test case and the explanatory cases. If the ANN makes different predictions for the test case and some of the explanatory cases, then those explanatory cases are not similar to the test case from the ANN's point-of-view.

Let CBE(test) be the set of cases the case-based explanation system returns for the test case. Let nn_pred(test) be the prediction the ANN makes for the test, and nn_pred(train$_i$) be the prediction the ANN makes for training case i. We define the Explanation_Error for the case-based explanation system on the test case as:

$$\text{Explanation\_Error(test)} = \sqrt{\sum_{i \in \text{CBE(test)}} (\text{nn\_pred(test)} - \text{nn\_pred(i)})^2 \Big/ \sum_{i \in \text{CBE(test)}} 1}$$

It measures the difference between the predictions the ANN makes for the test case and for the explanation cases. If the explanation error is small, the ANN makes similar predictions for the test and explanation cases. This suggests the explanation cases are similar to the test case from the ANN's point-of-view. If the explanation error is large, then the explanation cases are not similar to the test case according to the ANN, and thus probably are not a good explanation.

6 Empirical Results: Pneumonia Risk Prediction

In this section we use the metric proposed above to evaluate the performance of case-based explanation applied to an ANN trained to predict pneumonia risk. The ANN in Figure 1 is trained to predict pneumonia risk (0=lives, 1=dies) from the input features available before a patient is admitted to the hospital. We are not trying to determine if the patient has pneumonia; pneumonia has already been diagnosed from a chest x-ray. The goal is to determine which patients are at low/high risk from pneumonia to decide if they should be treated as out-patients/in-patients. The ANN is trained on 7500 pneumonia patients. The remaining 6699 cases are used as a test set.

The ANN predicts pneumonia risk well. It outperforms all other machine learning and statistical methods applied to this problem [4]. Unfortunately, it is so complex that we hesitate to use it. Instead, a more intelligible, though less accurate, logistic regression model was chosen for clinical evaluation. Our hope is that methods such as case-based explanation will allow us to use more accurate ANN models clinically.

We compare the explanation_error of kNN on hidden unit activations, with the explanation_error of explanations generated two other ways: using kNN on input features, and (as a baseline) explanation cases selected randomly.

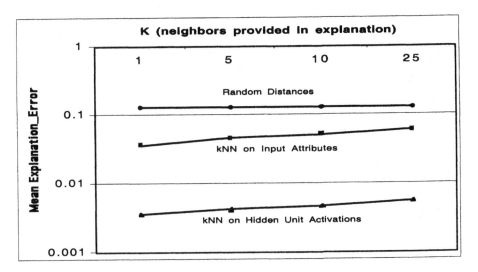

Figure 2. Mean Explanation_Error vs. Number of Cases in Explanation

Figure 2 shows the mean explanation error on the test set (6699 cases) as the number of cases in the explanations, varies from 1 to 25. The vertical axis is a log scale; small differences in height represent substantial differences in performance. Explanations generated by kNN on input features outperforms randomly generated explanations by a factor of 3–5. Explanations generated by kNN on hidden unit activations outperforms kNN on input features by a factor of 10.

7 Discussion

One might consider using prediction_error, instead of NeuralNetDistance, to select explanation cases. This is not a good approach because it does not consider *how* the ANN makes predictions. For example, the ANN probably models an 18 year old male with pneumonia acquired while hiking differently from an 85 year old diabetic woman with ischemic heart disease, though both patients could have similar predicted mortality. kNN on hidden unit activations takes into account *how* the ANN models different cases that have similar predicted risk.

One issue is how many cases to include in explanations. Different users may wish to view different numbers of cases, and the number of cases they want to view may depend on the test case being explained. We provide the user a list of explanatory cases sorted by hidden unit activation distance so the user can examine as many cases as they want. For each case we provide a score to help the user know how close that case is to the test case. We also determine what value of k would yield maximum predictive accuracy if we were to use kNN on hidden unit activations to make predictions for cases instead of the ANN. We "gray out" cases beyond this vaue of k to further suggest to the user which cases are less similar to the test case.

Low explanation error suggests that the explanation system finds explanatory cases that are similar to the test case from the ANN's point-of-view. But low explanation error does not indicate how useful these explanations are to users. For example, users might also want to see cases that are different from the test case in order to contrast the cases. We are planning an evaluation of user satisfaction and acceptance for case-based explanations to see what kinds of cases users consider most useful.

We are using case-based explanation with other opaque learning methods such as boosted/bagged decision trees, mixtures of experts, and Bayes nets. Our hope is that case-based explanation will allow us to use whatever learning method performs best on each medical application, even though the best models may not be intelligible.

References

[1] Kahn CE Jr. Artificial intelligence in radiology: decision support systems. RadioGraphics 1994; 14:849–861.

[2] Atkeson CG, Schaal SA, Moore AW. Locally weighted learning. AI Review 1997; 11:11-73.

[3] Craven M, Shavlik, J. Using sampling queries to extract rules from trained neural nets. 11th International Conference on Machine Learning 1994; 37–45.

[4] Cooper G, Aliferis C, Ambrosino CF, Aronis J, Buchanan, B, Caruana R, Fine M, Glymour C, Gordon G, Hanusa BH, Janosky J, Meek C, Mitchell T, Richardson T, Spirtes P. An evaluation of machine learning methods for predicting pneumonia mortality. AI in Medicine 1997; 9: 107–138.

Double Growing Neural Gas
for Disease Diagnosis

Guojian Cheng, Andreas Zell

University Tübingen

Wilhelm-Schickard-Institut for Computer Science

Dept. Computer Architecture

Köstlinstr. 6, D-72074 Tübingen, Germany

{cheng,zell}@informatik.uni-tuebingen.de

Abstract

We present here a variant of B. Fritzke's GNG (Growing Neural Gas) — Double GNG (DGNG). In each insertion step of the GNG only one new cell is inserted in the middle of the edge connecting Maximum Resource Vertex (MRV) and a MRV in its direct topological neighbourhood. But in our DGNG two new cells are inserted at the same time. Our goal is to speed up the convergence of the learning process. Although multiple growing cell mechanism can reduce the required number of learning epochs, it can also lead to an increased network structure. Simulation results on some neural network benchmarks indicate that, for many data sets, when the number of new cells in each insertion step is within three, the total performance of networks is at best, but that increasing the number of new cells beyond three has very little benefit and sometimes degrades performance. In this paper we consider only the DGNG. With two disease diagnosis benchmarks (Wisconsin breast cancer and soybean disease diagnosis) we tested the DGNG and indicated that DGNG performs better than the original GNG, measured by the required number of epochs and CPU times.

1 Introduction

Kohonen's Self-Organizing Maps (SOM) [4] generate mappings from high dimensional signal spaces to lower dimensional topological structures. Their main features are formation of topology-preserving feature maps and approximation of the input probability distribution. However, SOMs have some disadvantages: their fixed number of neural units makes them impractical for applications where the optimal number of units is not known in advance. Also, a topology of fixed dimensionality results in problems if this dimensionality does not match the dimensionality of the feature manifold. Incremental artificial neural networks grow as they learn and shrink as they forget. Based on SOM and Competitive Hebbian Learning (CHL) [5], B. Fritzke introduced an incremental self-organizing network with variable topology, known as Growing Neural Gas (GNG) [1]. The GNG requires no determination of the network dimensionality in advance. Instead the algorithm starts with two units at random positions and inserts new nodes at positions with high errors or resource values (the accumulated errors play a nutrition-like role for the cell growing). These nodes are connected by edges with a certain age, old edges are removed, which results in a topology preserving map with an induced Delaunay triangulation. All parameters used by GNG are constant. It is, therefore, not necessary to define a decay schedule as for the SOM.

Through a combination of GNG with radial basis functions (RBF) the Supervised GNG (SGNG) was developed [2]. The SGNG has three main advantages over other traditional supervised neural networks: First, number, diameter and position of RBF units are determined automatically through the growth process, which can be stopped as soon as the network performance is good enough. Second, since positioning of RBF units and supervised training of connection weights is performed in parallel, the current classification error (used as resource value) can be used to determine where to insert new RBF units. Third, the networks can generalize well and have a relatively small size. A performance comparisons of GNG with other neural network models have been made by M. Kunze [7] and F. Hamker [3].

2 Double Growing Neural Gas

2.1 The Mechanism of DGNG

The cell growing mechanism of DGNG is nearly the same as in GNG. The only difference is that in DGNG two new cells are inserted simultaneously instead of only one new cell as in GNG. In other words, in each insertion step of GNG only one new cell is inserted in the middle of the edge connecting the Maximum Resource Vertex (MRV) and it's maximum source neighbour node. But in DGNG two new cells are inserted at the same time. The first new cell is inserted in the same way as in GNG. In addition, another new cell is inserted between the edge connecting the second MRV and it's maximum source neighbour node. When the second MRV is in the direct topological neighbourhood of the first MRV, we insert only one new cell between first MRV and second MRV. It is important that the new cell should be inserted in such way that the expected value of a certain error measure (resource value) in the following process becomes equal for all neurons.

2.2 The Algorithm of Supervised DGNG

The algorithm diagram of supervised DGNG is showed on the next page (Table 1).

3 Disease Diagnosis Simulation Results

We applied our supervised DGNG algorithm to two neural network disease diagnosis benchmarks, Wisconsin breast cancer study and soybean disease recognition problem, which are adopted from the PROBEN1[6]. For the application of the validation method the datasets were divided into three sets: training set, validation set and test set. For controlling the training process the early stopping method was used. The criterion can be expressed as: stop training as soon as the error on the validation set is higher than it was the last time it was checked.

Table 1: Supervised DGNG Algorithm

Start with a random edge of 2 connected cells at age zero.
Initialize the learning parameters: λ, ε_{win}, ε_{dtn}, η, α, β, θ.
Create m linear output units and a weighted connection $\mathbf{w}_{jc}$ from each competitive cell c to each output unit j, $(j \in \{1, 2, \ldots, m\})$.
Associate every competitive cell with a Gaussian function.
while (classification error is not zero or maximum epoch is not reached):

    λ-Times Adaptation:

Load I/O-pattern pair (ξ, ζ) from training data.
Determine the first winner C_{win_1} and the second C_{win_2}.
Are C_{win_1} and C_{win_2} connected by an edge ? Yes: reset the corresponding edge-age to zero; No: connect C_{win_1} with C_{win_2}.
Increment the age of all edges emanating from C_{win_1}.
Remove edges at a certain age θ as well as resulting non-connected cells.
Increase matching for C_{win_1} and its direct topological neighbour cells by learning rates ε_{win} and ε_{dtn} resp. ($\varepsilon_{win} \gg \varepsilon_{dtn}$).
Compute activation d_c for every competitive cell c according to the Gaussian function.
All output unit activations are calculated by the product of $\mathbf{w}_{jc}$ and d_c.
Perform one δ-rule step for the output weights by learning rate η.
Increase resource variable R_{win1}.

Determine the cell C_{max1} and C_{max2} with the first and second maximum resource resp.
Determine the cell C'_{max2} with the maximum resource in the direct topological neighbourhood of C_{max1}.
Determine the cell C_{mrn1} and C_{mrn2} with the maximum resource of direct topological neighbourhood of C_{max1} and C_{max2} resp.
$C'_{max2} = C_{max2}$? Yes: insert a new cell C_{new1} between C_{max1} and C_{max2}. No: insert the first new cell C_{new1} between C_{max1} and C_{mrn1}; then insert the second new cell C_{new2} between C_{max2} and C_{mrn2}.
Decrease the resource values of $C_{max1,2}$ and $C_{mrn1,2}$ by a fraction α.
Interpolate the resource variable and output weight vector of the new cells C_{new1} and C_{new2}.
Decrease the resource variable of all units by a factor β.

3.1 Wisconsin Breast Cancer Study

The breast cancer diagnosis is derived from the descriptions gathered by microscopic examination. These descriptions include the clump thickness, the uniformity of cell size and cell shape, the amount of marginal adhesion, and the frequency of bare nuclei. The dataset consists of 9 inputs and 699 examples. The training set, validation set and test set contained 350, 175 and 174 patterns, respectively. For this simulation, the two classes (benign or malignant tumor) were coded on one outputs neuron. The accumulated squared output error served as resource updating: $\Delta R_{win} = ||\zeta - o||^2, \zeta, o \in \{0, 1\}$.

Table 2: Comparison DGNG with SGNG for the Wisconsin cancer classification problem.

Database	Algorithm	Epochs	CPU Time (Seconds)	Training Set CR	Testing Set CR
Cancer1	SGNG	154	42.06	100%	96.55%
Cancer1	DGNG	96	28.02	100%	97.70%
Cancer2	SGNG	156	40.98	100%	95.98%
Cancer2	DGNG	93	27.64	100%	96.55%
Cancer3	SGNG	146	36.82	100%	98.28%
Cancer3	DGNG	80	21.77	100%	97.13%

Table 2 shows the difference between SGNG and DGNG for this problem. CR represents the Classification Rate. The IBM-RS/6000 computer was used for testing of CPU time. As it can be seen the number of epochs and CPU time required by the DGNG is between 45% and 33% smaller than for original SGCS, while the generalization performance for SGCS and DGNG is almost same.

3.2 Soybean Disease Recognition Problem

This benchmark problem concerns the recognition of 19 different diseases of soybeans. The discrimination is done based on a description of the bean (e.g. whether its size and color are normal) and the plant (e. g. the size of spots on the leafs, whether these spots have a halo, whether plant growth is normal, whether roots are rotted) plus information about the history of the plant's life (e. g. whether changes in crop occurred in the last year or last two years, whether seeds were treated, how the environment temperature is).

The soybean benchmark dataset consists of 82 inputs, 19 outputs and 683 samples. This is the problem with the highest number of classes in the PROBEN1 benchmark set. Most attributes have a significant number of missing values. For this simulation, the 19 classes were coded on 19 outputs neurons, each class being associated with a neuron. The accumulated squared output error served as resource updating: $\Delta R_{win} = ||\zeta - o||^2, \zeta, o \in \{0, 1\}$. The soybean dataset with 342 examples used for the training set, 171 examples used for the validation set. 170 examples used for the testing set.

Table 3 shows the difference between SGNG and DGNG for this problem. As it can be seen the number of epochs and CPU time required by the DGNG is between

Table 3: Comparison of DGNG and SGNG for the soybean disease recognition problem.

Database	Algorithm	Epochs	CPU Time (Seconds)	Training Set CR	Testing Set CR
Soybean1	SGNG	125	132.55	94.44%	89.41%
Soybean1	DGNG	70	65.95	94.44%	89.41%
Soybean2	SGNG	105	107.75	90.94%	82.35%
Soybean2	DGNG	70	74.40	91.81%	84.12%
Soybean3	SGNG	120	129.50	93.27%	90.00%
Soybean3	DGNG	80	99.29	92.40%	87.64%

50% and 23% smaller than for original SGCS. For soybean1, to achieve same classification rate on testing set SGNG takes 125 epochs and 132.55 CPU seconds, but DGNG takes only 70 epochs and 65.95 seconds. For soybean2 the generalization performance of DGNG is better than SGNG. The classification rate is increased from 82.35% (SGNG) to 84.12% (DGNG). For soybean3 the generalization performance of DGNG is not so good as of SGNG, but DGNG needs only 80 epochs and 99.29 seconds CPU time.

4 Conclusions

In this paper we have developed a novel variant of Fritzke's GNG network. The main advantage of GNG is that it can automatically find a network structure and size suitable for a given classification pattern through cell growing. The DGNG algorithm can speed up the convergence of the learning process. Comparative experimental results on two benchmark datasets revealed that DGNG is both effective and efficient as a learning algorithm for diesease diagnosis problems. The required epochs in DGNG are about half to two third of the epochs required by GNG. The number of cells required for DGNG and GNG are nearly equal.

Acknowledgments

We would like to thank B. Fritzke, M. Kunze, F. Hamker, J. Bruske and L. Prechelt for valuable and constructive discussions about their growing neural network algorithm and neural network benchmark problems. We also wish to thank our colleagues Igor Fischer, Jürgen Wakunda, Fred Rapp and Jutta Huhse for their help in programming and proof reading. The first author also thanks German KAAD (Katholischer Akademischer Ausländer-Dienst) very much for the scholarship.

314

References

[1] B. Fritzke. A growing neural gas network learns topologies. In G. Tesauro, D. S. Touretzky, and T. K. Leen, editors, *Advances in Neural Information Processing Systems 7*, pages 625–632. MIT Press, Cambridge MA, 1995.

[2] B. Fritzke. Automatic construction of radial basis function networks with the growing neural gas model and its relevance for fuzzy logic. In *Applied Computing 1996: Proceedings of the 1996 ACM Symposium on applied computing*, pages 624–627, Philadelphia, 1996. ACM.

[3] F. Hamker and D. Heinke. Implementation and comparison of growing neural gas, growing cell structures and fuzzy artmap. In *Technical report, Schriftenreihe des FG Neuroinformatik der TU Ilmenau*. Report 1/97, ISSN 0945-7518, 1997.

[4] Teuvo Kohonen. *Self-Organizing Maps*. Springer, Berlin, Heidelberg, 1995. (Second Extended Edition 1997).

[5] Thomas Martinetz. Competitive hebbian learning rule forms perfectly topology preserving maps. In Stan Gielen and Bert Kappen, editors, *Proc. ICANN'93, Int. Conf. on Artificial Neural Networks*, pages 427–434, London, UK, 1993. Springer.

[6] Lutz Prechelt. PROBEN1 — A set of benchmarks and benchmarking rules for neural network training algorithms. Technical report, Fakultät für Informatik, Universität Karlsruhe, Germany.

[7] M. Kunze R. Berlich and J. Steffens. A comparison between the performance of feed forward neural networks and the supervised growing neural gas algorithm. In *In 5th Artificial Intelligence in High Energy Physics workshop, Lausanne*, 1996.

The Use of a Knowledge Discovery Method for the Development of a Multi-layer Perceptron Network that Classifies Low Back Pain Patients

M.L.Vaughn, S.J.Cavill,
Cranfield University (RMCS)
Shrivenham, Swindon UK
e-mail: M.L.Vaughn@rmcs.cranfield.ac.uk

S.J.Taylor, M.A.Foy, A.J.B.Fogg
Princess Margaret Hospital
Swindon, UK

Abstract

Using a new method published by the first author this paper discovers the ranked class profiles of key inputs used by a multi-layer perceptron (MLP) network that classifies low back pain patients into three diagnostic classes. It is shown how the validation of the class profiles leads to the discovery of 4 mis-diagnosed training cases and 2 further cases which were not relevant. By interpreting the test cases mis-classified by the MLP and comparing them with the validated class profiles a number of test cases were also found to have been mis-diagnosed by the clinicians. It is shown how the class profiles were used to develop a more optimal network with approximately half the number of inputs and only a marginally reduced test performance.

1 Introduction

In domains such as medical diagnosis it is especially important for clinicians to understand and to have confidence in a system's prediction. Most approaches for extracting knowledge from trained neural networks have used a search based approach [1]. However, a limitation of these methods is that the computational complexity of the search is exponential in the number of input features.

Using a new method published by the first author [2-4], this paper shows how to discover the ranked class profiles of key inputs used by a MLP network that classifies low back pain patients into 3 diagnostic classes. It is shown how the class profiles are used in the development of a more optimal low back pain network.

2 The initial low back pain MLP network

For this study, low back pain is classified into three diagnostic classes: simple (or mechanical) low back pain (SLBP); root pain (ROOTP) and abnormal illness behaviour (AIB) which is mechanical low back pain with signs and symptoms magnified as a sign of distress in response to chronic pain.

A data set of 198 actual cases was collected from patient questionnaires, physical and clinical findings. The distribution of cases by class was 80 SLBP, 91 ROOTP and 27 AIB. An initial MLP network was developed with 92 binary input neurons corresponding to 39 patient attributes. Using sigmoidal activation and generalised delta learning rule the network was trained and tested with 99 randomly selected patient cases. The best generalised network was a 92-10-3 architecture with a training performance of 96% and a test performance of 65%.

3 The interpretation and knowledge discovery method

The knowledge discovery method defines the *significant* inputs as the key inputs in an MLP input case which *positively* activate the classifying output neuron. The method first discovers the hidden layer *feature detector* neurons which contribute a positive input to the classifying neuron. For sigmoidal activations, the feature detectors are connected to the classifying neuron with *positive* weights. The first author has shown [3] that the MLP finds sufficiently many feature detector neurons with activation >0.5 which positively activate the classifying output neuron.

3.1 Discovery of the significant inputs

The significant inputs positively activate the feature detectors and, for binary encoded inputs, are *positive* inputs connected to the feature detectors with *positive* weights. The significant inputs can be ranked in order of decrease in activation at the classifying neuron when the input is switched *off* at the MLP input layer.

3.2 Discovery of the negated significant inputs

The method defines the *negated* significant inputs in the MLP input case as the inputs which de-activate the hidden layer neurons which are *not* feature detectors. These are the *zero*-valued inputs connected to not feature detectors with positive weights. The negated inputs can be ranked in order of decrease in activation at the classifying neuron when the input is switched *on* at the MLP input layer.

3.3 Discovery of the ranked class profiles for the initial low back pain MLP network

The ranked significant inputs were discovered for the 99 training cases in the low back pain MLP training set. With the aim of aggregating this knowledge in a meaningful way the *average* of the ranking values of the significant inputs was made by class, resulting in a ranked significant input profile for each diagnostic class, as shown in Table 1. Similarly, the negated significant input class profiles are shown in Table 2. It should be noted that the profiles are exponentially decreasing with respect to the ranked key inputs and only the top ten inputs are shown.

4 Validation of the class profiles

The validation of the ranked class profiles was achieved by first examining the profiles for anomalies and then by undertaking a clinical evaluation of the class profiles. Following the validation process the class profiles were then used to investigate the mis-classified test cases. This is explained as follows.

4.1 Inspection of the ranked class profiles

In the inspection of the class profiles for anomalies it was expected that the same attribute would not rank highly in both profiles for the same class. In comparing the SLBP class profiles, shown in Tables 1 and 2, four attributes (in bold), *worse(BP)*, *slrr(≥70)*, *LP(no)* and *lumbarex(<5)*, were highly ranked in both profiles. The anomaly was traced to 2 cases in the SLBP negated profile which were re-evaluated by the domain experts. Based on the evidence of the individual case rankings the clinicians agreed that the MLP classifications of ROOTP and AIB were correct.

The ROOTP negated significant input class profile, as shown in Table 2, indicated an anomaly in two attributes (in bold), *duration(recur)* and *slrr(≤45)*, since domain knowledge considered these two attributes as positive criteria for the ROOTP class. These attributes were traced to 2 cases which were again re-evaluated by the domain experts. The experts agreed that both cases showed more characteristics of the SLBP classification, as made by the MLP, than the ROOTP class.

The AIB class profiles did not show any anomalies. A revised set of SLBP and ROOTP average ranked class profiles was produced by correctingthe targets for the 4 mis-diagnosed training cases, as shown in Tables 3 and 4.

4.2 Clinical evaluation of the ranked class profiles

The clinical evaluation of the profiles by the domain experts indicated that the low back pain MLP network had largely determined relevant attributes as typical characteristics of patients belonging to the 3 diagnostic classes, although not necessarily in the same ranked order. The ranked order of the class profiles, however, was supported by a frequency analysis of the training data which indicated that, in general, the highest ranked inputs were the more unique discriminators for a class.

For example, the 4th ranked significant input for the SLBP class, *'no leg pain symptoms'*, was unique to 22% of SLBP patients. As another example, the highest ranked significant input for the SLBP class, *'back pain worse than leg pain'*, was present for 73% of all SLBP patients compared with 29% of the ROOTP class in the training set.

In the evaluation, the 8th ranked positive SLBP attribute *'no back pain'* was seen to be contradictory to the top ranked attribute *'worse back pain'*. Further investigation revealed that (the only) two SLBP training cases with attribute *'no back pain'* had been incorrectly included in the training set. This demonstrates the sensitivity of both the MLP network and the knowledge extraction method.

4.3 Validation of the test cases

Of the 99 test cases, 35 were apparently mis-classified by the low back pain MLP network. By individually interpreting each of the mis-classified test cases using the interpretation and knowledge discovery method the top ranked significant and

R	All SLBP training cases
1	**back pain worse than leg pain**
2	lumbar flexion ≥45 °
3	**straight right leg raise ≥ 70 °**
4	**no leg pain symptoms**
5	pain brought on by bending over
6	minimal ODI score
7	**lumbar extension < 5 °**
8	no back pain symptoms
9	straight left leg raise (46 to 69 °)
10	straight right leg raise not limited

R	All ROOTP training cases
1	back pain aggravated by coughing.
2	leg pain worse than back pain
3	lumbar flexion < 30 °
4	lumbar extension (5 to 14 °)
5	normal DRAM
6	low MSPQ score
7	low Zung depression score
8	use of walking aids
9	leg pain aggravated by coughing
10	pain brought on by lifting

R	All AIB training cases
1	claiming invalidity/ disability benefit
2	straight left leg raise≤45 °
3	high Waddell's inappropriate signs
4	distressed depressive
5	pain brought on by bending over
6	straight left leg raise limited by hamstrings
7	pain brought on by falling over
8	back pain aggravated by standing
9	high Zung depression score
10	back pain worse than leg pain

R	All SLBP training cases
1	**back pain worse than leg pain**
2	back pain aggravated by coughing
3	lumbar flexion < 30°
4	**no leg pain symptoms**
5	**straight right leg raise ≥ 70 °**
6	**lumbar extension < 5 °**
7	equal back and leg pains
8	normal DRAM
9	lumbar extension (5 to 14°)
10	recurring back pain

R	All ROOTP training cases
1	back pain worse than leg pain
2	lumbar flexion≥45°
3	**recurring back pain**
4	no leg pain symptoms
5	pain brought on by bending over.
6	minimal ODI score
7	straight right leg raise ≥ 70°
8	lumbar extension < 5°.
9	acute back pain
10	**straight right leg raise ≤45 °**

R	All AIB training cases
1	straight left leg raise ltd by leg pain
2	low Waddell's inappropriate signs
3	normal DRAM
4	low Zung depression score
5	leg pain worse than back pain
6	leg pain aggravated by walking
7	chronic back pain
8	straight left leg raise (46° to 69°)
9	back pain aggravated by coughing.
10	straight right leg raise limited by back pain

Table 1: **Significant** input class profiles   Table 2: **Negated** input class profiles

negated inputs of *each mis-classified test case* were compared with the class profiles. Many of the mis-classified case test rankings indicated a high degree of commonality with the ranked class profiles. On further investigation it was agreed by the domain experts that 15 of the 35 mis-classified cases were likely to have been *correctly* classified by the low back pain MLP and *incorrectly* classified by the clinicians. Of these, 12 cases (out of 19) were correctly classified by the network as

R	All **SLBP** training cases
1	back pain worse than leg pain
2	lumbar flexion ≥45 °
3	straight right leg raise ≥ 70 °
4	no leg pain symptoms
5	pain brought on by bending over
6	minimal ODI score
7	lumbar extension < 5 °
8	**no back pain symptoms**
9	straight left leg raise (46 to 69 °)
10	straight right leg raise not limited

R	All **ROOTP** training cases
1	back pain aggravated by coughing.
2	leg pain worse than back pain
3	lumbar flexion < 30 °
4	lumbar extension (5 to 14 °)
5	normal DRAM
6	low MSPQ score
7	low Zung depression score
8	use of walking aids
9	leg pain aggravated by coughing
10	pain brought on by lifting

R	All **SLBP** training cases
1	back pain aggravated by coughing
2	lumbar flexion < 30°
3	lumbar extension (5 to 14°)
4	equal back and leg pains
5	smoker
6	normal DRAM
7	loss of reflexes
8	low Zung depression score
9	recurring back pain
10	leg pain worse than back pain

R	All **ROOTP** training cases
1	back pain worse than leg pain
2	lumbar flexion≥45°
3	pain brought on by bending over.
4	no leg pain symptoms
5	minimal ODI score
6	straight right leg raise ≥ 70°
7	lumbar extension < 5°.
8	acute back pain
9	high Zung depression score
10	high MSPQ score

Table 3:   Revised SLBP & ROOTP **significant** input profiles

Table 4:   Revised SLBP & ROOTP **negated** input profiles

class AIB, 2 (out of 11) were correctly classified by the network as class ROOTP and 1 (out of 5) was correctly classified by the network as class SLBP. The difficulty experienced by clinicians in diagnosing AIB patients has been observed in other studies [5].

The initial 92-10-3 low back pain network was re-trained with the corrected training and test sets. The best test performance of the re-trained network was 79% which is similar to the results of other researchers [6].

5   Development of a more optimal low back pain MLP

By using the validated ranked class profiles to find the least significant inputs in the initial MLP network a more efficient network was developed on removing these inputs from the input layer of the initial network. The least significant inputs were arbitrarily selected as the inputs ranked in 20th position or below in the *significant* input class profiles which resulted in the removal of 48 inputs from the initial MLP. The reduced network was trained and tested with the same (reduced) cases as the re-trained initial network when a 48-5-3 network was found to have a best test

320

performance of 77%. This is a marginally reduced performance compared with the 92-10-3 network result of 79% but with approximately half the number of inputs. This result appears to confirm the validity of the knowledge discovery method.

6 Summary and Conclusions

Using a new knowledge discovery method this paper has presented the ranked class profiles from an initial low back pain MLP. In validating the profiles it was discovered that 4 training cases were mis-diagnosed by the clinicians and two training cases incorrectly included in the training set. This demonstrates the sensitivity of both the MLP network and the knowledge discovery method. In clinically evaluating the class profiles the experts considered that the network had determined largely valid attributes as typical characteristics of each class.

By directly interpreting the 35 mis-classified test cases and comparing them with the class profiles it was agreed that 15 cases were likely to have been *correctly* classified by the low back pain MLP. The test performance of the initial network was re-assessed as 79%, a similar result to other researchers.

By using the ranked class profiles to find the least significant inputs in the initial network a reduced 48-5-3 network was developed with a comparable test performance of 77%. It is concluded that the knowledge discovery method is a powerful tool for MLP network development.

References

[1] Andrews R, Diederich J, and Tickle AB. Survey and Critique of Techniques for Extracting Rules from Trained Artificial Neural Networks. Knowledge-Based Systems 1995; 8(6): 373–389

[2] Vaughn ML. Interpretation and Knowledge Discovery from the Multilayer Perceptron Network: Opening the Black Box. Neural Computing & Applications 1996; 4(2):72–82

[3] Vaughn ML. Derivation of the Multilayer Perceptron Weight Constraints for Direct Network Interpretation and Knowledge Discovery. Neural Networks 1999; 12 (9): 1259-1271

[4] Vaughn ML, Cavill SJ, Taylor SJ, Foy MA, Fogg AJB. Direct Explanations and Knowledge Extraction from a Multilayer Perceptron Network that Performs Low Back Pain Classification. In: Wermter S, Sun R (eds) Hybrid Neural Symbolic Integration. Springer, to be published Spring 2000

[5] Waddell G, Bircher M, Finlayson D, Main C, Symptoms and Signs:Physical Disease or Illness Behaviour. BMJ 1984; 289: 739–741

[6] Bounds DG, Lloyd PJ, and Mathew B. A Comparison of Neural Network and Other Pattern Recognition Approaches to the Diagnosis of Low Back Disorders. Neural Networks 1990; 3 : 583–591

Kernel PCA Feature Extraction of Event-Related Potentials for Human Signal Detection Performance

Roman Rosipal, Mark Girolami

Department of Computing and Information Systems, University of Paisley
Paisley, PA1 2BE, Scotland

Leonard J. Trejo

Human Information Processing Research Branch, NASA Ames Research Center
Moffett Field, CA

Abstract

In this paper, we propose the application of the Kernel PCA technique for feature selection in high-dimensional feature space where input variables are mapped by a Gaussian kernel. The extracted features are employed in the regression problem of estimating human signal detection performance from brain event-related potentials elicited by task relevant signals. We report the superiority of Kernel PCA for feature extraction over linear PCA.

1 Introduction

In many safety-critical applications (e.g., air traffic control, power plant operation, military applications) control is based on the ability of human operators to detect and evaluate task-relevant signals in the presented visual data. Performance quality of operators varies over time, often falling below acceptable limits, and may result in errors with potentially serious consequences. The likelihood of such errors could be reduced if physiological methods for assessment of human performance were available.

A fundamental part in the development of such a method is to construct a model reflecting the dependence between selected physiological metrics of mental workload (e.g., Event-Related Potentials (ERPs)) and the performance characteristics of a human operator (reaction time, accuracy, and confidence). Prior research has demonstrated that linear regression and nonlinear neural networks can model the relationships between ERPs and performance (see [1, 2, 3] and ref. therein). However, when we attempt to develop such a model we are confronted with the curse of dimensionality, which arises from the complexity of physiological data. For example, 225 data samples (dimensions) are required to describe a single 1.5s segment of ERP data from three electrodes. To address this problem, we can accept two general assumptions about the real world data sets. First, there exist some correlations among input variables; thus dimensionality reduction or so-called *feature extraction* allows us to restrict the entire input space to a sub-space of lower dimensionality. Second, in many practical problems we can assume a smooth mapping from input to output space; thus we can infer the values of the output for points where no input data are available. This can be done by an appropriate regularization technique.

In this study, we have used the recently proposed Kernel PCA [4] method for feature selection in kernel space. This allows us to obtain features (nonlinear principal components) with higher-order correlations between input variables, and second, we can extract more components if the number of data points is higher than their dimensionality [4]. The idea behind Kernel PCA [4] is based on computation of the standard linear PCA in a high dimensional feature space $\mathcal{F}$ (with dimension $\leq \infty$), into which the input data $\mathbf{x} \in R^{\mathcal{N}}$ are mapped via some nonlinear function $\Phi(\mathbf{x})$. To this end, we compute a dot product in space $\mathcal{F}$ using a kernel function, i.e. $K(\mathbf{x}, \mathbf{y}) = (\Phi(\mathbf{x}).\Phi(\mathbf{y}))$. This 'kernel trick' allows us to carry out any algorithm, e.g Support Vector Regression (SVR) [5], that can be expressed in the terms of dot products in space $\mathcal{F}$. Next, we used selected features to train SVR and Kernel Principal Component Regression to estimate the dependency between ERPs and the performance of the individual subjects. The results suggest the superiority of (nonlinear) Kernel PCA for feature extraction over linear PCA in some cases.

2 Methods

2.1 Kernel PCA, Multi-Layer SVR and Kernel PCR

The PCA problem in high-dimensional feature space $\mathcal{F}$ can be formulated as the diagonalization of a n-sample estimate of the covariance matrix $\hat{C} = \frac{1}{n} \sum_{i=1}^{n} \Phi(\mathbf{x}_i)\Phi(\mathbf{x}_i)^T$, where $\Phi(\mathbf{x}_i)$ are centered nonlinear mappings of the input variables $\mathbf{x}_i \in \mathcal{R}^{\mathcal{N}}$, $i = 1, ..., n$. The diagonalization represents a transformation of the original data to new coordinates defined by orthogonal eigenvectors $\mathbf{V}$. We have to find eigenvalues $\lambda \geq 0$ and non-zero eigenvectors $\mathbf{V} \in \mathcal{F}$ satisfying the eigenvalue equation $\lambda \mathbf{V} = \hat{C}\mathbf{V}$. Realizing, that all solutions $\mathbf{V}$ with $\lambda \neq 0$ lie in the span of mappings $\Phi(\mathbf{x}_1), ..., \Phi(\mathbf{x}_n)$, Schölkopf et. al. [4] derived the equivalent eigenvalue problem $n\lambda \mathbf{v} = \mathbf{K}\mathbf{v}$, where $\mathbf{v}$ denotes the column vector with coefficients $v_1, ..., v_n$ such that $\mathbf{V} = \sum_{i=1}^{n} v_i \Phi(\mathbf{x}_i)$ and $\mathbf{K}$ is a symmetric $n \times n$ matrix with $\mathbf{K}_{ij} = (\Phi(\mathbf{x}_i).\Phi(\mathbf{x}_j)) := K(\mathbf{x}_i, \mathbf{x}_j)$. Normalizing the solutions $\mathbf{V}^k$ corresponding to the non-zero eigenvalues λ_k of the matrix $\mathbf{K}$, translates into the condition $\lambda_k(\mathbf{v}^k.\mathbf{v}^k) = 1$ [4]. Finally, we can compute the projection of $\Phi(\mathbf{x})$ onto the k-th nonlinear principal component by

$$q(\mathbf{x})_k := (\mathbf{V}^k.\Phi(\mathbf{x})) = \sum_{i=1}^{n} v_i^k K(\mathbf{x}_i, \mathbf{x}). \qquad (1)$$

We then select the first r nonlinear principal components, e.g. the directions which describe a desired percentage of data variance, and thus work in an r-dimensional sub-space of feature space $\mathcal{F}$. This allows us to construct multi-layer support vector machines [4], where a preprocessing layer extracts features for the next regression or classification task. In our study we focus on the regression problem.

Generally, the SVR problem (see e.g.[5]) can be defined as the determination of function $f(\mathbf{x}, \mathbf{w})$ which approximates an unknown desired function and has the form $f(\mathbf{x}, \mathbf{w}) = (\mathbf{w}.\Phi(\mathbf{x})) + b$, where b is an unknown bias term, $\mathbf{w} \in \mathcal{F}$ is a vector of unknown coefficients and $(\mathbf{w}.\Phi(\mathbf{x}))$ is a dot product in space $\mathcal{F}$. In [7] the following regularized risk functional is shown to compute the unknown coefficients b and $\mathbf{w}$:

$$R_{reg}(\mathbf{w}) = \frac{1}{n}\sum_{i=1}^{n}|Err|_\varepsilon + \frac{\eta}{2}\|\mathbf{w}\|^2, \tag{2}$$

where $Err = y_i - f(\mathbf{x}_i, \mathbf{w})$, $\eta \geq 0$ is a regularization constant to control the trade-off between complexity and accuracy of the regression model and $|Err|_\varepsilon$ is Vapnik's ε-insensitive loss-function [7]. In [7] it is shown that the regression estimate that minimizes the functional (2) has the form: $f(\mathbf{x}, \mathbf{a}, \mathbf{a}^*) = \sum_{i=1}^{n}(a_i^* - a_i)K_1(\mathbf{x}_i, \mathbf{x}) + b$, where $\{a_i, a_i^*\}_{i=1}^{n}$ are Lagrange multipliers [5].

Combining the Kernel PCA preprocessing step with SVR yields a multi-layer SVR in the following form [4]: $f(\mathbf{x}, \mathbf{a}, \mathbf{a}^*) = \sum_{i=1}^{n}(a_i - a_i^*)K_1(\mathbf{q}(\mathbf{x}_i), \mathbf{q}(\mathbf{x})) + b$, where components of vectors $\mathbf{q}(.)$ are defined by (1). However, in practice the choice of appropriate kernel function $K_1(.,.)$ can be difficult. In this study, a polynomial kernel of first order $K_1(\mathbf{x}, \mathbf{y}) = (\mathbf{x}.\mathbf{y})$ is employed. We are thus performing a linear SVR on the r-dimensional sub-space of $\mathcal{F}$. The advantage of linear SVR over ordinary linear regression is the possibility of using a large variety of loss functions to suit different noise models [5], e.g. the proposed Vapnik's ε-insensitive function is more robust for noise distributions close to uniform. However, in the case of Gaussian noise the best approximation to the regression provides a loss function of the form $L(y_i, f(\mathbf{x}_i, \mathbf{c})) = [y_i - f(\mathbf{x}_i, \mathbf{c})]^2$. Therefore, we used a Kernel Principal Component Regression[1] technique which minimizes the following risk functional

$$R_{pcr}(\mathbf{c}) = \frac{1}{n}\sum_{i=1}^{n}[y_i - f(\mathbf{x}_i, \mathbf{c})]^2.$$

The solution $f(\mathbf{x}, \mathbf{c})$ has the form

$$f(\mathbf{x}, \mathbf{c}) = \sum_{k=1}^{r} b_k q(\mathbf{x})_k + b_0 = \sum_{k=1}^{r} b_k \sum_{i=1}^{n} v_i^k K(\mathbf{x}_i, \mathbf{x}) + b_0 = \sum_{i=1}^{n} c_i K(\mathbf{x}_i, \mathbf{x}) + b_0,$$

where $\{c_i = \sum_{k=1}^{r} b_k v_i^k\}_{i=1}^{n}$ and $\{q(\mathbf{x})_k\}_{k=1}^{r}$ are again defined by (1). The coefficients $\{b_k\}_{k=0}^{r}$ can be found by solving the *normal equations* for least squares estimation.

2.2 Data Sample Construction

We have used ERPs and performance data from an earlier study [2]. Eight male Navy technicians experienced in the operation of display systems performed a signal detection task. Each technician was trained to a stable level of performance and tested in multiple blocks of 50–72 trials each on two separate days. Blocks were separated by 1-minute rest intervals. A set of 1000 trials were performed by each subject. Inter-trial intervals were of random duration with a mean of 3s and a range of 2.5–3.5s. The entire experiment was computer-controlled and performed with a 19-inch color CRT display. Triangular symbols subtending 42 minutes of arc and of three different luminance contrasts (0.17, 0.43, or 0.53) were presented parafoveally at a constant

---

[1] A more detailed description of a Principal Component Regression is given in [6].

eccentricity of 2 degrees visual angle. One symbol was designated as the target, the other as the non-target. On some blocks, targets contained a central dot whereas the non-targets did not. However, the association of symbols to targets was alternated between blocks to prevent the development of automatic processing. A single symbol was presented per trial, at a randomly selected position on a 2-degree annulus. Fixation was monitored with an infrared eye tracking device. Subjects were required to classify the symbols as targets or non-targets using button presses and then to indicate their subjective confidence on a 3-point scale using a 3-button mouse. Performance was measured as a linear composite of speed, accuracy, and confidence. A single measure, PF1, was derived using factor analysis of the performance data for all subjects, and validated within subjects. The computational formula for PF1 was

$$PF1 = 0.33*\text{Accuracy} + 0.53*\text{Confidence} - 0.51*\text{Reaction Time}$$

using standard scores for accuracy, confidence, and reaction time based on the mean and variance of their distributions across all subjects. PF1 varied continuously, being high for fast, accurate, and confident responses and low for slow, inaccurate, and unconfident responses.

ERPs were recorded from midline frontal, central, and parietal electrodes (Fz, Cz, and Pz), referred to average mastoids, filtered digitally to a bandpass of 0.1 to 25 Hz, and decimated to a final sampling rate of 50 Hz. The prestimulus baseline (200 ms) was adjusted to zero to remove any DC offset. Vertical and horizontal electrooculograms (EOG) were also recorded. Epochs containing artifacts were rejected and EOG-contaminated epochs were corrected. Furthermore, any trial in which no detection response or confidence rating was made by a subject was excluded along with the corresponding ERP.

Within each block of trials, a running-mean ERP was computed for each trial. Each running-mean ERP was the average of the ERPs over a window that included the current trial plus the 9 preceding trials for a maximum of 10 trials per average. Within this 10-trial window, a minimum of 7 artifact-free ERPs were required to compute the running-mean ERP. If fewer than 7 were available, the running mean for that trial was excluded. Thus each running mean was based on at least 7 but no more than 10 artifact-free ERPs. This 10-trial window corresponds to about 30s of task time. The PF1 scores for each trial were also averaged using the same running-mean window applied to the ERPs, excluding PF1 scores for trials in which ERPs were rejected. Prior to analysis, the running-mean ERPs were clipped to extend from time zero (stimulus onset time) to 1500 ms post-stimulus, for a total of 75 time points.

3 Results

The present work was carried out with Gaussian kernels; $K(\mathbf{x},\mathbf{y}) = e^{-(\frac{\|\mathbf{x}-\mathbf{y}\|^2}{L})}$, where L is the width of the Gaussian function. The desired output PF1 was linearly normalized to have a range of 0 to 1. We trained the models on 50% of the ERPs and tested on the remaining data. The described results, for each setting of the parameters, are an average of 10 runs each on a different partition of training and testing data. The validity of the models was measured in terms of the proportion of data for which

PF1 was correctly predicted with 10% tolerance, i.e ±0.1 in our case. The performance of a Regularized Gaussian RBF (rGRBF) network [8] and SVR trained on data pre-processed by linear PCA (LPCA) in input space was compared with the results achieved by multi-layer SVR (MLSVR) and the proposed Kernel Principal Component Regression (KPCR) on features extracted by Kernel PCA. In both cases we used features (principal components) describing 99% of data variance. We used $\varepsilon = 0.01$, $\eta = 0.01$ parameter values in the case of SVR. The results achieved on subject A (592 ERPs), B(614 ERPs) and C (417 ERPs) are depicted in Figure 1. On subjects A and B we can see consistently better results on features extracted by Kernel PCA (top and middle left graphs). These superior results achieved using Kernel PCA representation were also observed on the remaining five subjects. In addition, we can see that in all cases the SVR was superior to rGRBF on inputs extracted by linear PCA (right graphs). We have to note that on all subjects we achieved similar results with features describing 98% of data variance. Without the PCA preprocessing step in feature space $\mathcal{F}$ we did not increase the overall performance. On the contrary, on four subjects the performance was on average decreased by 0.5% on test proportion error.

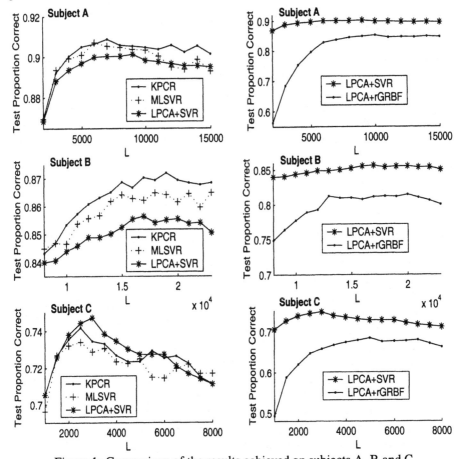

Figure 1: Comparison of the results achieved on subjects A, B and C.

4 Conclusions

The selection of appropriate features for regression has been investigated. On subjects A and B we demonstrated that (nonlinear) Kernel PCA provides a superior representation of the data set over that of linear PCA. However, on subject C the performance with features selected by linear PCA was slightly better. We have to note, that in this case the dimension of matrix $\mathbf{K}$ in feature space $\mathcal{F}$ is lower (209) than the input dimensionality (225), thus we can not exploit the advantage of Kernel PCA to improve overall performance by using more components in feature space than the number available in the input space. We used features describing 99% of the data variance that for different parameter L represents 70–90% of all nonlinear principal components and we showed that such a reduction of high-dimensional feature space $\mathcal{F}$ does not decrease the overall performance. Moreover, this can be seen as the denoising technique assuming that the noise is spread in directions with small variance. On all subjects, we demonstrated that the performance of SVR on features extracted by linear PCA was superior to Regularized Gaussian RBF.

Acknowledgments

The first author is funded by a research grant for the project "Objective Measures of Depth of Anaesthesia"; University of Paisley and Glasgow Western Infirmary NHS trust, and is partially supported by Slovak Grant Agency for Science (grants No. 2/5088/98 and No. 98/5305/468). Data were obtained under a grant from the US Navy Office of Naval Research (PE60115N), monitored by Joel Davis and Harold Hawkins. Dr. Trejo is supported by the Psychological and Physiological Stressors and Factors Project of the NASA Aerospace Operations Systems Program (RTOP 711-51-42), managed by J. V. Lebacqz.

References

[1] Trejo LJ, Shensa MJ. Feature Extraction of ERPs Using Waveletes: An Application to Human Performance Monitoring. Brain and Language 1999; 66:89–107.

[2] Trejo LJ, Kramer AF, Arnold JA. Event-related Potentials as Indices of Display-monitoring Performance. Biological Psychology 1995; 40:33–71.

[3] Koska M, Rosipal R, König A, Trejo LJ. Estimation of human signal detection performance from ERPs using feed-forward network model. In: Computer Intensive Methods in Control and Signal Processing, The Curse of Dimensionality. Birkhauser, Boston, 1997.

[4] Schölkopf B, Smola AJ, Müller KR. Nonlinear Component Analysis as a Kernel Eigenvalue Problem. Neural Computation 1998; 10:1299–1319.

[5] Smola AJ, Schölkopf B. A Tutorial on Support Vector Regression. Technical Report NC2-TR-1998-030, NeuroColt2 Technical Report Series, 1998.

[6] Jollife IT. Principal Component Analysis. Springer-Verlag, New York, 1986.

[7] Vapnik V. The Nature of Statistical Learning Theory. Springer, New York, 1998.

[8] Chen S, Chng ES, Alkadhimi K. Regularised orthogonal least squares algorithm for constructing RBF networks. Int Journal of Control 1996; 64.

Particle Swarm Optimisation in Feedforward Neural Network*

Chunkai Zhang

Automation Department, Shanghai Jiao Tong University
Shanghai, P. R. China, 200030
cgzhang812@mail1.sjtu.edu.cn

Huihe Shao

Automation Department, Shanghai Jiao Tong University
Shanghai, P. R. China, 200030

Abstract

The paper describes a new evolutionary system for evolving artificial neural networks (ANN's) called PSONN, which is based on the particle swarm optimisation (PSO) algorithm. The PSO algorithm is used to evolve both the architecture and weights of ANN's, this means that an ANN's architecture is adaptively adjusted by PSO algorithm, then the nodes of this ANN's are also evolved by PSO algorithm to evaluate the quality of this network architecture. This process is repeated until the best ANN's is accepted or the maximum number of generations has been reached. In PSONN, a strategy of evolving added nodes and a partial training algorithm are used to maintain a close behavioural link between the parents and their offspring, which improves the efficiency of evolving ANN's. PSONN has been tested on two real problems in the medical domain. The results show that ANN's evolved by PSONN have good accuracy and generalisation ability.

1 Introduction

A number of techniques exist with which ANN's can be trained. Most applications use the backpropagation (BP) algorithm or other training algorithms in the feedforward ANN's. But all these training algorithms assume a fixed ANN's architecture. There have been many attempts to design ANN's architectures automatically, such as various constructive and pruning algorithms [1, 2]. However, "Such structural hill climbing methods are susceptible to becoming trapped at structure local optima, and the result depends on initial network architectures." [3].

Design of a near optimal ANN architecture can be formulated as a search problem in the architecture space where each point represents an architecture. Miller et al. indicated that evolutionary algorithms are better candidates for searching the architecture space than those constructive and pruning algorithms mentioned above [4]. GA was used to evolve ANN's, but the evolution of ANN architectures often

* This work is supported by National 973 Fundamental Research Program of China.

suffers from the permutation problem [5]. This problem not only makes the evolution inefficient, but also makes crossover operators more difficult to produce highly fit offspring.

This paper proposes an evolutionary system for evolving feed-forward ANN's called PSONN, which combines the architectural evolution with weight learning. PSONN is based on the particle swarm optimisation (PSO) algorithm [6]. Compared with the GA based algorithm, it can avoid the permutation problem. And in PSONN a strategy of evolving added nodes (EAN) and a partial training (PT) algorithm are used to maintain the behavioural link between a parent and its offspring to prevent destruction of the behaviour already learned by the parent.

The rest of this paper is organised as follows. Section 2 describes the PSONN system and the motivations on how to evolve the entire ANN's. Section 3 presents experimental results on PSONN. The paper is concluded in Section 4.

2 PSONN system

2.1 Particle swarm optimisation (PSO)

In PSONN system, the learning algorithm is the PSO algorithm. PSO is a population based optimisation algorithm that is motivated from the simulation of social behaviour. Each individual in PSO flies in the search space with a velocity that is dynamically adjusted according to its own flying experience and its companions' flying experience. Compared with other evolutionary algorithms, such as GA, PSO algorithm possesses some attractive properties such as memory and constructive cooperation between individuals, so it has more chance to "fly" into the better solution areas more quickly and discover reasonable quality solution much faster [6]. In this paper we propose an improved PSO algorithm, which is as follows:

1) Initialise positions *Pesentx* and associated velocity v of all individuals (potential solutions) in the population randomly in the D dimension space.

2) Evaluate the fitness value of all individuals.

3) Compare the *PBEST[]* of every individual with its current fitness value. If the current fitness value is better, assign the current fitness value to *PBEST[]* and assign the current coordinates to *PBESTx[][d]* coordinates. Here *PBEST[]* represents the best fitness value of the nth individual, *PBESTx[][d]* represents the dth component of an individual.

4) Determine the current best fitness value in the entire population and its coordinates. If the current best fitness value is better than the *GBEST*, then assign the current best fitness value to *GBEST* and assign the current coordinates to *GBESTx[d]*.

5) Change velocities and positions using the following rules:

$$v[][d] = W * v[][d] + C1 * rand * (PBESTx[][d] - Pesentx[][d])$$
$$+ C2 * rand * (GBESTx[d] - Pesentx[][d])$$
$$Pesentx[][d] = Pesentx[][d] + v[][d] \tag{1}$$
$$W = W_\infty + (W_0 - W_\infty)[1 - \tfrac{1}{K}]$$

where $C1 = C2 = 2.0$, t and K are the number of current iterations and total generation. The balance between the global and local search is adjusted through the parameter $W \in (W_0, W_\infty)$.

6)   Repeat step 2) - 6) until a stop criterion is satisfied or a predefined number of iterations is completed.

Because there is not a selection operator in PSO, each individual in an original population has a corresponding partner in a new population. From the view of the diversity of population, this property is better than GA, so it can avoid the premature convergence and stagnation in GA to some extent.

2.2   PSONN system

In PSONN, evolving ANN's architectures and weight learning are alternated. This process can avoid a moving target problem resulted from the simultaneous evolution of both architectures and weights [7]. And the network architectures are adaptively evolved by PSO algorithm, starting from the parent's weights instead of randomly initialised weights, so this can preferably solve the problem of the noisy fitness evaluation that can mislead the evolution [3]. In PSONN all the data sets are partitioned into three sets: a training set, a validation set, and a testing set. The training set is used to evolve the nodes of ANN's with a given network architecture, and the fitness evaluation is equal to the mean squared error E of ANN's. But in evolving the architecture of network, the fitness evaluation is different from previous work in evolving the nodes of ANN's since it is determined through a validation set which does not overlap with the train set. Such use of validation set in evolutionary learning system improves the generalisation ability of evolved ANN's and introduces little overhead in computation time.

The major steps of PSONN can be described as follows:

1)   Generate an initial population of M networks. The number of hidden nodes for each network is uniformly generated at random within certain ranges, and the initial weights are uniformly distributed inside a small range.

2)   Use the partially training (PT) algorithm to train each network in the population on the training set.

- Choose a network as a parent network, then randomly generate $N-1$ initial individuals as a population where each individual's initial weights uniformly generated at random within certain ranges, but their network architectures are the same as the parent network architecture, and then the parent network is added into the population. Here each individual of the population is to parameterise a whole group of g nodes in ANN's, this means that every component of each individual represents a connection weight.
- Employ the PSO algorithm to evolve this population until the best individual found is accepted or the maximum number of generations has been reached.
- The best individual that survived will join the network architecture evolution.

3)   All survived networks form a new population. Evaluate the fitness values of every individual in this population. Here the mean squared error value E of each network on the validation set represents the fitness evaluation of each individual.

4) If the best network found is accepted or the maximum number of generations has been reached, stop and go to step 7). Otherwise continue.

5) Employ the PSO algorithm to only evolve the network architecture of each individual. Here each individual represents the number of the hidden nodes of a network.

6) When the network architecture of an individual changes, employ the strategy of evolving added nodes (EAN) to decide how to use the PT algorithm to evolve its weights. There are two choice:

- If some hidden nodes need to be added to a network, under the strategy of EAN, the PT algorithm only evolves the new added nodes to explain as much of the remaining output variance as possible. In this case the cost function that is minimised at each step of algorithm is the residual sum squared error that will remain after the addition of the new nodes, and the existing network nodes of the hidden layer are left unchanged during the search for the best new added nodes. Compared with the existing hidden nodes, the added nodes will represent or explain the finer details of this mapping that the entire network is trying to approximate between the inputs and outputs of the training data. This strategy can decrease the computation time for evolving the entire network and prevent destruction of the behaviour already learned by the parent.
- If some hidden nodes need to be deleted from a network, EAN strategy can remove the nodes in the reverse order in which they were originally added to the network, then the PT algorithm evolves the connection weights of the entire network, but sometimes a few of jump in fitness from the parent to the offspring is not avoided.

Then go to Step 3).

7) After the evolutionary process, train the best ANN's further with the PT algorithm on the combined training and validation set until it "converges".

In step 7), the generalisation ability of ANN's can be further improved by training the best ANN's with the PT algorithm on the combined training and validation set.

After evolving the architecture of networks every time, the strategy of evolving added nodes (EAN) and the partial training (PT) algorithm are used to optimise the weights of nodes with a given network architecture which has been evolved by the PSO algorithm. In other words, this process can evaluate the quality of this given network architecture and it is important to maintain the behavioural link between a parent and its offspring [11].

In the PT algorithm, each individual of the population in the PSO algorithm is to parameterise a whole group of g nodes in ANN's, this means that every component of each individual represents a connection weight. Compared with the encoding scheme that each individual represents a single node, and then the individuals are bundled together in the groups of g individuals each, this scheme is simple and easily implemented, and does not need a combinatorial search strategy. Although it possesses low efficiency, the fast convergent speed of the PSO algorithm can make up for this shortcoming.

3 Experimental studies

In order to evaluate the ability of PSONN in evolving ANN's, it was applied to two real problems in the medical domain, i.e., breast cancer and heart disease. All data sets were obtained from the UCI machine learning benchmark repository. These medical diagnosis problems have the following characteristics [12]:

● The input attributes used are similar to those a human expert would use in order to solve the same problem.

● The outputs represent either the classification of a number of understandable classes or the prediction of a set of understandable quantities.

● Examples are expensive to get. This has the consequence that the training sets are not very large.

The purpose of the breast cancer data set is to classify a tumour as either benign or malignant based on cell descriptions gathered by microscopic examination. It contains 9 attributes and 699 examples of which 485 are benign examples and 241 are malignant examples. The purpose of the heart disease data set is to predict the presence or absence of heart disease given the results of various medical tests carried out on a patient, it contains 13 attributes and 270 examples. For the breast cancer set, the first 349 examples of the whole data set were used for training, the following 175 examples for validation, and the final 175 examples for testing. For the heart disease set, the first 134 examples were used for training set, the following 68 examples for the validation set, and the final 68 examples for the testing set. Table 1 show the experimental results averaged 30 runs for the testing data set.

Table 1 Accuracy of the evolved ANN by PSONN

	Breast cancer data set			Heart disease data set		
	Mean	Min	Max	Mean	Min	Max
Error	1.322	0.153	3.362	11.724	9.958.	13.637
Error rate	0.011	0.000	0.051	0.146	0.112	0.186

We compare the results of PSONN with other works. For the breast cancer problem, Setiono and Hui have proposed a new ANN constructive algorithm called FNNCA [13]. Prechelt also reported results on manually constructed ANN using HDANN [12]. For the heart disease problem, Bennet reported a testing result with their MSM1 method [14], and the best manually designed ANN's also gave a result. The results are as following:

Table 2 Comparison with other work

	Breast cancer data set			Heart disease data set		
	FNNCA	HDANNS	PSONN	MSM1	HDANNS	PSONN
Error rate (Mean)	0.0145	0.0115	0.0112	0.1653	0.1478	0.1460

From the results, we know that PSONN is able to evolve both the architecture and weights of ANN's, and the ANN's evolved by PSONN has better accuracy and generalisation ability than the other algorithms.

4 Conclusion

This paper describes a new evolutionary system called PSONN for evolving artificial neural networks (ANN's). PSONN is based on the particle swarm optimisation (PSO) algorithm. The PSO algorithm is used to evolve both the architecture and weights of ANN's. It can effectively alleviate the noisy fitness evaluation problem and the moving target problem. PSONN has been tested on two real problems in the medical domain. The results shown the ANN's evolved by PSONN has good accuracy and generalisation ability.

References

[1]   N.Burgess. A constructive algorithm that converges for real-valued input patterns. Int. J. Neural Syst., vol 5, no. 1, pp. 59-66, 1994.

[2]   R. Reed. Pruning algorithms-A survey, IEEE trans. Neural Networks, vol 4, pp. 740-747, 1995.

[3]   D. B. Fogel. Evolutionary computation: toward a new philosophy of machine intelligence. New York: IEEE Press, 1995.

[4]   G. F. Miller, P. M. Todd, and S. U. Hegde. Designing neural networks using genetic algorithms. In proc. 3rd Int. Conf. Genetic Algorithms Their Applications, CA: Morgan Kaufmann, 1989, pp. 379-384.

[5]   R. K. Belew, J. MchInerney and N. N. Schraudolph. Evolving networks: Using GAs with connectionist learning. Computer Sci. Eng. Dept., Univ. California-San Diego, Tech. Rep. CS90-174 revised, Feb. 1991.

[6]   Kennedy, J., and Eberhart, R. C. Particle swarm optimization. Proc. IEEE International Conference on Neural Networks, IEEE Service Center, Piscataway, NJ, pp. 39-43, 1995.

[7]   X. Yao. A review of evolutionary artificial neural networks. Int. J. Intell. Syst., vol. 8, no. 4, pp. 539-567, 1993.

[8]   J. R. McDonnell and D. Waagen. Evolving recurrent perceptrons for time-series modeling. IEEE Tran. Neural Networks, vol. 5, no. 1, pp. 24-38, 1994.

[9]   V. Maniezzo. Genetic evolution of the topology and weight distribution of neural networks. IEEE Trans. Neural Networks, vol. 5, no. 1, pp. 39-53, 1994.

[10] Fahlman, S.E., and Lebiere, C.. The cascade-correlation learning architecture. In D.S.Touretzky. Advances in neural information processing systems 2, pp. 524-532, San Mateo, CA: Morgan Kaufmann, 1990.

[11] P.J. Angeline, G.M.Sauders, and J.B. Pollack. An evolutionary algorithm that constructs recurrent neural networks. IEEE Trans. Neural Networks, vol. 5, pp. 54-65. 1994.

[12] L. Prechelt. Some notes on neural learning algorithm benchmarking. Neurocomputing, vol. 9, no. 3, pp. 343-347. 1995.

[13] R. Setiono and L. C. K. Hui. Use of a quasinewton method in a feedforward neural network construction algorithm. IEEE Trans. Neural Network, vol. 6, pp. 740-747, 1995.

[14] K. P. Bennett and O. L. Mangasarian. Robust linear programming discrimination of two linearly inseparable sets. Optimization Methods Software, vol. 1, pp. 23-34, 1992.

Author Index